Rhodiola rosea

Traditional Herbal Medicines for Modern Times

Each volume in this series provides academia, health sciences, and the herbal medicines industry with in-depth coverage of the herbal remedies for infectious diseases, certain medical conditions, or the plant medicines of a particular country.

Series Editor: Dr. Roland Hardman

Volume 1
Shengmai San, edited by Kam-Ming Ko

Volume 2
Rasayana: Ayurvedic Herbs for Rejuvenation and Longevity, by H.S. Puri

Volume 3
Sho-Saiko-To: (Xiao-Chai-Hu-Tang) Scientific Evaluation and Clinical Applications, by Yukio Ogihara and Masaki Aburada

Volume 4
Traditional Medicinal Plants and Malaria, edited by Merlin Willcox, Gerard Bodeker, and Philippe Rasoanaivo

Volume 5
Juzen-taiho-to (Shi-Quan-Da-Bu-Tang): Scientific Evaluation and Clinical Applications, edited by Haruki Yamada and Ikuo Saiki

Volume 6
Traditional Medicines for Modern Times: Antidiabetic Plants, edited by Amala Soumyanath

Volume 7
Bupleurum *Species: Scientific Evaluation and Clinical Applications,* edited by Sheng-Li Pan

Volume 8
Herbal Principles in Cosmetics: Properties and Mechanisms of Action, by Bruno Burlando, Luisella Verotta, Laura Cornara, and Elisa Bottini-Massa

Volume 9
Figs: The Genus Ficus, by Ephraim Philip Lansky and Helena Maaria Paavilainen

Volume 10
Phyllanthus Species: Scientific Evaluation and Medicinal Applications edited by Ramadasan Kuttan and K. B. Harikumar

Volume 11
Honey in Traditional and Modern Medicine, edited by Laïd Boukraâ

Volume 12
Caper: The Genus Capparis, Ephraim Philip Lansky, Helena Maaria Paavilainen, and Shifra Lansky

Volume 13
Chamomile: Medicinal, Biochemical, and Agricultural Aspects, Moumita Das

Traditional Herbal Medicines for Modern Times

Rhodiola rosea

Edited by
Alain Cuerrier
Kwesi Ampong-Nyarko

CRC Press
Taylor & Francis Group
Boca Raton London New York

CRC Press is an imprint of the
Taylor & Francis Group, an **informa** business

CRC Press
Taylor & Francis Group
6000 Broken Sound Parkway NW, Suite 300
Boca Raton, FL 33487-2742

First issued in paperback 2019

ISBN-13: 978-1-4398-8840-7 (hbk)
ISBN-13: 978-0-367-37807-3 (pbk)

Visit the Taylor & Francis Web site at
http://www.taylorandfrancis.com

and the CRC Press Web site at
http://www.crcpress.com

Contents

Foreword vii
Preface xi
Editors xiii
Contributors xv

Chapter 1   Taxonomy of *Rhodiola rosea* L., with Special Attention to
Molecular Analyses of Nunavik (Québec) Populations 1

*Alain Cuerrier, Mariannick Archambault, Michel Rapinski,
and Anne Bruneau*

Chapter 2   Ethnobotany and Conservation of *Rhodiola* Species 35

Alain Cuerrier, Youri Tendland, and Michel Rapinski

Chapter 3   Phytochemistry of *Rhodiola rosea* 65

Fida Ahmed, Vicky Filion, Ammar Saleem, and John T. Arnason

Chapter 4   Cultivation of *Rhodiola rosea* in Europe 87

Bertalan Galambosi

Chapter 5   *Rhodiola rosea* Cultivation in Canada and Alaska 125

Kwesi Ampong-Nyarko

Chapter 6   Diseases of Wild and Cultivated *Rhodiola rosea* 155

*Sheau-Fang Hwang, Stephen E. Strelkov, Kwesi Ampong-
Nyarko, and Ron J. Howard*

Chapter 7   Biotechnology of *Rhodiola rosea* 173

Zsuzsanna György

Chapter 8   Pharmacological Activities of *Rhodiola rosea* 189

Fida Ahmed, Steffany A.L. Bennett, and John T. Arnason

Chapter 9   Evidence-Based Efficacy and Effectiveness of *Rhodiola* SHR-5
Extract in Treating Stress- and Age-Associated Disorders..............205

Alexander Panossian and Georg Wikman

Chapter 10 *Rhodiola rosea* in Psychiatric and Medical Practice.......................225

Patricia L. Gerbarg, Petra A. Illig, and Richard P. Brown

Chapter 11 Toxicology and Safety of *Rhodiola rosea*..253

Hugh Semple and Brandie Bugiak

Chapter 12 Commercialization of *Rhodiola rosea*..265

Nav Sharma and Raimar Loebenberg

Index..275

Foreword

In the late 1990s, I was invited to contribute an article to a mainstream medical journal outlining the clinical relevance of various herbs and phytomedicines for which I believed there was ample published clinical trial data to warrant *consideration* for inclusion in modern evidence-based clinical practice.

One of the herbs I suggested was Asian ginseng root (*Panax ginseng*), due to its increasingly documented immunological, "adaptogenic," and other potentially salubrious properties. A peer reviewer of the draft article objected to the use of the term "adaptogen" and "adaptogenic," stating that such terms were not recognized in modern (Western) pharmacology and medicine. When I attempted to explain their meaning and to defend their appropriate use in the context of my article, the journal editor, obviously siding with the reviewer and demonstrating his (and the reviewer's) unfortunate lack of familiarity with modern botanical and phytomedicinal literature, threatened to reject my invited article. I had a choice: drop the word *adaptogen* to help ensure that my article might be published (there were, of course, other areas of concern raised by what were most likely unqualified peer reviewers), or stand up for adaptogens, both as an appropriate term and as a category of natural therapeutic agents. I chose the latter, and, predictably, my article was not accepted.

During that time, more than 15 years ago, there was already a significant increase in the use of herbal dietary supplements and other so-called alternative medicine modalities by millions of Americans, Canadians, Europeans, and others, and this movement was just beginning to gain attention from the mainstream medical community. Of course, since then, the growth of the use of herbs in self-medication and in clinical practice has increased dramatically.

In a sense, I have paid some personal and/or professional dues in trying to champion the rational use of adaptogenic herbs and the language that describes them and their potential therapeutic benefits.

So, what is an *adaptogen*, and what is *adaptogenic*? In short, and as will be much more eloquently and technically described in Chapter 9 of this impressive book, adaptogens, as coined by Soviet pharmacologists in the late 1940s and early 1950s, are natural products (e.g., plants and plant extracts) that have the ability to safely support an organism in adapting to various types of nonspecific stressors.

There are a variety of herbs that are generally considered adaptogenic by knowledgeable experts. My focus at the time I was writing the above-mentioned, never-published article was Asian ginseng. Now, as I write this foreword, it is *Rhodiola rosea* root and its extracts.

This is the second time I have had the honor to write a foreword to a book that is devoted exclusively to the herb, *Rhodiola*. In 2005, I produced the foreword to *Rhodiola Revolution*, a book written by psychiatrists Richard Brown, MD, and Pat Gerbarg, MD. This consumer-oriented book published by Rodale Press helped to pave the path for enhanced consumer and professional acceptance of *R. rosea* in the

United States. (The book has the curious subtitle that was obviously a marketing effort by the publisher: "Transform Your Health with the Herbal Breakthrough of the 21st Century.") I had become acquainted with Dick and Pat (and their coauthor, the late pharmacologist Zakir Ramazanov*) in 2002 when they wrote an extensive literature review of *R. rosea*, which was published as the cover story in issue number 56 of *HerbalGram* (the peer-reviewed journal of the American Botanical Council).

When referring to "rhodiola," I believe it is essential to ensure that the appropriate species be clearly mentioned. As noted in the first chapter of this book (taxonomy), many botanists agree that there are 90 species in the genus *Rhodiola* (with some taxonomic systems noting a range from as low as 60 to as many as 200 species). *Rhodiola rosea* root is the subject of the most extensive level of pharmacological and clinical research to document its safety and potential adaptogenic health benefits.

There are concerns in the herb marketplace that some herbal dietary supplement products (called "natural health products" in Canada) potentially are being sold that do *not* contain *R. rosea*, but other, possibly lower cost and surely less well researched, species of *Rhodiola* instead. For the sake of this foreword, and throughout this book—except where specifically noted—the common name *"Rhodiola"* refers to the species *R. rosea*.

Rhodiola is increasing in popularity among consumers as well as knowledgeable health-care practitioners, particularly in Scandinavian countries, Eastern Europe and Russia, the United States, Canada, and other developed nations. With respect to the United States—where some market data are available to document the increased retail sales of herbal dietary supplements made with *Rhodiola* as the single or primary ingredient—the herb enjoyed a hugely significant increase of 165% in mainstream stores (grocery stores, drug stores, mass market retailers, etc.) in 2013 compared to sales in 2012, totaling about US$1.3 million. In natural and health foods stores—where *Rhodiola* sales have been more established for about a decade or more (i.e., compared to lower sales in the mainstream market channel)—retail sales of *Rhodiola* dietary supplements rose an impressive 26% in 2013 over the previous year, ranking *Rhodiola* as 34th in total herbal single/primary sales in natural foods stores (about US$2.3 million, without including sales from retail natural foods giant Whole Foods Market, whose sales are not included in econometric reports, the estimated sales from which might possibly double the sales figures to over US$4 million in the natural channel).

As is appropriate, this book contains a highly compelling chapter reviewing the extensive pharmacological and clinical research conducted on the world's most clinically tested *R. rosea* commercial product, SHR-5, the proprietary extract pioneered by the Swedish Herbal Institute, which, despite its name's implication, is actually a for-profit business in Göteborg, Sweden, owned by my old friend and colleague, Georg Wikman. He probably has done more than anyone to popularize this important medicinal plant via his company's pioneering and long-term investments in

---

* The author also respectfully acknowledges the tireless and pioneering work of this late Georgian researcher and entrepreneur, Zakir Ramazanov, PhD, who made available English translations of numerous Russian-language research papers on *R. rosea* and who helped increase knowledge and awareness about its health benefits.

pharmacological and clinical research on his proprietary extract. Such credit also must be shared with another friend, Dr. Alexander Panossian, Chief Science Officer at SHI, who has worked tirelessly over the past decades in conducting clinical pharmacological research, clinical trials, and review articles on *Rhodiola* (and other adaptogens).

It warrants mention that the information in Chapter 9 is specific to the SHI *Rhodiola* extract, and may or may not be applicable to other proprietary or generic preparations made from *R. rosea* root. This raises the important consideration of *phytoequivalence*, an issue of considerable concern and interest in the botanical medicine community. That is, to what extent, if any, can positive clinical trial results based on a proprietary phytomedicinal formulation be extrapolated to other formulations, which—although all may appear somewhat similar in the marketplace to the uninformed consumer or even to the health professional or researcher—may or may not reflect the safety and/or activity of the generic formulation? Or, to express this matter in the reverse, when, if ever, is it appropriate for the marketer of a "generic," nonclinically tested *Rhodiola* product to attempt to justify such a product's potential benefits or actual market claims by "borrowing" the science from the leading clinically tested proprietary product? In most cases, it is not appropriate, unless, perhaps, the generic product is able to demonstrate a chemical profile similarity and biological equivalence via some form of appropriate *in vitro* and/or *in vivo* studies. If this were the case, many phytomedicinal experts would suggest that such a generic product be subjected to appropriate human clinical trials to substantiate its market claims.

The scope of this book is both extensively broad and detailed, covering virtually all aspects of *R. rosea* as a modern phytomedicinal agent. The book is clearly the most detailed, technical, and up-to-date publication on *Rhodiola*, which most scientific and market indicators suggest will become a widely used adaptogen. Eventually—just as the common name/genus name *Echinacea* has become commonplace—*Rhodiola* is clearly destined to become a household word in industrialized nations, and beyond, in the coming decade.

This book will become the touchstone and *the* essential reference for anyone serious about *R. rosea* as a future medicine. And, whether or not *Rhodiola* becomes the "Herbal Breakthrough of the 21st Century," as hyperbolically stated in the subtitle of the above-mentioned consumer book, this volume clearly will provide researchers, health professionals, industry members, and others with a wide perspective and deep level of reliable information on this increasingly popular and beneficial medicinal plant.

Mark Blumenthal

Founder & Executive Director
American Botanical Council
Editor-in-Chief, HerbalGram *& HerbClip*
Austin, Texas

Preface

This book has been written to provide a critical review and analysis of *Rhodiola rosea* suitable for scientists, herbalists, physicians, naturopaths, homeopaths, students, growers, and others with an interest in traditional herbal medicine. The idea of the book was born at the 7th Natural Health Product Research Society of Canada Conference May 2010 in Halifax, Nova Scotia, Canada, where a session was devoted to *Rhodiola rosea*. We want to thank Taylor & Francis Group for making this idea a reality.

R. rosea has increasingly become important in recent years, mainly because of its medicinal and adaptogenic properties. There are gaps in the literature regarding the basic biology of *Rhodiola*, taxonomy at the subspecies level, pharmacological effects, phytochemistry, and clinical trials. Information on *R. rosea* has increased as more papers are being published every year, but studies are not always based upon similar extracts or concentrations of metabolites as well as divergent posologies, making comparisons difficult among assays. Our purpose in writing this book was to pull together and analyze valid available information on *R. rosea*. Readers and researchers can use this book as a starting point for new ideas and projects regarding this adaptogenic plant.

Chapters 1 and 2 cover botany, taxonomy, ethnobotany, and classification of *Rhodiola*. One chapter deals with the chemistry and phytochemistry of *R. rosea*. Four chapters are devoted to *R. rosea* agronomy, biotechnology, *in vitro* culture, and pests and diseases in Europe, Canada, and Alaska. The section on *Rhodiola* pharmacology covers pharmacological bioassays, meta-analysis of clinical trials, toxicology, and uses of *R. rosea* in clinical practice. The final chapter uses a model to illustrate the cultivation of *R. rosea* as an industrial crop from field to medicine cabinet.

As many of our contributors observed, writing their chapter was not straightforward. Being a new crop, we discovered that more scientific information has been published in recent years. Some references were obtained from previously untranslated Russian studies on *R. rosea*. All chapters were subjected to a peer-review process. At the end, the authors breathed a long sigh of relief and took great pride in their chapters.

We thank the authors included in this book and the reviewers we consulted for their contributions. The reviewers included Dr. Dagmara Head, Dr. Alphonsus Utioh, Dr. Michael Harding, Dr. Robert Conner, Dr. Steffany Bennett, Dr. Julie Daoust, Dr. Anja Hohtola, Dr. Denis Charlebois, Dr. Brian Foster, Dr. Annamaria Mészáros, and Dr. Gary Martin. We are grateful to Calli Stromner for her editorial help. A special word of thanks should go to Dr. Susan Lutz for putting together the workshop and AVAC Ltd., an Alberta-based investment company who identifies promising early stage commercial ventures in value-added agribusiness and who enabled the attendance of some of the participants. Also, we wish to thank Dr. Roland Hardman, John Sulzycki, and other members of the Taylor & Francis Group for their continuing support and help throughout the production of this book.

Alain Cuerrier
Montreal, Canada

Kwesi Ampong-Nyarko
Edmonton, Canada

Editors

Dr. Alain Cuerrier, researcher at the Montreal Botanical Garden and writer, earned his PhD in plant systematics (University of Montreal with one year at Harvard University) before switching to ethnobotany in 2001. He participated in the creation of the First Nations Garden in Montreal as well as in Laquenexy (France). Since then, he has started ethnobotanical and ethnoecological projects with the Inuit, Innu, Naskapi, and Cree people. As a member of the Canadian Institute Health Research Team in Aboriginal Antidiabetic Medicines, he has been active in traditional medicine since 2003. Dr. Cuerrier is a member of the Plant Biology Research Institute, an adjunct professor at the University of Montreal, and a member of ArcticNet and Quebec Centre for Biodiversity Science. He has been vice president of the Natural Health Product Research Society of Canada from 2010 to 2013 and is now the president of the International Society of Ethnobiology. Dr. Cuerrier has published more than seven books on plant uses by First Nations and Inuit of Canada.

Dr. Kwesi Ampong-Nyarko is a research scientist at Alberta Agriculture and Rural Development Government of Alberta, Canada. He earned his PhD in agricultural botany from the University of Reading, England. From 2004 to 2010, Dr. Ampong-Nyarko developed the technology for growing *Rhodiola rosea* and made great strides in popularizing and laying the foundation for the its cultivation and commercialization in Alberta. He has multidisciplinary background with demonstrated expertise in crop production, weed science, and International Agriculture Development. He worked at the International Rice Research Institute, Los Baños, Philippines; International Centre of Insect Physiology and Ecology, Kenya; International Maize and Wheat Improvement Center, (CIMMYT) Mexico. Dr. Ampong-Nyarko has worked as a consultant for the FAO of the United Nations. He is the author of *A Handbook for Weed Control in Rice*—a comprehensive textbook that provides practical information on weed control in rice worldwide. His current research focus is on crop diversification in the field and greenhouse crops. He currently resides in Edmonton, Alberta.

Contributors

Fida Ahmed
Ottawa-Carleton Institute of Biology
University of Ottawa
Ottawa, Ontario, Canada

Kwesi Ampong-Nyarko
Alberta Agriculture and Rural
    Development
Edmonton, Alberta, Canada

Mariannick Archambault
Département de Sciences Biologiques
Université de Montréal
Montréal, Québec, Canada

John T. Arnason
Ottawa-Carleton Institute of Biology
University of Ottawa
Ottawa, Ontario, Canada

Steffany A.L. Bennett
Department of Biochemistry,
    Microbiology and Immunology
University of Ottawa
Ottawa, Ontario, Canada

Richard P. Brown
Department of Psychiatry
Columbia University College of
    Physicians and Surgeons
New York, New York

Anne Bruneau
Département de Sciences Biologiques
Université de Montréal
Montréal, Québec, Canada

Brandie Bugiak
Alberta Innovates Technology Futures
Vegreville, Alberta, Canada

Alain Cuerrier
Institut de Recherche en Biologie
    Végétale
Université de Montréal
Montréal, Québec, Canada

Vicky Filion
Ottawa-Carleton Institute of Biology
University of Ottawa
Ottawa, Ontario, Canada

Bertalan Galambosi
Plant Production Research
Agrifood Research Finland
Mikkeli, Finland

Patricia L. Gerbarg
Department of Psychiatry
New York Medical College
Valhalla, New York

Zsuzsanna György
Department of Genetics and Plant
    Breeding
Corvinus University of Budapest
Budapest, Hungary

Ron J. Howard
Crop Diversification Centre South
Alberta Agriculture and Rural
    Development
Brooks, Alberta, Canada

Sheau-Fang Hwang
Crop Diversification Centre North
Alberta Agriculture and Rural
    Development
Edmonton, Alberta, Canada

Petra A. Illig
Anchorage, Alaska

Raimar Loebenberg
Faculty of Pharmacy and
    Pharmaceutical Sciences
University of Alberta
Edmonton, Alberta, Canada

Alexander Panossian
Research and Development
Swedish Herbal Institute
Vallberga, Sweden

Michel Rapinski
Institut de Recherche en Biologie
    Végétale
Université de Montréal
Montréal, Québec, Canada

Ammar Saleem
Ottawa-Carleton Institute
    of Biology
University of Ottawa
Ottawa, Ontario, Canada

Hugh Semple
Defence Research and Development
    Canada
Suffield Research Centre
Medicine Hat, Alberta, Canada

Nav Sharma
Alberta Innovation and Advanced
    Education
Edmonton, Alberta, Canada

Stephen E. Strelkov
Department of Agricultural, Food
    and Nutritional Science
University of Alberta
Edmonton, Alberta, Canada

Youri Tendland
Institut de Recherche en Biologie
    Végétale
Université de Montréal
Montréal, Québec, Canada

Georg Wikman
Research and Development
Swedish Herbal Institute
Vallberga, Sweden

1 Taxonomy of *Rhodiola rosea* L., with Special Attention to Molecular Analyses of Nunavik (Québec) Populations

Alain Cuerrier, Mariannick Archambault,
Michel Rapinski, and Anne Bruneau

CONTENTS

1.1 Introduction 2
    1.1.1 *Rhodiola*: Origin, Distribution, and Description................................3
    1.1.2 *R. rosea* in Nunavik (Québec) 5
1.2 Materials and Methods 5
    1.2.1 Sampling...5
    1.2.2 DNA Extraction 9
    1.2.3 ITS and *trn*L-F Sequences..9
    1.2.4 Parsimony Analyses...15
    1.2.5 Amplified Fragment Length Polymorphism....................................15
    1.2.6 AFLP Analyses..16
1.3 Results..17
    1.3.1 ITS and *trn*L-F Sequence Results 17
    1.3.2 AFLP Results..18
1.4 Discussion..21
    1.4.1 Taxonomic Distinction and Medicinal Implications 21
    1.4.2 Infraspecific Taxonomy and Confusion..25
    1.4.3 Phylogeography, Environment, and Population Structure................27
1.5 Conclusion 28
Acknowledgments...29
References...29

1.1 INTRODUCTION

The medicinal plant *Rhodiola rosea* L., known as roseroot, golden root, or Arctic root, has a long history of traditional medicinal use in Eurasia (Kelly 2001; Brown et al. 2002; Alm 2004) and North America (Blondeau et al. 2010; Cuerrier and Elders of Kangiqsualujjuaq 2011). Although the species has been highlighted as a promising Canadian medicinal crop (Small and Catling 2000), North American populations of *R. rosea* have not been studied as extensively as European populations (Olfelt et al. 2001; Filion et al. 2008; Avula et al. 2009) and remain the subject of little research. This is of particular concern as *R. rosea* has a complex taxonomic history. It has been given more than 20 names and in North America, the western roseroot, *Rhodiola integrifolia* Raf., has sometimes been considered a subspecies or variety of *R. rosea* (Table 1.1), and this infraspecific nomenclature has been used in some regional floras (Polunin 1940; Scoggan 1978; Hultén and Fries 1986; Gleason and Cronquist 1991; Roland and Zinck 1998; Cody 2000; Aiken et al. 2003; Dodson and Dunmire 2007). This taxonomic confusion can lead to unconscious adulteration of commercialized products, which, in this case, is of special interest as, to date, no medicinal properties have been reported for *R. integrifolia*. In this chapter, first we will give an overview of the taxonomy (and distribution) of the genus *Rhodiola*, and second we will elucidate the genetic diversity of *R. rosea* populations in Nunavik (low Arctic region of Northern Québec, Canada), and refer to the taxonomy proposed by Ohba (2002).

TABLE 1.1

Synonyms for *R. integrifolia* and *R. rosea*

Species	Synonym
Rhodiola integrifolia Raf.	*Rhodiola rosea* subsp. *integrifolia* (Rafinesque) H. Hara
	Sedum integrifolium (Rafinesque) A. Nelson
	Sedum integrifolium (Rafinesque) A. Nelson subsp. *integrifolium*
	Rhodiola rosea var. *integrifolia* (Rafinesque) Jepson
	Sedum rosea var. *integrifolia* (Rafinesque) A. Berger
	Tolmachevia integrifolia (Rafinesque) Á. Löve and D. Löve
	Sedum rosea subsp. *integrifolium* (Rafinesque) Hultén
Rhodiola rosea L.	*Sedum rosea* (L.) Scopoli
	Sedum rhodiola de Candolle
	Rhodiola roanensis (Britton) Britton
	Sedum roanensis (Britton)
	Sedum rosea var. *roanense* (Britton) A. Berger
	Sedum rosea (L.) Scopolli subsp. *rosea*
	Sedum rosea (L.) Scopolli var. *rosea*

Source: Brouillet, L. et al., *VASCAN, the Database of Vascular Plants of Canada*, 2010+.

1.1.1 *Rhodiola*: Origin, Distribution, and Description

The Angiosperm Phylogeny Group lists 90 species belonging to the genus *Rhodiola* (Stevens 2001 onward). Though this estimate is commonly supported in the scientific literature (Fu and Ohba 2001; Guest 2009; Hermsmeier et al. 2012; Liu et al. 2013), numbers are nonetheless extremely variable, varying from as low as 60 (Mayuzumi and Ohba 2004; Gontcharova et al. 2009; Zhang et al. 2014) to as high as 200 species (Elameen et al. 2008; Tasheva and Kosturkova 2011, 2012). The genus is thought to have originated in the mountainous regions of southwest China and the Himalayas (Brown et al. 2002), where it is mainly distributed (Mayuzumi and Ohba 2004). Not surprisingly, the flora of China lists 55 species, 16 of which are endemic to the country (Fu and Ohba 2001). *Rhodiola rosea*, the most studied species of its genus, is thought to have originated in the southern Siberia highlands (Kozyrenko et al. 2011).

R. rosea is an amphi-Atlantic species found in Asia (from Russia to Japan), Europe (Scandinavia, United Kingdom, most of the mountains of Central Europe, including the Pyrenees to those found in southern Bulgaria), Iceland, Greenland, and in North America (Figure 1.1; Polunin 1969; Komarov 1971; Hultén and Fries 1986; Ohba 2002; Galambosi 2006). Altogether, it is reported in 28 countries in the northern hemisphere (Kylin 2010). It is considered an Arctic-Alpine species even though its distribution includes some southern localities. The species occurs in tundra, borders of brooks and river banks, slopes, cliffs, often in isolated rock crevices, or in association with moss and other local vegetation like sedges and grasses; it can tolerate low organic substratum, but dwells well in richer ones (Aiken et al. 2003; Kylin 2010; Tasheva and Kosturkova 2012). In North America, *R. rosea* occurs in a few locations in Nunavut, but mostly along the East coast of Nunavik to North Shore (Québec), Labrador, Newfoundland, New Brunswick and Nova Scotia, Maine and in some isolated populations south to Pennsylvania (Polunin 1940; Clausen 1975; Roland and Zinck 1998; United States Department of Agriculture 1999; Hinds 2000; Moran 2009).

FIGURE 1.1 Global distribution of *Rhodiola rosea* (in dark gray).

According to the Flora of North America (Moran 2009), *R. rosea* (Crassulaceae) is a perennial, succulent herbaceous plant with a characteristic oblong or cylindrical, thick (0.5–2.5 cm in diameter), fleshy, fragrant rhizome characteristic of rose. Leaves are pale green, ovate, obovate, elliptic-oblanceolate or oblong (1–5 cm × 0.4–1.5 cm), fleshy, glabrous, and usually glaucous; the leaf margin is entire or dentate, and the apex acute or obtuse. The flowering stems produce dense terminal inflorescences in corymbose cymes containing up to 150 flowers and reaching 6.5 cm in diameter. The flower buds are yellow to red, and the petals turn yellow before anthesis. The species is diploid ($2n = 22$) (Amano et al. 1995) and has been reported to be generally dioecious, though monoecious plants and hermaphroditic flowers are also known (Small and Catling 2000). The plant can thus be seen as functionally dioecious. The flowers are 4(-5)-merous, with petals that are oblong, 1–3.5 mm, and shorter than stamens; the petals are erect on pistillate flowers, and they are spreading, 0.7- to 1.1-mm-wide, on staminate flowers. Seeds, born from 4- to 9-mm-long follicles with spreading beaks, are slightly winged at both ends, 1.7- to 2.2-mm long.

Three *Rhodiola* species are recognized in the Flora of North America (Moran 2009): *R. rhodantha* (A. Gray) H. Jacobsen, *R. rosea*, and *R. integrifolia*. The latter two have historically been included together as subspecies or varieties of *R. rosea*, and have hence been considered as one and the same species (Table 1.1). Yet, the two taxa are most distinctly distinguished by genetic differences, with *R. integrifolia* possessing a chromosome number of $2n = 36$, instead of $2n = 22$ (Clausen 1975; Amano et al. 1995; Moran 2000). Morphologically, *R. rosea* is known to be quite variable (Serebryanaya and Shipunov 2009; Kylin 2010) and difficult to differentiate from *R. integrifolia* during the vegetative phase. After flowering, seed coat morphology is not considered a reliable tool for identification (Gontcharova et al. 2009). Although Clausen (1975) found few absolute morphological distinctions between the two taxa, his best key characters, notably, petal width of staminate flowers and flower color, remain those used by the Flora of North America to differentiate the two species. *Rhodiola integrifolia*, has purple to reddish flowers and wider petals in staminate flowers (1.3–1.7 mm).

Despite an overlapping range for the two species in Siberia (Amano et al. 1995), *R. integrifolia* is an amphi-beringian species. Its distribution is limited to western North America, from Alaska to New Mexico (Hermsmeier et al. 2012), and a general geographic dichotomy exists between both species. This attribute, however, cannot entirely be relied on to distinguish the taxa as isolated populations of *R. integrifolia* have been confirmed, by chromosomal counts, in the state of Minnesota and New York (Moran 2009; Hermsmeier et al. 2012). The presence of *R. rosea* in western North America also remains ambiguous. Though these have not been confirmed by chromosomal counts, the Flora of North America reports the presence of isolated populations far into western Alaska (Moran 2009). Genetic studies of herbarium specimen from Alaska and the Bering Sea have since found them all to be *R. integrifolia*, including a specimen with yellow inflorescences from western Alaska (Hermsmeier et al. 2012). Nonetheless, herbaria still report, among their collections, occurrences of *R. rosea* in western North America (GBIF 2013).

The current classification of *R. rosea* does not include any subspecies or varieties in North America (Moran 2009; Brouillet et al. 2010+), but the Flora of China lists two varieties, *R. rosea* var. *rosea* and *R. rosea* var. *microphylla* (Fu and Ohba 2001). In a revision of Asiatic *Rhodiola* species, Ohba (1981) clearly demonstrated the taxonomic challenges associated with *R. rosea* due to its extremely high morphological variability. Although multiple varieties had been proposed, Ohba (1981) did not establish infraspecific taxa because the variation, which was thought to show geographical consistency, was no longer sufficient to support such divisions. Though there is no formally accepted infraspecific classification for the species, some botanists continue to recognize subspecific taxa. Gontcharova et al. (2009) recognize three subspecies: *R. rosea* ssp. *rosea*, *R. rosea* ssp. *arctica*, and *R. rosea* ssp. *sachalinensis*. A recent study, which noted differences among sequences of the internal transcribed spacer (ITS) regions between *R. rosea* and other species, found that the *R. rosea* sequences could not be differentiated from those of *R. iremelica* Boriss., suggesting that this species, endemic to the southern Urals (Yanbaev et al. 2007), may be a subspecies of *R. rosea* (György et al. 2012). Indeed, high genetic variability exists at the species level (Elameen et al. 2008, 2010; Soni et al. 2010; Kozyrenko et al. 2011; György et al. 2012), and an analysis of its genetic structure from populations ranging from Poland to the Kamchatka Peninsula, in the Russian Far East, support the existence of at least two distinct evolutionary lines (Kozyrenko et al. 2011).

1.1.2 *R. rosea* in Nunavik (Québec)

Exacerbating the taxonomic confusion regarding *R. rosea*, almost no genetic work has been done that includes samples from Eastern Canadian populations (see Guest 2009 who includes a few samples in an extensive molecular study of *R. integrifolia*). Here, we present some molecular analyses that compare the Nunavik populations to European ones. We sequenced the plastid *trn*L-F region and the nuclear ribosomal ITS, which are known for their high rate of nucleotide substitution and their usefulness in resolving infrageneric relationships in other plant taxa (Taberlet et al. 1991; Muschner et al. 2003). In addition, we evaluate regional genetic variation to examine phylogeographic relationships among regions. We are particularly interested in differences among regions along Ungava Bay in Québec. We also verified whether the North American populations had differentiated in the short period since the last glaciations and whether geographic and genetic distances were correlated. This information will help to better understand the relationship between Nunavik and other populations and to elaborate dispersal hypotheses for this species. To achieve this, we used the amplified fragment length polymorphism (AFLP) technique elaborated by Vos et al. (1995). This technique is relatively easy, fast, reliable, and useful in genetic variation analyses below the species level (Mueller and Wolfenbarger 1999; Boninet al. 2007; Meudt and Clarke 2007).

1.2 MATERIALS AND METHODS

1.2.1 Sampling

The sampled taxa included *R. rosea* and three other species used as outgroups: *R. integrifolia*, *R. kirilowii*, and *R. seminowii* (Table 1.2). Figure 1.2a shows an overview of the sampling in North America. Fieldwork was undertaken principally around the shore of Ungava Bay in Nunavik, Québec (Figure 1.2b). Each of the collected

TABLE 1.2
Rhodiola **Samples Sequenced for the Nuclear Ribosomal ITS and *trn*L-F Chloroplast Regions**

Species	Voucher Information	Origin Location, Country	Latitude, Longitude or Origin Specifications		ITS	*trn*L-F
R. rosea (440)	Archambault and Filion (RR2006-MM)	Quaqtaq (Cape Hope Advance), Canada	N 61 07 816	W 069 55 179	GQ374182	GQ374209
R. rosea (42)	Archambault, Cuerrier and Filion (RR2005-C)	Kangirsualujjuaq, Canada	N 58 30 969	W 065 57 962	GQ374184	GQ374211
R. rosea (240)	Archambault and Filion (RR2006-CC)	Killiniq, Canada	N 60 42 201	W 064 83 934	GQ374181	GQ374208
R. rosea (140)	Archambault and Filion (RR2005-M)	Kuujjuaq, Canada	N 58 06 697	W 068 19 541	GQ374185	GQ374212
R. rosea (196)	Charest et al. (96-578)	Swale Island, Newfoundland, Canada	N 48 61 534	W 053 69 197	GQ374183	GQ374210
R. rosea (670)	Archambault (RR2006-NB-cap)	Cap Enragé, New Brunswick, Canada	N 45 35 43	W 064 46 40	GQ374189	GQ374216
R. rosea (197)	Charest et al. (NDC99-391)	Limit Québec-Labrador, Canada	N 51 41 743	W 057 10 510	GQ374186	GQ374213
R. rosea (182)	Parks Canada (RR2005-S)	Mingan Archipelago (Grande Île), Canada	N 50.241	W 63.914	GQ374188	GQ374215
R. rosea (189)	Sunniva Aagaard (RR-2005-Nw)	Baktifjellet, Selbu, Norway	N 63 26 44	E 10 19 52	GQ374191	GQ374218
R. rosea (787)	Galambosi (RR2008-G17)	Altäi, Russia	Seedlings from Tomsk Botanical Garden		GQ374201	GQ374228
R. rosea (706)	Galambosi (RR2008-G1)	Kilpisjärvi, Jeähkäjärvi, Finland	From natural populations		GQ374192	GQ374219
R. rosea (704)	Galambosi (RR2008-G18)	Hirvas, Finland	Plants from Kaloti nursery		GQ374193	GQ374220
R. rosea (779)	Galambosi (RR2008-G15)	Adanello, Italy	Seedlings from seeds from Toronto		GQ374195	GQ374222
R. rosea (780)	Galambosi (RR2008-G16)	Bondelo, Italy	Seedlings from seeds from Toronto		GQ374196	GQ374223
R. rosea (781)	Galambosi (RR2008-G13)	Alps, Austria	Plants from nature from Obertauern		GQ374197	GQ374224
R. rosea (782)	Galambosi (RR2008-G12)	Germany	Seed from Kiel Botanical Gardens		GQ374198	GQ374225

R. rosea (783)	Galambosi (RR2008-G11)	Sweden	Commercial seed from the Swedish company Impecta	GQ374199	GQ374226
R. rosea (807)	Galambosi (in cultivation)	Norway	Plants from Särkä nursery	GQ374190	GQ374217
R. rosea (784)	Galambosi (in cultivation)	Mattark, Switzerland	Seedlings from RAC	GQ374200	GQ374227
R. rosea (792)	Thompson (SLT#2007-003)	Peggy's Cove, Nova Scotia, Canada	N 44 29 31　　W 063 55 5	GQ374187	GQ374214
R. rosea (701)	Thompson (SLT#2007-001)	Tatrzanski Park Narodowy, Poland	N 49 15 04　　E 19 54 04	GQ374194	GQ374221
R. rosea (808)	Kwesi (in cultivation)	Russia	Cultivated plants in Alberta (Kwesi)	GQ374181	GQ374229
R.integrifolia	Cody and Ginns (34697)	Reindeer Mountain, Yukon, Canada	N 63 37　　W 139 22	GQ374205	GQ374232
R.integrifolia	Guest, Heidi (046085) (UVIC)	La Plata Canyon, Colorado, USA	N 37 25 938　　W 108 02 124	GQ374204	GQ374231
R.integrifolia	Jorgenson (RI2007-001)	Mt. Chiginagak, Alaska, USA	N 57 10 707　　W 157 06 786	GQ374203	GQ374230
R.kirilowii	Archambault (MA-2008-01)	Montreal Botanical Garden (371-2007)	Seeds from cultivated materials of Botanischer Garten des Landes Karnten, Klagenfurt, Austria	GQ374206	GQ374233
R.semenovii	Archambault (MA-2008-02)	Montreal Botanical Garden (372-2007)	Seeds from cultivated materials of Botanischer Garten des Landes Karnten, Klagenfurt, Austria	GQ374207	GQ374234

The number in parentheses corresponds to the sample DNA extraction number. For each sample, locality, collector, and voucher information are given. Specimens are deposited at Marie-Victorin herbarium (MT), except one specimen of *R. integrifolia* from Colorado, which is at UVIC. Genbank accession number are given for ITS and *trn*L-F. Samples from Galambosi were cultivated by Bertalan Galambosi, Agrifood Research Finland.

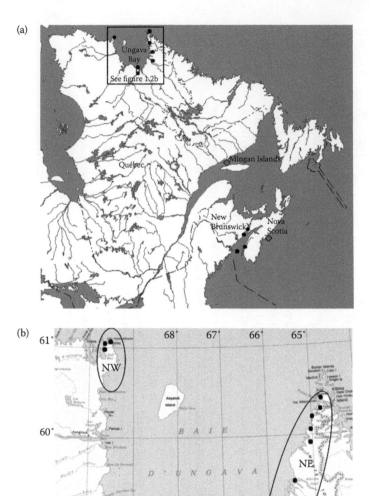

FIGURE 1.2 Sampling of *Rhodiola rosea* in (a) North America and (b) along the shore of Ungava Bay, Nunavik, Québec. (a) Circles represent populations that we collected and lozenges show populations collected by collaborators. (b) Empty circles represent populations that were collected in 2005 and full circles are populations collected in 2006. Three regions are defined relative to their proximity: NW for West Nunavik, NE for East Nunavik and NS for South Nunavik. This map is modified from the regional map of Québec MCR 42 of Energy, Mines and Resources Canada.

sites was considered a distinct population, bearing in mind that most populations occurred on separate islands or were separated by several kilometers. All populations were coastal, mostly found on the east shore of Ungava Bay, except for the three populations found at Quaqtaq. We collected leaves in silica gel from a total of 32 populations, with either 10 or 20 samples per population. Three populations from New Brunswick and two from Nova Scotia were also collected. Eight samples from Mingan Islands and those from Finland, Italy, Norway, Sweden, Austria, Germany, Switzerland, Poland, and Russia were obtained with the help of collaborators for a wide representation of *R. rosea*. The sampling for the sequence analyses of the ITS and *trn*L-F regions included 27 individuals, 22 of which were from different populations of *R. rosea* and five individuals representing the three outgroup species (Table 1.2). Over 700 samples were amplified for AFLP analyses; a breakdown of the number of samples and populations per geographic region is presented in Table 1.3.

1.2.2 DNA EXTRACTION

DNA extractions were performed on leaves dried in silica gel using a protocol based on the work of Xia et al. (2005) and modified from the CTAB protocol of Doyle and Doyle (1987). We added 1% polyvinylpyrrolidone and 40 µg of RNAse A to the extraction buffer. The chloroform-isoamyl alcohol extraction was performed twice and the precipitation was followed by a 20-minute centrifugation at 14,000 rpm at 4°C. The pellets were washed using 500 µL of 70% ETOH or wash buffer (76% ETOH, 10 mM ammonium acetate) twice, dried, and suspended again in 100 µL $TE_{0.1}$ (10 mM Tris–HCl, 1 mM EDTA disodium, pH 7.4), and then incubated at 65°C for at least 10 minutes to dissolve the DNA.

1.2.3 ITS AND *TRN*L-F SEQUENCES

Two regions were sequenced, the nuclear ribosomal ITS region of the 18S–26S and the chloroplast *trn*L-F region (containing the *trn*L intron, *trn*L 3' exon, and *trn*L-*trn*F intergenic spacer). The polymerase chain reaction (PCR) was performed in a Perkin-Elmer GeneAmp PCR System 9700 Thermocycler (Applied Biosystems [ABI], Foster City, California) or a 2720 Thermal Cycles (ABI). Amplification of the ITS region was done in a total volume of 25 µL containing 1× PCR buffer (Roche Diagnostics, Laval, Canada) (with a total 1.5 mmol/L $MgCl_2$ concentration), 200 µmol/L of each dNTP (MBI Fermentas, Burlington, Canada), 0.1 µg/µL bovine serum albumin (BSA) (New England Biolabs, Pickering, Canada), 6% of DMSO, 0.06% of Tween-20, 0.4 µmol/L of each primer and two units of *Taq* polymerase with 1 µL of total DNA, using the primers AB101 and AB102 (Douzery et al. 1999). The PCR protocol consisted of an initial denaturation step (3 minutes at 94°C) followed by 40 cycles of denaturation (30 seconds at 94°C), annealing (30 seconds at 50°C) and elongation (1 minute at 72°C), with a final extension (7 minutes at 72°C) step.

    PCR for the *trn*L-F region was done in a total volume of 25 µL containing 1× PCR buffer, a total 2 mmol/L $MgCl_2$ concentration, 200 µmol/L of each dNTP (MBI Fermentas), 0.4 µmol/L of each primer and two units of *Taq* polymerase with 1 µL of total DNA, using the primers *trn*L-c and *trn*L-f (Taberlet et al. 1991). The PCR

TABLE 1.3
Rhodiola rosea Samples Used for AFLP Analysis

Rg	Pop	No	H	L	H-Rg	L-Rg	Locality	Latitude, Longitude	Voucher Information
NE	a	10	10	48	258	74	Canada, Québec, Ungava Bay, Kangirsualujjuaq (around)	58.69258 65.95013	Archambault et al. (RR2005-A)
	b	9	9	47			Canada, Québec, Ungava Bay, Kangirsualujjuaq (around)	58.67742 66.01660	Archambault et al. (RR2005-B)
	c	8	8	32			Canada, Québec, Ungava Bay, Kangirsualujjuaq (around)	58.51160 65.96603	Archambault et al. (RR2005-C)
	d	8	8	41			Canada, Québec, Ungava Bay, Kangirsualujjuaq (around)	58.55162 65.93748	Archambault et al. (RR2005-D)
	e	7	7	45			Canada, Québec, Ungava Bay, Kangirsualujjuaq (around)	58.59725 65.89087	Archambault et al. (RR2005-E)
	f	10	10	37			Canada, Québec, Ungava Bay, Kangirsualujjuaq (around)	58.67680 66.00843	Archambault et al. (RR2005-F)
	g	6	6	32			Canada, Québec, Ungava Bay, Kangirsualujjuaq (around)	58.69422 65.94262	Archambault et al. (RR2005-G)
	aa	19	19	43			Canada, Québec, Ungava Bay, Kangirsualujjuaq (around)	58.74270 66.02954	Archambault and Filion (RR2006-AA)
	bb	17	17	51			Canada, Québec, Ungava Bay, Kangirsualujjuaq (around)	58.75073 66.03168	Archambault and Filion (RR2006-BB)
	cc	15	15	47			Canada, Québec, Ungava Bay, Killiniq	60.42201 64.83934	Archambault and Filion (RR2006-CC)
	dd	16	16	50			Canada, Québec, Ungava Bay, shore between Kangirsualujjuaq and Killiniq	60.35905 64.85030	Archambault and Filion (RR2006-DD)
	ee	18	18	52			Canada, Québec, Ungava Bay, shore between Kangirsualujjuaq and Killiniq	60.22697 64.96510	Archambault and Filion (RR2006-EE)

	ID						Lat	Long	Locality	Source
	ff	18	18	48			60.09253	65.07035	Canada, Québec, Ungava Bay, shore between Kangirsualujjuaq and Killiniq	Archambault and Filion (RR2006-FF)
	gg	17	17	52			59.92101	65.02571	Canada, Québec, Ungava Bay, shore between Kangirsualujjuaq and Killiniq	Archambault and Filion (RR2006-GG)
	hh	16	16	36			59.30652	65.66747	Canada, Québec, Ungava Bay, shore between Kangirsualujjuaq and Killiniq	Archambault and Filion (RR2006-HH)
	ii	15	15	48			59.49055	65.51294	Canada, Québec, Ungava Bay, shore between Kangirsualujjuaq and Killiniq	Archambault and Filion (RR2006-II)
	jj	18	18	49			59.05420	65.74654	Canada, Québec, Ungava Bay, shore between Kangirsualujjuaq and Killiniq	Archambault and Filion (RR2006-JJ)
	kk	18	18	49			58.93758	65.97757	Canada, Québec, Ungava Bay, shore between Kangirsualujjuaq and Killiniq	Archambault and Filion (RR2006-KK)
	ll	17	17	49			59.06207	65.72964	Canada, Québec, Ungava Bay, shore between Kangirsualujjuaq and Killiniq	Archambault and Filion (RR2006-LL)
NS	h	7	7	32	96	65	58.02645	68.44675	Canada, Québec, Ungava Bay, Kuujjuaq (along Koksoak river)	Archambault and al. (RR2005-H)
	i	9	9	34			58.14810	68.33710	Canada, Québec, Ungava Bay, Kuujjuaq (along Koksoak river)	Archambault and Filion (RR2005-I)
	j	10	10	43			58.35753	68.19277	Canada, Québec, Ungava Bay, Kuujjuaq (along Koksoak river)	Archambault and Filion (RR2005-J)
	k	10	10	34			58.44615	68.19597	Canada, Québec, Ungava Bay, Kuujjuaq (along Koksoak river)	Archambault and Filion (RR2005-K)
	l	9	9	28			58.27043	68.28210	Canada, Québec, Ungava Bay, Kuujjuaq (along Koksoak river)	Archambault and Filion (RR2005-L)
	m	8	8	32			58.11162	68.32568	Canada, Québec, Ungava Bay, Kuujjuaq (along Koksoak river)	Archambault and Filion (RR2005-M)
	n	6	6	31			58.10328	68.10328	Canada, Québec, Ungava Bay, Kuujjuaq (along Koksoak river)	Archambault and Filion (RR2005-N)
	pp	17	17	44			58.53354	68.12910	Canada, Québec, Ungava Bay, Kuujjuaq (along Koksoak river)	Archambault and Filion (RR2006-PP)

(Continued)

TABLE 1.3 *(Continued)*
Rhodiola rosea Samples Used for AFLP Analysis

Rg	Pop	No	H	L	H-Rg	L-Rg	Locality	Latitude, Longitude	Voucher Information
	qq	14	14	44			Canada, Québec, Ungava Bay, Kuujjuaq (along Koksoak river)	58.55163 68.18744	Archambault and Filion (RR2006-QQ)
	rr	6	6	36			Canada, Québec, Ungava Bay, Kuujjuaq (along Koksoak river)	58.32655 68.25588	Archambault and Filion (RR2006-RR)
NW	mm	15	15	35	46	58	Canada, Québec, Ungava Bay, Quaqtaq	61.07816 69.55179	Archambault and Filion (RR2006-MM)
	nn	17	17	49			Canada, Québec, Ungava Bay, Quaqtaq	61.04693 69.63371	Archambault and Filion (RR2006-NN)
	oo	14	14	46			Canada, Québec, Ungava Bay, Quaqtaq	61.07090 69.60801	Archambault and Filion (RR2006-OO)
N	All				400	77			
Mg	p	4	4	31	40	64	Canada, Québec, Mingan Archipelago, Île Nue de Mingan	50.211 64.117	Parks Canada (RR2005-P)
	q	4	4	34			Canada, Québec, Mingan Archipelago, Petite île au Marteau	50.216 63.563	Parks Canada (RR2005-Q)
	r	3	3	39			Canada, Québec, Mingan Archipelago, Île du Havre	50.213 63.608	Parks Canada (RR2005-R)
	s	7	7	36			Canada, Québec, Mingan Archipelago, Grande île	50.241 63.914	Parks Canada (RR2005-S)
	ss	9	9	36			Canada, Québec, Mingan Archipelago, Île du Havre (2)	50.20803 63.6047	Parks Canada (RR2006-SS)
	tt	5	5	34			Canada, Québec, Mingan Archipelago, Île du Havre (1)	50.20767 63.60477	Parks Canada (RR2006-TT)

Section	Code					Latitude	Longitude	Collector	Locality
	uu	4	4	28		50.20945	63.55788	Parks Canada (RR2006-UU)	Canada, Québec, Mingan Archipelago, Petite île au Marteau
	vv	4	4	27		50.22098	63.68042	Parks Canada (RR2006-VV)	Canada, Québec, Mingan Archipelago, Île du Fantôme
Ma	GM	4	4	18	37 66	44.76417	66.73465	Archambault (RR2006-NB-GM)	Canada, New Brunswick, Grand Manan Island
	WQ	6	6	31		45.31513	65.55387	Archambault (RR2006-NB-WQ)	Canada, New Brunswick, West Quaco
	SC	8	8	28		44.49767	66.10502	Archambault (RR2006-NS-SC)	Canada, Nova Scotia, Sandy Cove
	CE	10	10	53		45.59390	64.77993	Archambault (RR2006-NB-cap)	Canada, New Brunswick, Cap Enragé
	NE	9	9	32		44.49674	63.91913	Thompson (SLT#2007-003)	Canada, Nova Scotia, Peggy's Cove
Sč	FK1	5	5	33	71 65	69.04647	20.79530	Galambosi (RR2008-G1)	Finland, Kilpisjärvi (Jeáhkájärvi)
	FK2	8	8	37		69.04647	20.79530	Galambosi (RR2008-G2)	Finland, Kilpisjärvi (Saana)
	FK3	5	5	38		69.04647	20.79530	Galambosi (RR2008-G3)	Finland, Kilpisjärvi (Saananmaja)
	FK4	7	7	29		69.04647	20.79530	Galambosi (RR2008-G4)	Finland, Kilpisjärvi (Tsahkaljärvi South)
	FK5	7	7	34		69.04647	20.79530	Galambosi (RR2008-G5)	Finland, Kilpisjärvi (Tsahkaljärvi North)
	FK6	7	7	36		69.04647	20.79530	Galambosi (RR2008-G6)	Finland, Kilpisjärvi (Marjajärvi)
	FH1	5	5	23		69.30733	21.26964	Galambosi (RR2008-G7)	Finland, Halti (Jogasjärvi)
	FH2	5	5	36		69.30733	21.26964	Galambosi (RR2008-G8)	Finland, Halti (Valtijoki)
	FH3	6	6	40		69.30733	21.26964	Galambosi (RR2008-G9)	Finland, Halti (Somaslompolo)
	FU	5	5	30		68.90486	27.02605	Galambosi (RR2008-G10)	Finland, Utsjoki (Goahppelasjoki)
	Norw	10	11	41		63.44568	10.33111	Sunniva Aagaard (RR2006-Nw)	Norway, Trøndelag
	Norw	1				60.47203	8.46895	Galambosi (in culture)	Norway (nursery)

(Continued)

TABLE 1.3 *(Continued)*
Rhodiola rosea **Samples Used for AFLP Analysis**

Rg	Pop	No	H	L	H-Rg	L-Rg	Locality	Latitude, Longitude	Voucher Information
Eur	Eur	1	6	51	6	51	Austria, Alps	*47.55483 14.74436*	Galambosi (RR2008-G13)
	Eur	1					Germany (Botanical Garden)	*54.32268 10.12775*	Galambosi (RR2008-G12)
	Eur	1					Italy, Valle d'Aperta	*46.60396 11.45397*	Galambosi (RR2008-G14)
	Eur	1					Italy, Bondolo	*46.13926 8.33197*	Galambosi (RR2008-G16)
	Eur	1					Poland, Tatrzanski Park Narodowy	*49.25111 19.90111*	Thompson (SLT#2007-001)
	Eur	1					Russia, Altaï	*50.83239 86.95778*	Galambosi (RR2008-G17)

For each population, number of samples, locality, collector, and voucher specimen deposited at Marie-Victorin herbarium (MT) are given. The localities where we estimated the latitude and longitude are in italic. Pop, population code; No, number of samples used for AFLP analysis; H, number of haplotypes per populations; L, number of polymorphic loci on a possibility of 88; Rg, geographic region (NE, Ungava Bay-East; NS, Ungava Bay-South; NW, Ungava Bay-West; Mg, Mingan Islands; Ma, Maritimes; Sc, Scandinavia; Eur, Eurasia); H-Rg, number of haplotypes per region; L-Rg, number of polymorphic loci per region.

conditions followed the ITS protocol except for the annealing temperature that was set to 48°C and only 35 cycles were used. The PCR products were purified using the polyethylene glycol (PEG) purification procedure as described by Joly et al. (2006) with the following modifications for centrifugation times. For the first precipitation, centrifugation was for 20 minutes rather than 15 minutes, and the centrifugation for the two washes was increased to 7 minutes.

Sequencing reactions using the same primers as above were performed with BigDye terminator (v. 1.1) (ABI) following the manufacturer's protocols. Sequences were run on a 3100-*Avant* automated sequencer (ABI). Sequencher (version 4.7) (GeneCodes, Ann Arbor, Michigan) was used to visualize and edit the sequences. Sequence alignment was performed manually with BioEdit v.1.6.0 (Hall 1999). Indels were coded by Seqstate v.1.32 software (Müller 2006) using the simple indel coding model (Simmons and Ochoterena 2000).

1.2.4 PARSIMONY ANALYSES

We used the parsimony criterion for the analysis, but maximum likelihood yielded similar results in preliminary analyses. The analysis was performed on the combined matrix (ITS + *trn*L-F), which included only individuals for which both sequences were available. The matrix with 27 sequences was analyzed with PAUP* (v.4.0; Swofford 2002) using a heuristic search strategy with 1,000 random addition replicates, with tree-bisection-reconnection (TBR) branch swapping and one tree held at each step during stepwise addition with a maximum of 10,000 trees in memory. A second analysis was performed using all trees in memory as starting trees with TBR branch swapping. Nodal support was estimated using 1000 bootstrap replicates in a heuristic search with TBR branch swapping. Strict consensus trees were obtained. Before combining the data, separate analyses were conducted on each matrix, and the two datasets were tested for incongruence with the incongruence length difference (ILD) test implemented in PAUP.*

1.2.5 AMPLIFIED FRAGMENT LENGTH POLYMORPHISM

Initial AFLP reactions were conducted with the restriction enzymes *Eco*RI and *Mse*I but because these lacked variation, we used the restriction enzymes *Mse*I and *Taq*αI, which are both four nucleotides and thus more frequent base cutters (Table 1.4).

TABLE 1.4
Adapters and Primers Used in AFLP Analyses with the Restriction Enzymes *Taq*I

Adapters *Taq*I	Top Strand	5′-GACGATGAGTCCTGAG
	Bottom strand	5′-CGCTCAGGACTCAT
Primers *Taq*I	Taq + 1	5′-GACGATGAGTCCTGAGCGA **C**
	Taq + CAG*Hex	5′-GATGAGTCCTGAGCGA **CAG**
	Taq + CTC*Fam	5′-GATGAGTCCTGAGCGA **CTC**
	Taq + CGC*	5′-GATGAGTCCTGAGCGA **CGC**

The reactions were done using a Perkin–Elmer GeneAmp PCR System 9700 thermocycler (ABI). Primers were ordered from Alpha DNA (Montréal). AFLP analyses were conducted following the protocol described by Soltis and Gitzendanner (1999). More than 700 samples were amplified in the AFLP analyses, but samples that did not work were discarded in one or both primer combinations in the final combined matrix (Table 1.3). A total of 24 individuals were arbitrarily chosen as replicates, and these were analyzed twice for calculation of the error rate.

Total DNA was quantified by optical density at 260 nm. The first digestion step was done in a volume of 7 µL, in a mix that contained 0.7 µL buffer 2 (New England Biolabs), 0.7 µL of BSA (1 mg/mL), and 5 U *Mse*I for 300 ng of total DNA. The reaction was performed for at least 3 hours at 37°C. The second step was the digestion with the restriction enzyme *Taq*αI (New England Biolabs). A volume of 3 µL, containing 4 U of *Taq*αI and 0.5 µL buffer 3 (New England Biolabs), was added to the first digestion tube. The digestion was performed for at least 3 hours at 65°C.

The ligation mix, in a total volume of 5 µL, contained 2 µL of T4 DNA ligase buffer (New England Biolabs), 1 µL of *Mse*I and *Taq*αI adapters (50 µmol/L), and 0.25 U of T4 DNA ligase (New England Biolabs). A total of 5 µL of the digestion was added to the ligation mix and placed for at least 3 hours or overnight at room temperature.

The preselective amplification was performed using the protocol described by Soltis and Gitzendanner (1999). The PCR master mix was modified by using a total 3 mmol/L $MgCl_2$ concentration and 1.6 U of *Taq* polymerase in a total volume of 20 µL. An electrophoresis of 10 µL of the PCR product followed the PCR + 1. Samples where we visualized a smear were subsequently diluted with 100 µL H_2O.

For the selective amplification two combinations were chosen because they were the most variable: *Taq*αI+CAG (green chromophore 5′-HEX) and *Mse*I + CAC, *Taq*αI+CTC (blue chromophore 5′-FAM) and *Taq*αI + CGC. The PCR + 3 mix containing 1× PCR buffer for a total 3 mmol/L $MgCl_2$, 200 µmol/L of each dNTP, 1.5 mmol/L $MgCl_2$, 100 pmol/L of selective primer, and 0.8 U of *Taq* polymerase with 1.5 µL of PCR + 1, in a total volume of 10 µL. The condition for selective amplification was modified only by decreasing the number of cycles to 20 in the second step of the amplification.

Following the selective amplification, samples were diluted with 10 µL of water, 0.5 µL of each combination (green and blue chromophore) were pooled together with 12 µL Hi-Di Formamide (ABI) and 0.2 µL GeneScan-500 ROX size standard (ABI) and denatured at 95°C for 5 minutes. Samples were run on an ABI 3100-*Avant* automatic sequencer. Because of time and cost constraints, we used only two combinations of primers, but the most variable were chosen.

1.2.6 AFLP Analyses

The data from the automatic sequencer were imported and aligned using the software Genographer v. 1.6 (Benham 2001). AFLP bands were scored using the same software for fragments between 50 and 500 bp. AFLP fragments were scored as present or absent, only for the unambiguous ones and were verified manually.

The binary matrix of presence/absence from Genographer was transferred to Excel. Only the individuals present in both combinations were retained in the final matrix. The data matrix was analyzed using Jaccard's distance that considers shared presence but not shared absence to maximize homology (Joly and Bruneau 2007). The conversion of the presence/absence matrix into a distance matrix was done for the Mantel test with the software R Package version 4.0 (Casgrain and Legendre 1999) and with the statistic software R (R Development Core Team 2012) for the redundancy analysis and clustering.

The distance matrix was computed using the "binary" method, which is equivalent to Jaccard's distance in the statistic software R. A cluster analysis was implemented with R using Ward's cluster algorithm in the "agnes" function of the "cluster" package. Ward's (1963) algorithm was chosen because it is useful for large-scale analyses ($n > 100$) and it forms hierarchical groups that are mutually exclusive.

To test the correlation between genetic and geographic distances, we performed a Mantel test (Mantel 1967) with the R Package version 4.0. Significance was based on 1000 permutations. Geographic coordinates were mostly obtained directly in the field, but some were estimated with Google Earth (http://earth.google.com/, 2008) (Table 1.3). Different tests were performed using different geographical scales of sampling (i.e., using different subsets of samples): all samples, North American samples (without estimated coordinates for the European samples) and Nunavik samples.

To study allelic variation among populations and regions, we defined different geographical groups. The definition was based on the proximity of populations: we divided the Nunavik samples into three groups depending on which side of the Ungava Bay shore they were collected (Figure 1.2b), we grouped all populations from Mingan Islands together, and the samples from New Brunswick and Nova Scotia were labeled as Maritime populations. Although all these populations are from Canada, we refer to them collectively as the North American populations. All samples from Eurasian countries were divided into two main groups: Norway and Finland were grouped as Scandinavia samples and the rest were considered as the Eurasian samples. To evaluate the allelic variation, an analysis of molecular variance (AMOVA) was performed using the Arlequin software (Excoffier et al. 2005) with 10,000 permutations to test for significance. All samples were considered as a different haplotype. The analyses were designed to test for variation at the geographic level. First, we checked to see if our Nunavik subdivision was suitable or not. Then, variation between our defined regions was tested to see if these could stand as individual groups. Finally, variation within and among different populations was elucidated.

We also evaluated the experimental effect of the laboratory through two different analyses. First, we defined laboratory groups based upon the running AFLP plate in the sequencer because we observed some alignment differences between plates. We then did an AMOVA on these groups to see how variation was distributed among or within the groups. We also used the function "rda" and "varpart" in the "vegan" package in R to estimate the impact of different variables (AFLP plates, geographical region, populations), which could explain data variability.

1.3 RESULTS

1.3.1 ITS AND *trn*L-F SEQUENCE RESULTS

For each locality, only a single sample was retained in the analysis because no sequence variation was detected within a locality. The combined matrix contained

27 sequences, including five as outgroups. The *trn*L-F matrix contained 883 bp plus 11 indels, and the ITS matrix contained 805 bp plus nine indels for a total of 1708 bp with the insertion of 20 indels for the combined matrix. We found 5% and 14.6% parsimony informative characters for *trn*L-F and ITS, respectively, for a total of 9.5% in the combined matrix including outgroups. In the combined matrix we had a total of 1.6% missing data. The three matrices (*trn*L-F, ITS, and *trn*L-F+ITS) showed similar results, but with different levels of resolution. In all these analyses, outgroup species are clearly separated from the *R. rosea* clade, even though the placement of *R. integrifolia* samples from different regions differs among analyses.

The *trn*L-F and ITS matrices were combined in a single analysis because the ILD test did not detect any significant incongruence ($p = .001$) between the two data sets. Maximum parsimony analysis of the combined matrix found 190 trees of 189 steps with a consistency index (CI) of 0.96 and a RI of 0.98. The strict consensus tree is shown (Figure 1.3) with bootstrap values noted on branches. This analysis shows approximately the same pattern as that from the *trn*L-F matrix but the *R. rosea* clade is better supported with a bootstrap of 100% rather than 95%. The *R. rosea* clade is separated into three main groups: a nonresolved mix composed of Eurasian samples, a subclade with Poland and Russian samples, and a larger subclade supported by a bootstrap value of 95%. This larger subclade contains all the samples from North America, including the Nunavik populations, and samples from Norway and Finland (North Scandinavia). They all share two important duplications of 23 and 19 bp in the *trn*L-F marker. The sample from Sweden (Scandinavia) does not have these duplications.

Preliminary analyses (not shown) of our sequences, in combination with sequences from NCBI, also distinguish a *R. rosea* clade, but with lower support, because the NCBI sequences are shorter, thus adding missing data to our matrix.

1.3.2   AFLP Results

Only unambiguous AFLP fragments were scored as presence/absence for 553 *R. rosea* samples. A total of 60 fragments were scored for the blue combination between 107 and 491 bp in length and 28 fragments for the green combination between 117 and 420 bp in length. All samples were exemplified by a different haplotype, as shown in Table 1.3 where the number of haplotypes and the number of polymorphic loci is given for each population and region. The overall patterns from the outgroup *Rhodiola* species were easily distinguished from the *R. rosea* pattern.

All our tests confirm a geographical component in our data. The Mantel test, at the largest scale with all samples of *R. rosea*, reveals a positive correlation ($r_M = 0.412$, $p = .001$) between genetic and geographic distances. The Ward dendrogram shows geographic clustering of Scandinavian, Eurasian, and Maritime samples, but none at the population level even in these well-defined regions. The tree was simplified to see the main clusters and the origin of the samples (Figure 1.4). Three main clusters are defined and supported by PCoA analysis with the same samples (data not shown). This also reveals the presence of these same three groups (1-2-3) on the first two principal coordinates.

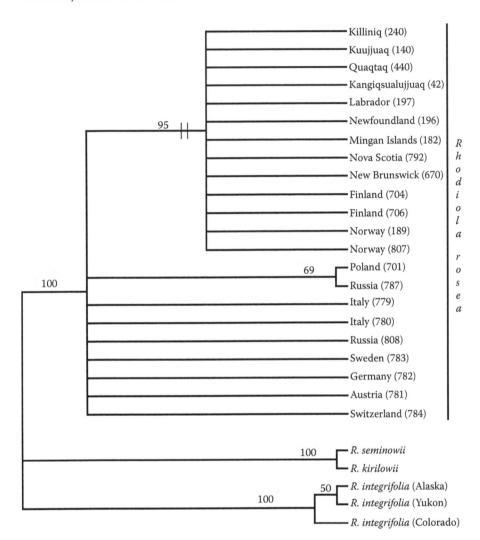

FIGURE 1.3 Strict consensus of the combined matrix (ITS and *trn*L-F) of *R. rosea* and outgroups (*R. integrifolia, R. seminowii, R. kirilowii*) analyzed by maximum parsimony. Bootstrap support values and indels (represented by vertical marks) are added on the branches.

The relationship between the differently defined groupings found in the dendrogram identifies some unexpected relationships such as the distinction between groups 1 and 2 in the dendrogram. There is some overlap between the two groups. Group 1 is only composed of samples from Nunavik and Mingan. Within group 2, Scandinavian samples cluster into three closely related groups. Two of them cluster together and the other cluster with a group formed by Mingan and Nunavik samples. These form a sister group to a cluster comprising Mingan and Nunavik samples. Nunavik and Mingan samples can be found randomly in groups 1 and 2, suggesting a lack of resolution at a regional level for these two regions. There is no apparent

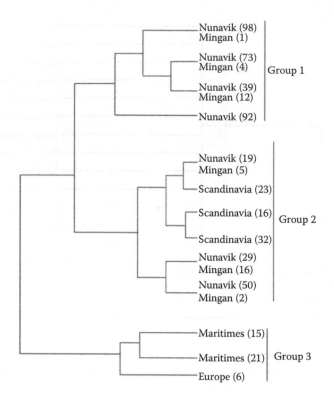

FIGURE 1.4   A simplified dendrogram of *Rhodiola rosea* clustering using Ward's algorithm with AFLP data. The main geographic regions are followed by the number of samples for each in parentheses.

differentiation between Nunavik and Mingan samples. Group 3, sister to the group 1 and 2 dichotomy, indicates a relationship between two groups of Maritime samples and six samples from Eurasia.

Other Mantel tests were performed on subsets of the samples and confirmed the similarity of geographic and genetic distances with the same threshold ($p = .001$). The strongest positive correlation was found between geographic and genetic distances at the North American level ($r_M = 0.642$, $p = .001$). When we removed the Maritime samples, the correlation found in the Mantel tests was weaker ($r_M = 0.339$, $p = .001$). When samples from both the Maritimes and Mingan were removed, with only Nunavik samples remaining, the correlation observed decreased ($r_M = 0.267$, $p = .001$).

Partitions of variance for groupings are represented in Table 1.5. The AMOVA reveals that more than a third of the allelic variation is distributed between populations, and that most of the variation is from the population itself. The lowest allelic variance (11.76%) among subregions was found in the Nunavik region. These subregions are situated in the South, East, and West around Ungava Bay. The highest allelic variance 69.35% was, however, found within these Nunavik populations. In

TABLE 1.5

Summary of the Partition of Variance (AMOVA) among and within Different Groups of *Rhodiola rosea*

N	Group and Level of Variation	d.f.	Sum of Squares	Variance Components	Percentage of Variation
	Ungava Bay (Nunavik)				
3	3 subregions: (NE) (NW) (NS)				
	Among subregions	2	340.05	1.30	11.76
	Among populations within subregions	29	979.19	2.08	18.88
	Within populations	372	2,851.59	7.67	69.35
	North America				
3	3 subregions: (N-Mg-Ma)				
	Among subregions	2	627.59	4.21	28.81
	Among populations within subregions	42	1,582.89	2.79	19.06
	Within subregions	436	3,323.79	7.62	52.13
	Geographic regions				
5	5 regions: (N-Mg-Ma-Sc-Eur)				
	Among regions	4	1,027.16	3.67	26.27
	Among populations within regions	52	1,759.16	2.63	18.84
	Within populations	501	3,839.74	7.66	54.89
	Populations				
1	57 populations				
	Among populations	56	2786.32	4.32	36.04
	Within populations	501	3839.74	7.66	64.07

Geographic region (NE, Ungava Bay-East; NS, Ungava Bay-South; NW, Ungava Bay-West; N, Ungava Bay (Nunavik); Mg, Mingan Islands; Ma, Maritime; Sc, Scandinavia; Eur, Europe). *N*, number of groups in the analysis; d.f., degrees of freedom. All P-values (P (rand ≥ obs)) were <0.001.

each analysis, most of the allelic variance is found within populations or regions instead of among populations or regions.

1.4 DISCUSSION

1.4.1 Taxonomic Distinction and Medicinal Implications

Using samples from Eastern Canada, we were looking to determine if and how *R. rosea* populations from Nunavik are related to Eurasian populations. Thus, we have compared the Nunavik populations with those from Russia that have been studied for their medicinal properties for several decades.

Although the strong support for the monophyly of this species with respect to the outgroup taxa suggests all our *R. rosea* samples are the same taxon, infraspecific variation remains important. The strict consensus topologies for the ITS and *trn*L-F resolved one subclade of *R. rosea* that includes North American and Scandinavian, but not Eurasian samples (Figure 1.3). Samples in this subclade share two short duplications of 23 and 19 bp in the *trn*L-F region that are absent in Eurasian samples. The rest of the *R. rosea* samples from Central Europe and Russia form a polytomy, but are grouped together in the AFLP cluster analysis (Figure 1.4). These results suggest the possibility for differences between coastal and Alpine populations of *R. rosea*, whether this distinction needs to be recognized using subspecies or varieties is debatable. However, ethnopharmacological studies in progress (see Ahmed et al. 2013) have shown differential activity in addressing *in vivo* Alzheimer bioassays with samples of indigenous Eastern Canada and Russian origins, a result which lends support to this molecular analysis as well as to their phytochemical profiles (Avula et al. 2009; Rumalla et al. 2011). Based on seed coat ornamentation, Gontcharova et al. (2009) considered *R. rosea* ssp. *rosea* to hide cryptic species or, at least, subspecies. There is a need for more comprehensive molecular analyses before revising the subspecies level.

The AFLP dendrogram also reveals clustering of some regions within *R. rosea* such as the grouping of Maritime and Scandinavia samples. We observed morphological differences between Finland and Nunavik specimens, which have thicker and shorter leaves, and specimens from Central Europe or Russia, where the leaves can be more lanceolate and dentate (unpublished observations). Similarly, Asdal et al. (2006) noted morphological differences between Finnish and Central European samples.

When populations from across the range are taken into account, our analyses clearly suggest the presence of at least two evolutionary lineages within the species, one of which is at the origin of the Scandinavian and North American populations, the other embraces the Alpine populations in Eurasia. Because of the coastal nature of these populations (Figure 1.2), there is strong evidence supporting the presence of a subspecies or variety, genetically characterized by the presence of two short duplications in the *trn*L-F plastid marker and an affinity for coastal environments and habitats. This is in contrast to the Alpine populations, in which duplications are yet to be found. Not surprisingly, a study on the genetic diversity of *R. rosea* from Sweden, Greenland, and Faroe Islands using four microsatellite (SSR) and four interspecific sequence repeat (ISSR) markers found no population-specific primers (Kylin 2010). All populations from Greenland and Faroe Islands were coastal, whereas those from Sweden were collected in proximity to the Norwegian border, within 200 km from the coast. Although different markers were used, this nonetheless brings further support to a genetically distinct lineage of *R. rosea*, which would be the origin of populations from North American and Scandinavia.

In an analysis of the genetic structure of *R. rosea* from Russia and Poland, Kozyrenko et al. (2011) also came to the conclusion that there were distinct evolutionary lineages in the species. The clustering of these populations based on the genetic distances of 252 ISSR markers showed a highly supported cluster comprising populations from Poland and Zabaikal'e in Russia, and a weaker one comprising all

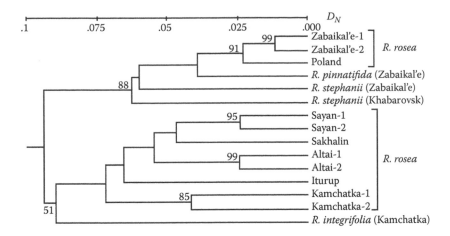

FIGURE 1.5 UPGMA dendrogram of eleven *Rhodiola rosea*, two *R. stephanii*, *R. integrifolia* and *R. pinnatifida* populations constructed based on Nei's (1973) genetic distances calculated from 252 ISSR markers. The numbers on the branches indicate the bootstrap support (only those above 50% are shown). (Adapted from Kozyrenko, M.M. et al., *Flora—Morphology, Distribution, Functional Ecology of Plants*, 206 (8), 691–96, 2011.)

other Russian populations (Figures 1.5 and 1.6). Clustering based on nuclear ribosomal ITS and *trn*L-F chloroplast regions also indicate a close relationship between populations from the Polish Tatra Mountains and the Russian Altaï Mountains (Figure 1.3, Table 1.1). Their affinity to other European populations from mountainous regions (Figure 1.3 and 1.4) suggests an evolutionary lineage of *R. rosea* confined to Alpine environments and habitats. The study of Kozyrenko et al. (2011) comprises a relatively large Eurasian sample size; all of which were collected at or above 1000 m latitude though some were also of coastal origin. The evolutionary story among Alpine populations of *R. rosea* may be more complex. This would be better understood by a study on the genetic structure of the species from populations covering its entire range, from eastern North America and Europe to eastern Asia.

Phytochemical analyses also identified some regional differences. Studies by Filion et al. (2008), Avula et al. (2009, 2010), and Rumalla et al. (2011) on four key phytochemical markers (salidroside, rosavin, rosarin, rosin) for a similar geographic sampling showed that Nunavik populations possess the key compounds, but in lesser amounts than the Siberian populations. It thus confirms the close phytochemical relationship between Siberian and Nunavik *R. rosea* populations. New compounds, which are yet to be identified, were also found in the Nunavik populations, but are absent in the Siberian ones (Filion et al. 2008). Avula et al. (2009, 2010) evaluated 14 compounds, of which four were not detected in *R. rosea* from Nunavik but were present in Eurasian populations. Our data emphasize that Nunavik populations belong to *R. rosea* despite the regional variation observed.

The intraspecific variability observed among the regions (morphology, molecular, or phytochemical) is likely the result of their independent evolution. For instance, Serebryanaya and Shipunov (2009) found that short-scale evolutionary processes

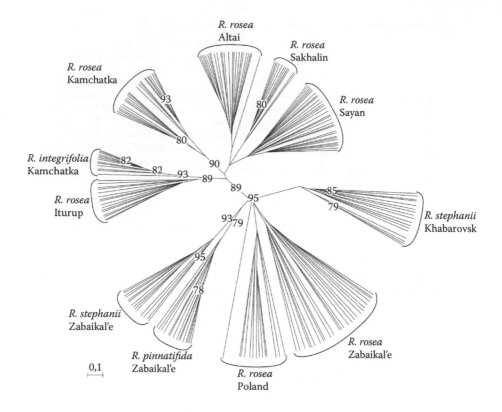

FIGURE 1.6 Unrooted UPGMA dendrogram showing genetic relationships between the individuals in the population studied constructed based on Nei and Li's (1979) genetic distances from 252 ISSR markers. The number on the branches indicated the bootstrap support (≥50%). (Adapted from Kozyrenko et al., *Flora—Morphology, Distribution, Functional Ecology of Plants* 206 (8), 691–96, 2011.)

best explained morphological variations of *R. rosea* between islands and mainland populations. Variability within a species is dependent on different evolutionary processes such as mutation, gene flow, genetic drift, and natural selection, which may differ among regions. Morphological and phytochemical variations may also be related to environmental variables. Clausen (1975) found that significant morphological differences among wild populations were lost once grown together under uniform conditions. Also, the quantity of phytochemical compounds was reported to vary depending on the distribution and the age of the plants (Furmanowa et al. 1995). György (2006) also noted that the detection of compounds depends on the origin of the samples, the extraction and analysis method used. Filion et al. (2008) also found variations according to the plant gender (pistillate, staminate, hermaphroditic), physiological conditions, and age. Thus, despite infraspecific and regional variation, we consider Nunavik populations of *R. rosea* to be good candidates for future locally sourced cultivation, following further phytochemical and pharmacological assays and investigations.

1.4.2   INFRASPECIFIC TAXONOMY AND CONFUSION

Over its wide distribution, *R. rosea* has received more than 20 different Latin names (Ohba 2002). Despite the recent and thorough taxonomic account of the genus *Rhodiola* by Ohba (2002), phylogenetic studies have focused mainly on Asian species (Figure 1.7a). In North America, *R. integrifolia* has historically been included as a subspecies of *R. rosea* (Table 1.1), many authors, therefore, still use the old nomenclature. Clausen (1975) refers to them as two species, and studies of chromosome counts have revealed differences in chromosome numbers between *R. integrifolia* and *R. rosea* (Uhl 1952; Amano et al. 1995). Nonetheless, confusion persists in the taxonomic determination of plant material. The nomenclature remains crucial for correct identification, but often depends on the regional flora used, where subspecies and varieties are sometimes recognized.

Our maximum parsimony analysis (Figure 1.3) of nuclear ribosomal ITS and *trn*L-F chloroplast regions reveals a well-supported clade composed of *R. rosea* samples from Europe, Russia, and eastern North America with a clear distinction from *R. integrifolia*, *R. kirilowii*, and *R. seminowii*. Our AFLP patterns also allow us to distinguish between *R. rosea* and other members of the genus. Clustering analysis with AFLP data including the outgroup species (data not shown) clearly separate them from all samples of *R. rosea* in the dendrogram. This distinction between this species and outgroup species suggests that samples of *R. rosea*, independently of their origin, are more closely related to each other than to other members of the genus. Similar conclusion can be made from a phylogenetic study of 18 *Rhodiola* species by Guest (2009), including *R. rosea* samples from New Brunswick, Greenland, and Japan (Figure 1.7b). In addition, the clear distinction of *R. integrifolia* from *R. rosea* in Figures 1.3, 1.5, 1.6, and 1.7b provides further support to the differentiation between these two species and to the idea that western roseroot, *R. integrifolia*, is more closely related to *R. rhodantha* and other Asian species, such as *R. algida* and *R. semenovii* (Guest 2009; Hermsmeier et al. 2012; DeChaine et al. 2013; Zhang et al. 2014).

Our sequence analysis revealed misidentification of some Canadian *R. rosea* plants under cultivation. Some cultivated samples originating from wild western North American populations were identified as *R. rosea*, but their origin as well as our analysis, suggest they are *R. integrifolia*. In Yukon, herbalists have used and sold what they thought was *R. rosea*, but is instead *R. integrifolia* (Cuerrier, personal communication). This illustrates the important need for molecular as well as phytochemical markers. This is in line with Chan et al. (2007) and the Smithsonian Institution who have begun to develop a barcode library for medicinal plants, using a variety of markers (Pennisi 2007; CBOL Plant Working Group 2009). A consensus for the compilation of medicinal plant DNA sequences is expected, so that other groups may add sequences to this database and others (Hollingsworth et al. 2011; see also Sharma and Sarkar 2013). As a first step, our publicly available ITS and *trn*L-F sequences generated for *R. rosea* can be used as an initial barcode to taxonomically validate samples of this medicinal plant. This type of information could be useful for future commercial exploitation of *R. rosea*, to prevent misidentification before market distribution. Furthermore, with the two informative duplications in the chloroplast *trn*L-F region, it is possible to distinguish if plants are of Eurasian or of Scandinavian-North American origin.

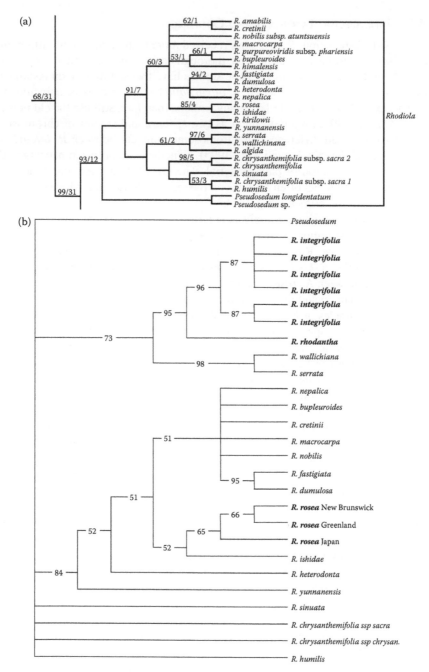

FIGURE 1.7 Phylogenetic tree for (a) 23 *Rhodiola* species of eastern Asia and (b) 18 species including the three North American species *R. integrifolia*, *R. rhodantha*, and *R. rosea*. (a) The strict consensus tree of 874 MP trees based on ITS seQuénces. (Adapted from Mayuzumi, S., and Ohba, H., *Systematic Botany*, 29 (3), 587–98, 2004.) (b) Maximum parsimony 50% majority rule consensus tree based on ITS sequences. (Adapted from Guest, H.J., Systematic and phylogeographic implications of molecular variation in the western North American roseroot, *Rhodiola integrifolia* (Crassulaceae), MSc thesis, University of Victoria, 2009.)

1.4.3 PHYLOGEOGRAPHY, ENVIRONMENT, AND POPULATION STRUCTURE

The plastid *trn*L-F region reveals two duplications in the *R. rosea* subclade comprising North American (Maritimes, Nunavik, Mingan) and Scandinavian (Norway and Finland) samples (Figure 1.3). This cluster is similar to one observed in *Arabis alpina* where only one haplotype of *trn*L-F was found for the samples from Canada, North Europe, Greenland, Iceland, and Svalbard (Koch et al. 2006). In the case of *R. rosea*, the subclade is geographically and environmentally distinct. In contrast to the Eurasian samples (Russia and Central Europe) that are from high altitude regions, such as the Alps and Altaï Mountains, this North American–Scandinavian subclade comprises coastal samples on both sides of the Atlantic Ocean.

From its emergence in the southern Siberian highlands during the Pliocene, it is hypothesized that *R. rosea* migrated westward to the southern Urals during the Pleistocene maximal glaciation (Kozyrenko et al. 2011). From there, two possible lines of evolution may have diverged early in the species when populations migrated northward toward the Arctic along the Ural Mountains and those of eastern Siberia (Kozyrenko et al. 2011). More extensive genetic studies are required to readily dissect the historical movements of the species into Western Europe.

Our analyses of ITS, *trn*L-F, and AFLP markers enable us to propose several dispersal hypotheses to explain the present distribution of this species in North America and Scandinavia.

The first hypothesis implies a long-distance dispersal from Scandinavia to North America and the colonization of the northern part following glacial retreat by the populations that survived south of the ice-sheet. The second hypothesis is a variant of the first, and implies northern and southern refugia during the last glaciations. The third hypothesis suggests a first dispersal event from Europe to North America during early Quaternary or late Tertiary period via the North Atlantic land bridge and a second event from North America to Scandinavia during or after the last glaciations. This hypothesis is similar to that proposed for the amphi-atlantic *Saxifraga paniculata* (Reisch 2008). Other regions could have played a role in the dispersal of *R. rosea*, but more extensive sampling is needed to confirm any of these hypotheses.

Our results (Figure 1.4) show more geographic structure among regions than population structure within regions. Our Mantel results pinpoint to some geographic differentiation in our North American sampling. This contrasts with the lack of resolution observed for Nunavik and Mingan samples in the dendrogram. Only a third of the variation is found among the three subregions of our North American sampling (Nunavik, Mingan, Maritimes). A Mantel test on a more restrictive part of the sampling, excluding the Maritime samples that are strongly grouped in the cluster analysis, reduced considerably the correlation. For this reason, most of the support for the Mantel test comes from the Maritime group, which is the only differentiated group on the dendrogram for the North American continent. In a similar manner, György et al. (2013a,b) demonstrated in an analysis of four SSR and eight ISSR markers from *R. rosea* populations in northern Norway that the more geographically distant these were from one another, the higher the observed genetic difference.

The Maritimes is the first of the areas sampled that was deglaciated around 12,000 years BP and was occupied by tundra (10,000 years BP) (Delcourt and Delcourt 1993), compared to Nunavik (Ungava Bay), which was completely ice-free 7,000 years BP, except for some eastern parts of Ungava Bay that were ice-free around 10,000 years BP (Brouillet and Whetstone 1993). The earlier retreat of the ice-sheet from the Maritimes allows more time for regional differentiation after the colonization. Maritime samples are differentiated by the AFLP dendrogram but morphological variation was also observed mostly in the etiolation of the plant, the leaf color and their height. These variations can be due to acclimatizing to environmental conditions that are different from subarctic or high-altitude regions. Our AFLP analyses (Figure 1.4) also indicate a close relationship between the two groups of Maritime samples and the samples from Eurasia. These results contrast with those of the ITS and *trn*L-F sequence analysis, where the Maritime samples group with the North America–Scandinavia clade. This pattern could be explained by the second and third dispersal hypothesis described above.

Our subdivision of the Nunavik populations on each side of the Ungava Bay was not confirmed by the partitioning of the allelic variance. Most of the variance is found within populations instead of among the subregions (Table 1.5). The dendrogram also reveals a random distribution of the Nunavik samples (Figure 1.4). The Mantel test based only on Nunavik samples indicates the lower correlation between geographic and genetic distances. Nothing indicates that these populations are distinct (or differentiated) in different subgroups. The colonization of northern regions, such as Nunavik and Mingan, is too recent to show regional differentiation by clustering together, but the genetic uniformity also could be the result of gene flow during postglacial colonization (Reisch 2008).

Overall with the present sampling, even in the regional groupings, there is no population structure in the dendrogram. The allelic analysis of populations also reported that two-thirds of the variation occurs within populations rather than among them (Table 1.5). Similar results were obtained by Elameen et al. (2008), Soni et al. (2010), Kozyrenko et al. (2011), and György et al. (2012, 2013a) on *R. rosea* populations from India, Norway, Poland, and Russia. It is thought that this is characteristic for species with wide distribution ranges and susceptible to short-scale evolutionary processes (Serebryanaya and Shipunov 2009; Kozyrenko et al. 2011). The lack of population differentiation is probably due to high gene flow (Elameen et al. 2008; Soni et al. 2010) and the short time since their colonization.

1.5 CONCLUSION

Our molecular results corroborate the phytochemical analyses done on the same samples (Filion et al. 2008; Avula et al. 2009, 2010; Rumalla et al. 2011): both indicate some level of difference with Russian populations of *R. rosea*. It thus highlights the need for further pharmacological analyses of these Canadian *R. rosea* populations (Cayer et al. 2013) (see also Chapter 7). The sequences obtained in this study could be useful for verifying the identity of the species under cultivation or collected before processing for market utilization. Scandinavian and North American populations are genetically more similar, sharing two duplications in the chloroplast *trn*L-F sequence. The distribution of this amphi-Atlantic species and the observed phylo-geographic pattern suggest a common dispersal history for populations on either side of the ocean. Geographic differences were observed in our analyses, but no differentiation was found at the population level. In North America, only the Maritime samples form a clearly distinct geographic entity. Our analyses suggest that the Nunavik region is not genetically differentiated from the rest of the North American populations, perhaps due to the short time since the last glaciations.

In addition, our results point to possible differences between coastal and Alpine populations of *R. rosea*, at least at the genetic level. More research should be done that involves better sampling, with a focus on the differentiation of populations stemming from these two habitats. Also, additional molecular data will be necessary to understand the phylogeography of the species. Once we establish a robust phylogeny, it will be possible to combine phytochemical profiling with the history of the species and explore rigorously the medicinal properties in relation with specific genomes. This may lead to natural products with specific claims.

ACKNOWLEDGMENTS

Many thanks are due to Nunavik Biosciences, Inc. (Makivik Corporation) for financial support to Alain Cuerrier. The project was also financially supported by an NSERC grant (to Anne Bruneau) and a bursary from Fondation Museums Nature Montréal received by Mariannick Archambault. The Inuit communities of Kangirsuk and Kangirsualujjuaq are acknowledged for welcoming us and helping us with the logistics regarding fieldtrips to the eastern and western coasts of Ungava Bay. Further, we recognize Bill Doidge from Makivik Research Station and Josée Brunelle from Kativik Regional Government for their help during our stay in Kuujjuaq. Finally, we thank Bernard Angers and Simon Joly for their help with the AFLP.

REFERENCES

Ahmed, F., C. Cieniak, H. Xu, A. Saleem, A. Cuerrier, D. Figeys, J.T. Arnason, and S.A.L. Bennett. 2013. Evaluation of the neuroprotective effects of *Rhodiola rosea* (Crassulaceae) in the TgCRND8 mouse model of Alzheimer disease. In *7th Annual Meeting of the Canadian Association for Neuroscience 2013*. Toronto, Canada, May 21–24, 2013.

Aiken, S.G., M.J. Dallwitz, L.L. Consaul, C.L. McJannet, L.G. Gillepsie, R.L. Boles, G.W. Argus et al. 2003. *Flora of the Canadian Archipelago: Descriptions, Illustrations, Identification, and Information Retrieval*. Version 29.

Alm, T. 2004. Ethnobotany of *Rhodiola rosea* (Crassulaceae) in Norway. *SIDA* 21 (1): 324–44.

Amano, M., M. Wakabayashi, and H. Ohba. 1995. Cytotaxonomic studies of Siberian Sedoideae (Crassulaceae) I. Chromosomes of *Rhodiola* in the Altai Mountains. *Journal of Japanese Botany* 70: 334–38.

Asdal, Å., B. Galambosi, K. Olsson, K.W. Bladh, and E. Poorvaldsdóttir. 2006. *Spice- and Medicinal Plants in the Nordic and Baltic Countries*, Conservation of Genetic Resources. Report from a project group at the Nordic Gene Bank. Alnarp, pp. 94–104.

Avula, B., Y.-H. Wang, Z. Ali, T.J. Smillie, V. Filion, A. Cuerrier, J.T. Arnason, and I.A. Khan. 2009. RP-HPLC determination of phenylalkanoids and monoterpenoids in *Rhodiola rosea* and identification by LC-ESI-TOF. *Biomedical Chromatography* 23 (8): 865–72.

Avula, B., Y.-H. Wang, Z. Ali, T.J. Smillie, V. Filion, A. Cuerrier, J.T. Arnason, and I.A. Khan. 2010. Erratum: RP-HPLC determination of phenylalkanoids and monoterpenoids in *Rhodiola rosea* and identification by LC-ESI-TOF. *Biomedical Chromatography* 24 (6): 682.

Benham, J. 2001. Genographer. Bozeman: Montana State University. http://hordeum.msu.montana.edu/genographer.

Blondeau, M., C. Roy, and A. Cuerrier. 2010. *Plants of the Villages and the Parks of Nunavik.* Québec: Éditions MultiMondes.

Bonin, A., D. Ehrich, and S. Manel. 2007. Statistical analysis of amplified fragment length polymorphism data: a toolbox for molecular ecologists and evolutionists. *Molecular Ecology* 16 (18): 3737–58.

Brouillet, L., F. Coursol, S.J. Meades, M. Favreau, M. Anions, P. Bélisle, and P. Desmet. 2010+. *Rhodiola rosea* Linnaeus. In *VASCAN, the Database of Vascular Plants of Canada.* Accessed October 30, 2013. http://data.canadensys.net/vascan/.

Brouillet, L. and R. Whetstone. 1993. Climate and physiography. In *Flora of North America North of Mexico*, edited by F.o.N.A.E. Committee. Oxford: Oxford University Press.

Brown, R.P., P.L. Gerbarg, and Z. Ramazanov. 2002. *Rhodiola rosea*: A phytomedicinal overview. *HerbalGram* 56: 40–52.

Casgrain, P. and P. Legendre. 1999. *The R Package.* Montreal: Department of Biological Sciences, University of Montreal.

Cayer, F., F. Ahmed, V. Filion, A. Saleem, A. Cuerrier, M. Allard, G. Rochefort, Z. Merali, and J.T. Arnason. 2013. Characterization of the anxiolytic activity of Nunavik *Rhodiola rosea*. *Planta Medica* 79 (15): 1385–91.

CBOL Plant Working Group. 2009. A DNA barcode for land plants. *PNAS* 106 (31): 12794–7.

Chan, L., S.C.C. Chik, A.S. Lau, and C.C. Liu. 2007. Identification of multiple markers for phylogenetic and DNA barcoding studies for medicinal plants. In *Second International Barcode of Life Conference.* Taipei, Taiwan: Academia Sinica.

Clausen, R.T. 1975. *Sedum of North America North of the Mexican Plateau.* Ithaca, NY: Cornell University Press.

Cody, W.J. 2000. *Flora of the Yukon Territory.* Ottawa, ON, Canada: NRC Research Press.

Cuerrier, A. and Elders of Kangiqsualujjuaq. 2011. *The Botanical Knowledge of the Inuit of Kangiqsualujjuaq, Nunavik.* Inukjuak, QC, Canada: Avataq Cultural Institute.

DeChaine, E.G., B.R. Forester, H. Schaefer, and C.C. Davis. 2013. Deep genetic divergence between disjunct Refugia in the Arctic-Alpine King's Crown, *Rhodiola integrifolia* (Crassulaceae). *PLOS ONE* 8 (11): e79451.

Delcourt, P.A. and H.R. Delcourt. 1993. Paleoclimates, paleovegetation, and paleofloras during the late Quaternary. In *Flora of North America North of Mexico*, edited by F.O.N.A.E. Committee. Oxford: Oxford University Press.

Dodson, C. and W.W. Dunmire. 2007. *Mountain Wildflowers of the Southern Rockies.* Albuquerque: UNM Press.

Douzery, E.J., A.M. Pridgeon, P. Kores, H.P. Linder, H. Kurzweil, and M.W. Chase. 1999. Molecular phylogenetics of diseae (Orchidaceae): A contribution from nuclear ribosomal ITS sequences. *American Journal of Botany* 86 (6): 887–99.

Doyle, J.J. and J.L. Doyle. 1987. A rapid DNA isolation procedure for small quantities of fresh leaf tissue. *Phytochemical Bulletin* 19 (1): 11–5.

Elameen, A., S. Dragland, and S.S. Klemsdal. 2010. Bioactive compounds produced by clones of *Rhodiola rosea* maintained in the Norwegian germplasm collection. *Die Pharmazie* 65 (8): 618–23.

Elameen, A., S.S. Klemsdal, S. Dragland, S. Fjellheim, and O.A. Rognli. 2008. Genetic diversity in a germplasm collection of roseroot (*Rhodiola rosea*) in Norway studied by AFLP. *Biochemical Systematics and Ecology* 36 (9): 706–15.

Excoffier, L., G. Laval, and S. Schneider. 2005. Arlequin (Version 3.0): An integrated software package for population genetics data analysis. *Evolutionary Bioinformatics Online* 1: 47–50.

Filion, V.J., A. Saleem, G. Rochefort, M. Allard, A. Cuerrier, and J.T. Arnason. 2008. Phytochemical analysis of Nunavik *Rhodiola rosea* L. *Natural Product Communications* 3 (5): 721–26.

Fu, K.J. and H. Ohba. 2001. *Rhodiola* (Crassulaceae). In *Flora of China,* Vol. 8, edited by Z.Y. Wu and P. Raven, 251–68. Beijing: Science Press.

Furmanowa, M., H. Oledzka, M. Michalska, I. Sokolnicka, and D. Radomska. 1995. XXIII *Rhodiola rosea* L. (Roseroot): In vitro regeneration and the biological activity of roots. In *Biotechnology in Agriculture and Forestry, Vol. 33, Medicinal and Aromatic Plants VIII,* edited by Y.P.S. Bajaj, 412–26. Heidelberg, Germany: Springer-Verlag.

Galambosi, B. 2006. Demand and availability of *Rhodiola rosea* L. raw material. In *Medicinal and Aromatic Plants,* edited by R.J. Bogers, L.E. Craker, and D. Lange, 223–36. Wageningen: Springer.

GBIF. 2013. Biodiversity occurrence data published by: Field Museum of Natural History, Museum of Vertebrate Zoology, University of Washington Burke Museum, and University of Turku. Available through GBIF Data Portal, data.gbif.org. Accessed November 4, 2013.

Gleason, H.A. and A. Cronquist. 1991. *Manual of Vascular Plants of Northeastern United States and Adjacent Canada.* Bronx: NYBG Press.

Gontcharova, S.B., A.A. Gontcharov, V. V. Yakubov, and K. Kondo. 2009. Seed surface morphology in some representatives of the Genus *Rhodiola* sect. *Rhodiola* (Crassulaceae) in the Russian Far East. *Flora — Morphology, Distribution, Functional Ecology of Plants* 204 (1): 17–24.

Guest, H.J. 2009. Systematic and phylogeographic implications of molecular variation in the western North American roseroot, *Rhodiola integrifolia* (Crassulaceae). MSc thesis. Victoria: University of Victoria.

György, Z. 2006. Glycoside production by in vitro *Rhodiola rosea* cultures. Oulu: University of Oulu.

György, Z., E. Fjelldal, M. Ladányi, P.E. Aspholm, and A. Pedryc. 2013a. Genetic diversity of roseroot (*Rhodiola rosea*) in North-Norway. *Biochemical Systematics and Ecology* 50: 361–67.

György, Z., E. Fjelldal, A. Szabó, P.E. Aspholm, and A. Pedryc. 2013b. Genetic diversity of golden root (*Rhodiola rosea* L.) in northern Norway based on recently developed SSR markers. *Turkish Journal of Biology* 37: 655–60.

György, Z., M. Szabó, D. Bacharov, and A. Pedryc. 2012. Genetic diversity within and among populations of roseroot (*Rhodiola rosea* L.) based on molecular markers. *Notulae Botanicae Horti Agrobotanici Cluj-Napoca* 40 (2). Cluj-Napoca: University of Agricultural Sciences and Veterinary Medicine: 266–73.

Hall, T.A. 1999. BioEdit: A user-friendly biological sequence alignment editor and analysis program for Windows 95/98/NT. *Nucleic Acids Symposium Series*, 41:95–98.

Hermsmeier, U., J. Grann, and A. Plescher. 2012. *Rhodiola integrifolia*: hybrid origin and Asian relatives. *Botany* 90 (11): 1186–90.

Hinds, H.R. 2000. *Flora of New Brunswick: A Manual for the Identification of the Vascular Plants of New Brunswick.* 2nd ed. Fredericton: University of New Brunswick.

Hollingsworth, P.M., S.W., Graham, and D.P. Little. 2011. Choosing and using a plant DNA barcode. *PLOS ONE* 6 (5): e19254.

Hultén, E. and M. Fries. 1986. *Atlas of North European Vascular Plants (North of the Tropic of Cancer).* Königstein: Koeltz Scientific Books.

Joly, S. and A. Bruneau. 2007. Delimiting species boundaries in *Rosa* Sect. *Cinnamomeae* (Rosaceae) in eastern North America. *Systematic Botany* 32 (4): 819–36.

Joly, S., J.R. Starr, W.H. Lewis, and A. Bruneau. 2006. Polyploid and hybrid evolution in roses east of the Rocky Mountains. *American Journal of Botany* 93 (3): 412–25.

Kelly, G.S. 2001. *Rhodiola rosea*: a possible plant adaptogen. *Alternative Medicine Review* 6 (3): 293–302.

Koch, M.A., C. Kiefer, D. Ehrich, J. Vogel, C. Brochmann, and K. Mummenhoff. 2006. Three times out of Asia Minor: The phylogeography of *Arabis alpina* L. (Brassicaceae). *Molecular Ecology* 15 (3): 825–39.

Komarov, V.L. 1971. *Flora of the USSR. IX. Rosales and Sarraceniales*. Leningrad: The Botanical Institute of Science the USSR.

Kozyrenko, M.M., S.B. Gontcharova, and A.A. Gontcharov. 2011. Analysis of the genetic structure of *Rhodiola rosea* (Crassulaceae) using inter-simple sequence repeat (ISSR) polymorphisms. *Flora – Morphology, Distribution, Functional Ecology of Plants* 206 (8): 691–96.

Kylin, M. 2010. Genetic diversity of roseroot (*Rhodiola rosea* L.) from Sweden, Greenland and Faro Islands. Uppsala: Swedish University of Agricultural Sciences.

Liu, Z., Y. Liu, C. Liu, Z. Song, Q. Li, Q. Zha, C. Lu et al. 2013. The chemotaxonomic classification of *Rhodiola* plants and its correlation with morphological characteristics and genetic taxonomy. *Chemistry Central Journal* 7 (1). 118.

Mantel, N. 1967. The detection of disease clustering and a generalized regression approach. *Cancer Research* 27 (2): 209–20.

Mayuzumi, S. and H. Ohba. 2004. The phylogenetic position of eastern Asian Sedoideae (Crassulaceae) inferred from chloroplast and nuclear DNA sequences. *Systematic Botany* 29 (3): 587–98.

Meudt, H.M. and A.C. Clarke. 2007. Almost forgotten or latest practice? AFLP applications, analyses and advances. *Trends in Plant Science* 12 (3): 106–17.

Moran, R. 2000. *Rhodiola integrifolia* Raf.-a rambling account, with two new combinations. *Cactus and Succulent Journal* 72 (3): 137–39.

Moran, R. V. 2009. *Rhodiola*. In *Flora of North America North of Mexico*. 16+ vols, edited by Flora of North America Editorial Committee. 1993+. New York and Oxford: Flora of North America Association, Vol. 8, pp. 164–7.

Müller, K. 2006. Incorporating information from length-mutational events into phylogenetic analysis. *Molecular Phylogenetics and Evolution* 38 (3): 667–76.

Mueller, U.G. and L.L. Wolfenbarger. 1999. AFLP genotyping and fingerprinting. *Trends in Ecology & Evolution* 14 (10): 389–94.

Muschner, V.C., A.P. Lorenz, A.C. Cervi, S.L. Bonatto, T.T. Souza-Chies, F.M. Salzano, and L.B. Freitas. 2003. A first molecular phylogenetic analysis of *Passiflora* (Passifloraceae). *American Journal of Botany* 90 (8): 1229–38.

Nei, M. 1973. Analysis of gene diversity in subdivided populations. *Proceedings of the National Academy of Sciences* 70 (12): 3321–23.

Nei, M. and W.-H. Li. 1979. Mathematical model for studying genetic variation in terms of restriction endonucleases. *Proceedings of the National Academy of Sciences* 76 (10): 5269–73.

Ohba, H. 1981. A revision of the Asiatic species of Sedoideae (Crassulaceae). Part 2. *Rhodiola* (sugen. *Rhodiola* sect. *Rhodiola*). *Journal of the Faculty of Science, University of Tokyo, Section III* 13: 65–119.

Ohba, H. 2002. *Rhodiola*. In *Illustrated Handbook of Succulent Plants: Crassulaceae*, edited by U. Eggli. Berlin, Heidelberg: Springer-Verlag.

Olfelt, J.P., G.R. Furnier, and J.J. Luby. 2001. What data determine whether a plant taxon is distinct enough to merit legal protection? A case study of *Sedum integrifolium* (Crassulaceae). *American Journal of Botany* 88 (3): 401–10.

Pennisi, E. 2007. Wanted: a barcode for plants. *Science* 318: 190–91.

Polunin, N.V. 1940. *Botany of the Canadian Eastern Arctic. Part I: Pteridophyta and Spermatophyta*. Ottawa: National Museum of Canada.

Polunin, O. 1969. *Flowers of Europe: A Field Guide*. Toronto: Oxford University Press.

R Development Core Team. 2012. R: A language and environment for statistical computing Vienna, Austria: R Foundation for Statistical Computing. http://www.r-project.org/.

Reisch, C. 2008. Glacial history of *Saxifraga paniculata* (Saxifragaceae): molecular biogeography of a disjunct arctic-alpine species from Europe and North America. *Biological Journal of the Linnean Society* 93 (2): 385–98.

Roland, A.E. and M. Zinck. 1998. *Roland's Flora of Nova Scotia*. Halifax: Nimbus Publishing and Nova Scotia Museum.

Rumalla, C.S., B. Avula, Z. Ali, T.J. Smillie, V. Filion, A. Cuerrier, J.T. Arnason, and I.A. Khan. 2011. Quantitative HPTLC analysis of phenylpropanoids in *Rhodiola* species. *JPC – Journal of Planar Chromatography — Modern TLC* 24 (2): 116–20.

Scoggan, H.J. 1978. *The Flora of Canada. Part 1–4*. Ottawa: National Museums of Canada.

Serebryanaya, A. and A. Shipunov. 2009. Morphological variation of plants on the uprising islands of northern Russia. *Annales Botanici Fennici* 46 (2): 81–9.

Sharma, V. and I.N. Sarkar. 2013. Bioinformatics opportunities for identification and study of medicinal plants. *Briefings in Bioinformatics* 14 (2): 238–50.

Simmons, M.P. and H. Ochoterena. 2000. Gaps as characters in sequence-based phylogenetic analyses. *Systematic Biology* 49 (2): 369–81.

Small, E. and P.M. Catling. 2000. *Rhodiola rosea* (L.) Scop. In *Les Cultures Médicinales Canadiennes*. Ottawa: NRC Research Press.

Soltis, D. and M. Gitzendanner. 1999. Soltis Lab AFLP protocol.

Soni, K., S. Rawat, A. Gupta, K. Yangzom, S. Pandit, P.K. Naik, and H. Singh. 2010. Genetic characterisation of *Rhodiola rosea* using gene specific SSR and CAPS molecular markers. *Genetic Engineering and Biotechnology Journal* 2010: 1–10.

Stevens, P.F. 2001. Angiosperm Phylogeny Website. Version 12, July 2012 (and more or less continuously updated since). http://www.mobot.org/MOBOT/research/APweb/.

Swofford, D.L. 2002. *PAUP: Phylogenetic Analysis Using Parsimony (and Other Methods), Version 4.0b10*. Sunderland, MA: Sinauer Associates.

Taberlet, P., L. Gielly, G. Pautou, and J. Bouvet. 1991. Universal primers for amplification of three non-coding regions of chloroplast DNA. *Plant Molecular Biology* 17 (5): 1105–9.

Tasheva, K. and G. Kosturkova. 2011. *Rhodiola rosea* L. in vitro plants morphophysiological and cytological characteristics. *Romanian Biotechnological Letters* 16 (6):79–85.

Tasheva, K. and G. Kosturkova . 2012. The role of biotechnology for conservation and biologically active substances production of *Rhodiola rosea*: Endangered medicinal species. *The Scientific World Journal* 2012: 1–13.

Uhl, C.H. 1952. Heteroploidy in *Sedum rosea* (L.) Scop. *Evolution* 6. JSTOR: 81–6.

United States Department of Agriculture. 1999. Natural Resources Conservation Service: Plants Database. http://plants.usda.gov/.

Vos, P., R. Hogers, M. Bleeker, M. Reijans, T. van de Lee, M. Hornes, A. Friters et al. 1995. AFLP: A new technique for DNA fingerprinting. *Nucleic Acids Research* 23 (21): 4407–14.

Ward, J.H. 1963. Hierarchical grouping to optimize an objective function. *Journal of the American Statistical Association* 58 (301): 236–44.

Xia, T., S. Chen, S. Chen, and X. Ge. 2005. Genetic variation within and among populations of *Rhodiola alsia* (Crassulaceae) native to the Tibetan Plateau as detected by ISSR markers. *Biochemical Genetics* 43 (3/4): 87–101.

Yanbaev, Y.A., N.R. Bairamgulov, N.N. Redkina, and R.Y. Mullagulov. 2007. Differentiation among populations of the *Rhodiola iremelica* Boriss. (Grassulaceae) in the southern Urals. *Russian Journal of Genetics* 43 (11):1314–18.

Zhang, J.-Q., S.-Y. Meng, J. Wen, and G.-Y. Rao. 2014. Phylogenetic relationships and character evolution of *Rhodiola* (Crassulaceae) based on nuclear ribosomal ITS and plastid *trnL-F* and *psbA-trnH* sequences. *Systematic Botany* 39 (2): 441–51.

2 Ethnobotany and Conservation of *Rhodiola* Species

Alain Cuerrier, Youri Tendland, and Michel Rapinski

CONTENTS

2.1 Introduction 35
2.2 Ethnobotany of *Rhodiola* Species 36
    2.2.1 *Rhodiola* Species in Central Asia 36
    2.2.2 Circumpolar Ethnobotany of *R. rosea* and *R. integrifolia* 39
        2.2.2.1 Eurasia 41
        2.2.2.2 North America 43
    2.2.3 *Rhodiola* in the Modern Day 44
2.3 Conservation of *Rhodiola*: An Upcoming Challenge 45
    2.3.1 Conservation in the Old World 46
        2.3.1.1 Species Status and Protection 46
        2.3.1.2 Genetic Diversity 49
        2.3.1.3 Biotechnology 50
    2.3.2 Conservation in the New World 51
    2.3.3 Cultivation: The Future of *Rhodiola rosea*? 54
References 56

2.1 INTRODUCTION

The genus *Rhodiola* is a fascinating one in terms of its wide and diversified uses and also for its widespread distribution. Interest in medicinal plants worldwide and, more recently, in adaptogens has targeted and publicized such plants as ginseng, astragalus, and roseroot. A large number of natural health products can be found on the shelves of natural food stores or within different websites. Besides the extensive work done by Alm (2004) in Norway, no synthesis has been published about the uses of *Rhodiola* spp. on a larger scale. Similarly, the conservation issues related to *Rhodiola* wild-crafting are still largely unknown and no real project looking at its impact has been done. Nevertheless, we do have some information, in some cases more detailed than others, stemming from different countries where *Rhodiola* spp. are facing overharvesting. Conservation can only be accomplished with the help of local people, hence the importance of an ethnobotanical approach. First we deal with local/traditional knowledge and then with the conservation status of this important adaptogenic plant.

2.2   ETHNOBOTANY OF *RHODIOLA* SPECIES

2.2.1   *RHODIOLA* SPECIES IN CENTRAL ASIA

There are an estimated 90 species belonging to the genus *Rhodiola* (Stevens 2001; Mayuzumi and Ohba 2004), which thrive in the cold regions of the northern hemisphere (Lei et al. 2003). Although certain species display a circumpolar distribution, the genus is thought to have originated in the mountainous regions of southwest China and the Himalayas (Brown, Gerbarg, and Ramazanov 2002). There, the Qinghai-Tibetan plateau primarily covers parts of the Tibet Autonomous Region and Qinghai Province in western China, as well as northern India's Jammu and Kashmir state. Not surprisingly, this region contains a high diversity of *Rhodiola* species. The flora of China lists 55 species (Mayuzumi and Ohba 2004), a large number of which are employed in traditional Chinese and Nepalese medicine, the Sow rigpa, or Tibet's traditional medicine system, and those of the region's ethnic minorities. Altogether, over 20 species are used throughout Asia (Kelly 2001) (Table 2.1).

In China, the roots and rhizomes of *Rhodiola crenulata*, *da hua hong jing tian*, are popularly and widely used for treating a variety of health problems. Employed for activating blood circulation, thoracic obstruction, cardiodynia, apoplexy, hemiplegia, lassitude, and asthma, it is, nonetheless, the only species included in the official Chinese pharmacopoeia (Li and Zhang 2008b). Local communities, however, have made use of a variety of locally found *Rhodiola* species for the treatment of similar and some entirely different symptoms. In Yunnan Province, the Lisu People use *R. yunnanensis*, locally known as *mingleshi*, where the whole plant is made into a poultice for the treatment of fractures and rheumatoid arthritis, as well as infections such as mastitis or furuncles (Ji, Shengji, and Chunlin 2004), while the Yi people know it as *haisainai* and use it for laryngitis, dysentery, wounds, and fractures (Long et al. 2009).

Conversely, *R. crenulata,* called *suo-luo-ma-bu*, is well known and is present in the Sow rigpa, or Tibetan medicine, and used in similar fashion to that of Chinese medicine (Lei et al. 2003; Li and Zhang 2008a; Byg, Salick, and Law 2010). Although it is difficult to determine which traditional system has had the most influence on the other, it is undeniable that there exists a long history of knowledge transfer between both Tibetan and Chinese traditional medicine. Multiple *Rhodiola* species, and their uses, such as those of *R. kirilowii, R. yunnanensis, R. fastigiata, R. imbricata,* and *R. quadrifida*, are indeed common to both (Table 2.1).

The number of *Rhodiola* species used in traditional Chinese medicine may very well have been influenced by various cultures within and around its borders. Altogether, the *Encyclopedia of Traditional Chinese Medicines* provides specific uses for *R. crenulata, R. kirilowii, R. quadrifida, R. sacra,* and *R. yunnanensis* and makes mention of *R. atuntsuensis, R. algida, R. coccinea, R. himalensis,* and *R. subopposita* as well (Zhou, Xie, and Yan 2011). Much like in Tibetan medicine where species such as *R. alsia* and *R. chrysanthemifolia* have been used as a substitute to more popular *Rhodiola* (Xia et al. 2005, 2007), many of these are more often used as

TABLE 2.1
List of *Rhodiola* Species Used across Central Asia

Species	Use[a]	Country/Origin[b]	References
(1) *R. algida* (Ledeb.) Fisch. & C. A. Mey.	M	China (TCM)	Li et al. (2009), Zhou, Xie, and Yan (2011)
(2) *R. alsia* (Fröd.) S. H. Fu	M	China (TTM)	Xia et al. (2005)
(3) *R. atuntsuensis* (Praeger) S.H. Fu	M	China (TCM)	Zhou, Xie, and Yan (2011)
(4) *R. chrysanthemifolia* (H. Lév.) S.H. Fu	M	China (TTM)	Xia et al. (2005)
(5) *R. coccinea* (Royle) Boriss.	M	China (TCM)	Zhou, Xie, and Yan (2011)
(6) *R. crenulata* (Hook. f. & Thomson) H. Ohba	M	China (TCM, TTB)	Lei et al. (2003), Li and Zhang (2008b), Byg, Salick, and Law (2010), Zhou, Xie, and Yan (2011)
(7) *R. fastigiata* (Hook. f. & Thomson) S.H. Fu	M	China (TCM, TTB)	Hui et al. (2002), Lei et al. (2003), Li and Zhang (2008b), Liu et al. (2006)
	M	India (TTB)	Angmo, Adhikari, and Rawat (2012)
(8) *R. gelida* Schrenk	M	Tajikistan (Pamiri people)	Kassam et al. (2010)
	M	Afghanistan (Pamiri people)	Kassam et al. (2010)
(9) *R. heterodonta* (Hook. F. & Thomson) Boriss.	M	Uzbekistan China (TTB) India (TTB) Nepal	Yousef et al. (2006) Kumar et al. (2011) Sharma et al. (2011) Rokaya et al. (2012)
	F	India	Rana et al. (2011)
(10) *R. himalensis* (D. Don) S.H. Fu	M	Nepal (+, TTB) China (TCM) India (+, TTB)	Uprety et al. (2010), Humagain and Shrestha (2010), Ghimire and Aumeeruddy-Thomas (2009), Zhou, Xie, and Yan (2011) Kumar, Paul, and Anand (2009), Angmo, Adhikari, and Rawat (2012)
(11) *R. imbricata* Edgew.	M	China (TCM, TTB) India (+ TTB)	Kumar et al. (2011), Gupta et al. (2012) Angmo, Adhikari, and Rawat (2012); Shah, Abass, and Sharma (2012)
	F	India	Rana et al. (2011)
(12) *R. kirilowii* (Regel) Maxim.	M	China (TCM, TTB) Kyrgyzstan	Wójcik et al. (2009), Li and Zhang (2008b), Zhou, Xie, and Yan (2011), Zuo et al. (2007), Shang et al. (2012), Zaurov et al. (2013)
	V	China (TTB)	Shang et al. (2012)
(13) *R. pamiroalaica* Boriss.	M	Kyrgyzstan	Ismailov et al. (1998)

(Continued)

TABLE 2.1 (Continued)
List of *Rhodiola* Species Used across Central Asia

Species	Use[a]	Country/Origin[b]	References
(14) *R. pinnatifida* Boriss.	M	Russia	Hanelt (2001)
		Mongolia	Hanelt (2001)
(15) *R. quadrifida* (Pall.) Fisch. & C.A. Mey.	M	Mongolia China (TCM, TTB)	World Health Organization (2013) Li and Zhang (2008b), Zhou, Xie, and Yan (2011), Yoshikawa et al. (1995)
(16) *R. sachalinensis* Boriss.	M	China (TCM)	Tingli Li et al. (2007), Yu et al. (2011)
(17) *R. sacra* (Prain ex Raym.-Hamet) S.H. Fu	M	China (TCM)	Wang et al. (2004), Zhou, Xie, and Yan (2011), Wong et al. (2006), Ohsugi et al. (1999)
(18) *R. semenovii* (Rgl. Et Herd.) Boriss.	M	Kyrgyzstan	Yousef et al. (2006)
(19) *R. subopposita* (Maxim.) Jacobsen	M	China (TCM)	Zhou, Xie, and Yan (2011)
(20) *R. tangutica* (Maxim.) S.H. Fu	M	China (TCM, TTB)	Zhang, Ma, and Yuan (2010), Shang et al. (2012)
	V	China (TTB)	Shang et al. (2012)
(21) *R. wallichiana* (Hook.) S.H. Fu	M	Nepal	Rokaya et al. (2012), Manandhar (2002)
(22) *R. yunnanensis* (Franch.) S.H. Fu	M	China (+, TCM, TTB)	Li and Zhang (2008b), Zhou, Xie, and Yan (2011), Ji, Shengji, and Chunlin (2004), Long et al. (2009)

Names based on The Plant List, a collaboration between KEW, Royal Botanical Garden, and Missouri Botanical Garden (The Plant List 2010) and the Flora of China (eFloras 2008).

[a] F, Food; M, Medicine; V, Veterinary.

[b] + including tribal communities of different ethnic origin; TCM, tradition Chinese medicine; TTB, traditional Tibetan medicine (or Sow-rigpa).

a substitute to, and even sold as, *R. crenulata* in the Chinese market (Li and Zhang 2008b). Such accidental (or even intentional) substitutions pinpoint the importance of these *Rhodiola* species in the Chinese economy.

The Sow rigpa contains elements of Chinese medicine, but Ayurveda and the traditional healing systems of Nepal, Persia, and Mongolia are also thought to have contributed to its evolution (Kumar et al. 2011). Indeed, *R. quadrifida*, known as ***tsan*** in Tibet, goes by ***dorvolson mugez*, *altan gagnuur*, or *zerleg mugez*** in Mongolia and is used traditionally for enhancing strength and oral health (World Health Organization 2013). In India's Jammu and Kashmir state, the root of *R. imbricata*, known as ***dhodlli*** by the local communities of Gujjar and Bakerwal, is chewed for fatigue (Shah, Abass, and Sharma 2012), while the whole plant of *R. fastigiata*, known as ***rholo mukpo***, is used for coughs and congested chests (Angmo, Adhikari, and Rawat 2012). The root bark of *R. himalensis*, a species also used by the Tamang people in central Nepal to treat gastrointestinal disorders, fevers, and headaches (Uprety et al. 2010; Humagain and Shrestha 2010), is known as ***dand jari*** and is used for oral health by the people of Kishtwar district of India's Jammu and Kashmir state (Kumar, Paul, and Anand 2009).

Evidently not restricted to China and Tibet's pharmacopoeia, the ethnobotany of *Rhodiola* species extends itself not only to the surrounding of the Qinghai-Tibetan plateau, but beyond. *R. heterodonta*, a medicinal plant common to both Nepalese and Tibetan systems (Kumar et al. 2011; Rokaya et al. 2012), can be found in Uzbekistan's folk medicine (Yousef et al. 2006), while in Kyrgyzstan, an infusion of the underground parts of *R. kirilowii* is used to treat weariness, neurotic conditions, and decreased ability to work (Zaurov et al. 2013). In this country, *R. semenovii* is also known for its endurance enhancing and antihypoxic properties (Yousef et al. 2006), and other species, such as *R. pamiroalaica*, are also present within the country's traditional folk medicine (Ismailov et al. 1998). Nearby, another species, *R. gelida*, is used by the Pamiri people of Tajikistan's and Afghanistan's Pamir Mountains (Kassam et al. 2010). Further east, in Mongolia and Russia, *R. pinnatifida* is used medicinally like the more widespread *R. rosea* (Table 2.2) and is even taken into cultivation in southern Siberia (Hanelt 2001).

As can be seen by the variety of uses for *Rhodiola* species across Central Asia, it is apparent that they are prized mainly for their medicinal potential. Although these have been primarily for human use, *R. algida* has also been collected for veterinary purposes in Tibetan medicine (Shang et al. 2012). Local communities in the Indian Himalaya are also known to use *Rhodiola* species for ornamental purposes and fodder while the fresh leaves of *R. imbricata* and *R. heterodonta* are prepared for human consumption in a dish called ***tantur*** (Rana et al. 2011; Gupta et al. 2012). It appears, however, that the versatility of *Rhodiola* species in its traditional and modern uses, as a food source, in material culture and in medicine, is further understood when considering the widespread global distribution of *R. rosea* and *R. integrifolia* on the northern hemisphere.

2.2.2 CIRCUMPOLAR ETHNOBOTANY OF *R. ROSEA* AND *R. INTEGRIFOLIA*

R. rosea is one of the most studied species of the genus. Its interest in the medical literature worldwide may be attributed to its wide distribution, where it can be found anywhere in Low-Arctic to high-temperate regions of Asia, Europe, Greenland, and North America (Scoggan 1978). Commonly called roseroot, the plant, however, is known under a multitude of names depending on its ethnobotanical origin. It may be known as arctic root, in reference to its distribution among arctic regions, or golden root, perhaps as a reference to the perceived value of the root. The Mongolian name is ***altan gagnuur***, meaning golden soldier (Shatar, Adams, and Koenig 2007), or ***yagaan mugez*** (World Health Organization 2013). In Russia, it is known as ***rodiola rozovaia*** (Hanelt 2001; Davydov 2011), a direct translation of its Latin binomial name, but also goes by ***zolotoi koren***, which translates to golden root (Hanelt 2001; Metzo 2011). Within Russia, it is also locally known to the Ural's Komi people as ***dzhurtan turun*** (Ivanitskiy 1890), and to Siberia's Evenkis as ***uildyn*** or ***uildyun*** (Davydov 2011). In Canada, Inuit from Nunavik call it ***tallirunnaq*** or ***tullirunnaq***, and ***uqaujatuinnait*** or ***utsuqammat*** when referring specifically to the leaves or roots, respectively (Cuerrier and Elders of Kangiqsualujjuaq 2011). Inuit from Nunatsiavut, on the other hand, refer to the roseroot plant as ***tulligunnak***, and the

TABLE 2.2
Circumpolar Use of *R. integrifolia* and *R. rosea*

Use[a]	Country/Origin[b]	References
R. integrifolia		
M	USA/Alaska (Cup'it, Yup'ik)	Griffin (2001), Ager and Ager (1980), Lantis (1959), Smith (1973)
F	USA/Alaska (Cup'it, Yup'ik, Iñupiat)	Griffin (2001), Ager and Ager (1980), Anderson (1939), Heller (1953), Lantis (1946), Hughes (1960), Jones (2010)
***R. rosea*[c]**		
M	Russia/Siberia (+, Evenkis)	Alm (2004), Davydov (2011)
	Georgia	Brown, Gerbarg, and Ramazanov (2002)
	China	Brown, Gerbarg, and Ramazanov (2002)
F	Russia/Siberia (Chukchi)	Porsild (1953), Nordenskiöld (1882)
MC	Russia/Siberia (Evenkis)	Davydov (2011)
R. rosea		
M	Sweden	Sparschuch (1775)
	Iceland	Alm (2004), Hjaltalin (1830)
	Denmark	Alm (2004)
	Norway (+, Sámi, Russians)	Alm (2004), Galambosi (2006), Ryvarden (1993)
	France	Virey (1811)
	Italy	Alm (2004)
	Bulgaria	Nedelcheva (2012)
	Russia (+, Komi)	Mamedov (2005), Mamedov, Gardner, and Craker (2005), Shatar, Adams, and Koenig (2007), Ivanitskiy (1890), Popov (1974), Iljina (1997)
	China (TCM, TTB)	Shatar, Adams, and Koenig (2007), Yu et al. (2011), Li (2002)
	Mongolia	Shatar, Adams, and Koenig (2007), Brown, Gerbarg, and Ramazanov (2002), World Health Organization (2013)
	Canada (Inuit)	Cuerrier and Hermanutz (2012)
V	Britain/Ireland	Allen and Hatfield (2004)
	Norway	Alm (2004)
F	Mongolia	Shatar, Adams, and Koenig (2007)
	Norway (+, Sámi, Russians)	Alm (2004), Galambosi (2006), Kylin (2010), Dragland (2001)
	Greenland (Inuit)	Birket-Smith (1928), Høygaard (1941), Porsild (1953), Rink (1974), Egede (1818)
	Canada (Inuit)	Blondeau (2009), Cuerrier and Elders of Kangiqsualujjuaq (2011); Cuerrier and Hermanutz (2012)
MC	Norway	Kylin (2010), Hanelt (2001), Alm (2004)

[a] F, Food; M, Medicine; MC, Material Culture; V, Veterinary.
[b] + including uses by people not belonging to the country's ethnic minorities.
[c] May be *R. integrifolia* due to its location or history.

rhizome as ***utsuKammak*** (Cuerrier and Hermanutz 2012). Finally, there are over 50 different names in Norwegian or Sámi (Alm 2004).

In North America and Asia, there are two species, *R. integrifolia* and *R. rosea*, that were once thought to be subspecies of *R. rosea*. As a consequence, both species have been commonly referred to as roseroot and information about these in the literature can be confusing. Although their ranges overlap in Siberia, a geographical dichotomy exists between the two in North America. *R. rosea* is found on the eastern part of the continent, and in the alpine habitats of Europe and Asia, while the other is found in western North America (Hermsmeier, Grann, and Plescher 2012). As a result, it is now fair to assume that references to roseroot from western North America are in fact *R. integrifolia*, instead of *R. rosea*, though this is not necessarily true for those from Siberia. Regardless of this, both species have a long tradition of use in North America and Asia. The succulent young stems and leaves have been eaten raw or cooked (Porsild 1953). Among ethnic groups in Alaska and Siberia, they have been one of the 20 most used plants where parts have been soured, something similar to sauerkraut (Källman 1997; Jones 2010). Although little is known about the pharmacology and phytochemistry of *R. integrifolia*, herbalists suggest that it may be used similarly to *R. rosea* for medicinal applications (Gray 2011). Indeed, these two species have been traditionally used as medicine and food across North America, Europe, and Asia (Table 2.2).

2.2.2.1 *Eurasia*

The first written reference of roseroot's medicinal use dates back to AD 77, when the Greek physician and botanist Dioscorides wrote of *rodia riza* in *De Materia Medica* (Brown, Gerbarg, and Ramazanov 2002; Hedman 2000). Along with Roman legions, knowledge of this plant, under the name of *R. riza,* was carried to various parts of Europe, including England. It was later sold, during the Middle Ages, in pharmacies under the name *lignum rhodium* (Hedman 2000). Adding it to his 1749 book, *Materia Medica*, noting its astringent and styptic properties, and its use in the treatment of leucorrhoea, hysteria, dysentery, hernia, and headaches, the Swedish botanist, Carl Linnaeus, gave the plant its Latin name, *R. rosea*, with its specific epithet referring to the rose-like fragrance of the fresh or dried root (Brown, Gerbarg, and Ramazanov 2002). Not long after was it included in the first official Swedish pharmacopeia (Sparschuch 1775).

Including Sweden, *R. rosea* has appeared in the literature of many European countries. It is included in early collections on herbal remedies such as Fuchs' 1543 *New Kreüterbüch* (Fuchs 1964). Noted for its rose-like taste, it has appeared in the French *Materia medica*, where it has been described as being very cephalic and astringent (Virey 1811). In the north of Italy, a decoction prepared in milk was used to promote pregnancy (Alm 2004). Uses of *R. rosea*, however, seem to find their way in the traditional medicines of many places of Norse origin and Alm (2004) has written an extensive review pertaining to its uses in Norway. In Norwegian traditional medicine, roseroot was also considered an astringent; it was used to treat wounds, swollen limbs, lung disorders and ***mosott***, a frequently diagnosed, yet ill-defined disease of the folkloric medicine, while the Sámi made a decoction to treat urinary disorders (Alm 2004). Furthermore, in Iceland and Denmark's Faroe

Islands, a decoction of the roots was used on wounds and as a hair wash (Alm 2004). The Icelandic botanist Hjaltalin (1830) describes in his *Islands botanik* its use for several illnesses, at a dose of two teaspoonful daily from a rhizome and root decoction.

Various ethnic groups from Russia have been historically known to employ rose-root for medicinal purposes. In the Ural Mountains, its health benefits were known to the Komi people, also known as Zyrian (Ivanitskiy 1890; Popov 1974), who collected and kept the dry roots in small birch bark boxes (Iljina 1997). Used as a general fortifying remedy for weariness and nervous diseases, the root tincture also found a more general use as a panacea (Iljina 1997). It was particularly prized by the hunters who, living far from home for months in the extreme conditions of the northern taiga, valued the root for giving new vigor and stimulating the organism (Iljina 1997). In Siberia, the Evenkis people have used it to treat flatulence. It is now used as a universal medicinal tea and its name *uildyun*, meaning liver, reflects the beliefs of the plant's benefits toward that organ (Davydov 2011).

Siberians, in general, used it to ensure long life and as an aphrodisiac (Alm 2004). Newly married couples in Siberia were presented with a bouquet of roseroot to symbolize and encourage a fruitful union (Alm 2004). A similar custom is still observed in the mountain villages of the Republic of Georgia (Brown, Gerbarg, and Ramazanov 2002), where Siberians exported the root via the Caucasian Mountains to trade for wine, honey, fruits, and garlic (Brown, Gerbarg, and Ramazanov 2002).

Indeed, it would seem that roseroot was the subject of much ancient trade. It is said that in ancient China, emperors, in their search for longevity and superior health, would send expeditions to Siberia to trade for the valuable medicinal plant (Brown, Gerbarg, and Ramazanov 2002), where roseroot was considered a "gift of the spirits." Indeed, its uses reached into traditional Tibetan medicine and beyond (Shatar, Adams, and Koenig 2007). In Mongolia, roseroot was used in the treatment of tuberculosis and cancer (Brown, Gerbarg, and Ramazanov 2002), as a mouthwash for bad breath, treating inflammation of the lung, fever, strengthening of the body (World Health Organization 2013), as well as long-term illness and weakness due to infection (Shatar, Adams, and Koenig 2007). While on the other hand, the uses of *R. rosea* in Bulgarian folk medicine are thought to have been influenced by Asian medicinal systems (Nedelcheva 2012).

In addition to its traditional medicinal uses, roseroot is also an important traditional food source. Every part of the plant is edible, from its leaves and shoots down to the rhizome and root. In 1599, when the Danish-Norwegian king, Kristian IV, visited Norway, he was served *R. rosea* rhizomes (Dragland 2001; Galambosi 2006; Kylin 2010), where it was reportedly used by both the Sámi and the Russians inhabiting the area to treat scurvy (Alm 2004; Galambosi 2006). In Norway, traditional uses vary from eating the leaves raw and cooked to mixing them grounded into bread dough (Alm 2004). On the Kola peninsula, the Sámi and Russians are reported to have also used the roots to flavor beer (Alm 2004) and in Siberia, the Chukchi not only preserved roseroot's stalks and leaves in seal-skin sacks (Nordenskiöld 1882), but also favored roseroot's flowering stems as a delicacy (Porsild 1953).

Roseroot is also used in spiritual and material culture. Siberia's Evenkis made rosaries from cut-up pieces of the rhizome, a use for which the name *uildyn*, meaning connection or bundle, may be derived from Davydov (2011). In Norway, it was historically employed in the planting of turf roofs, as it was traditionally believed to protect against fire by averting the anger of the Norse god Thor, who controlled thunder and lighting (Alm 2004). For its insulating properties and while being low-maintenance and fire-resistant, it was endorsed by the Norwegian government in 1762 (Hanelt 2001; Alm 2004; Kylin 2010). Roseroot was also used to dye woolen textiles, imparting a green hue when prepared with alum (Kylin 2010). In cosmetic applications, roseroot was traditionally used in Norway as a hair wash, both for its pleasant fragrance, now coveted by the perfume industry (Dragland 2001), and to encourage hair growth (Ryvarden 1993; Alm 2004).

Roseroot has also been used as forage or fodder for cattle, as indicated by several of its traditional Norwegian names, such as **kalvegror** or "calf growth" (Alm 2004). Its vitamin C content and antiscorbutic properties may have been important in maintaining healthy livestock, particularly at winter's end, when other vitamin C–rich foods in scarce supply would be reserved for human consumption (Alm 2004). Furthermore, it is reported in folk Norwegian veterinary medicine to be used as "horse medicine" and given to cattle to treat intestinal parasites (Alm 2004), while in Britain and Ireland, it was mixed with *Ligusticum scoticum* L. (Scotch lovage) to be used as a purge for calves (Allen and Hatfield 2004). In Southern Greenland, roseroot can be observed where sheep cannot get access to it as they are highly appreciated and preferentially grazed by animals (Kylin 2010).

2.2.2.2 *North America*
In the Arctic, in early springtime, the shoots of roseroot often emerge prior to other vegetation. At that time, the fleshy, tuberous rhizomes are eaten (MacKinnon et al. 2009), where the taste is said to vary from sweet to bitter (Jones 2010). The uses of *R. integrifolia* and *R. rosea* in North America appear mostly among nations of the Eskimo-Aleut linguistic family, although they have been reported in books of edible plants seemingly geared toward the general population. In one such book for Alaska, *R. integrifolia* is reported (Heller 1953), and for Newfoundland and Labrador, *R. rosea* appears (Scott 2010). In both, the leaves and stems are written to be good raw in mixed salads or cooked as a green vegetable. This knowledge most likely seems to have stemmed from the local knowledge of people inhabiting the Arctic. Indeed, written records show these two species to be most popularly used as food.

2.2.2.2.1 *Rhodiola integrifolia*
In western North America, western roseroot, *R. integrifolia,* is given various names by diverse groups of the Eskimo-Aleut linguistic family. In Alaska, the Iñupiat call it **eveeahkluk** (Heller 1953), **eluaklak** (Anderson 1939), or **iviaqtuk** and **ikutaq** (Jones 2010). The Yup'ik call it **cuqlamcaraat** (Ager and Ager 1980), and the Cup'it call it **megtat neqiat** or **ca'klax** (Griffin 2001). The whole plant, including inflorescences and underground parts, have been used as a food source by all three groups (Anderson 1939; Heller 1953; Hughes 1960; Ager and Ager 1980; Griffin 2001). Furthermore, the Iñupiat of Northwest Alaska and the Yup'ik from St. Lawrence Island ate the aerial parts and fermented them in water to preserve them for later usage (Anderson 1939; Hughes 1960; Jones 2010). The resulting juice and fermented

plants could be consumed together with any kind of blubber or oil, such a walrus (Hughes 1960; Jones 2010).

As a green vegetable, it is an important food source. Hughes (1960) considered *R. integrifolia* to be the most important of the "greens" collected by the Yup'ik throughout the summer. It is thus not surprising that the Iñupiat actively spread the plant by digging out the roots and planting them elsewhere so that they could be harvested at a later time, in case of hard time and scarcity of food (Jones 2010). On top of providing a natural source of niacin, iron, phosphorus, protein, carbohydrates, calcium, and vitamin A, *R. integrifolia* is particularly rich in vitamin C (Kershaw 2000; Jones 2010), conferring important health benefits, as an antiscorbutic, for example, after a long winter, at a time and place when few other green vegetables are available.

Naturally, the people of Alaska also traditionally employ it as a medicine. Of Nelson Island, the Yup'ik chewed the rhizomes for mouth sores (Ager and Ager 1980). While on Nunivak Island, the Cup'it used the leaves of *R. integrifolia* (Griffin 2001) and its inflorescences in infusion with *Rhododendron tomentosum* Harmaja for medicine (Smith 1973), although such a tea was also simply enjoyed for its taste (Lantis 1946). Such a decoction of the inflorescences was employed for stomach or intestinal discomforts as well, and the plant was eaten raw for tuberculosis (Lantis 1959).

2.2.2.2.2   Rhodiola rosea

In eastern North America, consumption of *R. rosea* as a food is just as common as it is for its western counterpart, *R. integrifolia*. The Inuit from Nunavik, in northern Québec, prepare the rhizomes of *R. rosea* in **suvak** or with seal oil (Blondeau, Roy, and Cuerrier 2010; Cuerrier and Elders of Kangiqsualujjuaq 2011). Further east, the Greenlandic Inuit store the aerial parts with seal blubber in skin bags called **imigarmît** and eat it with dried blood (**akâq**) (Høygaard 1941). In general, the whole plant is consumed, raw or cooked, such as a potherb, by the Greenlandic Inuit (Egede 1818; Birket-Smith, 1928; Porsild, 1953; Rink, 1974) and by the Inuit of Nunatsiavut, in northern Labrador (Cuerrier and Hermanutz 2012) and of Nunavik (Blondeau 2009; Cuerrier and Elders of Kangiqsualujjuaq 2011). Like in Europe and Asia, where the roseroot is prized primarily for medicinal purposes, it is an important medicinal plant to the Inuit of Nunatsiavut who extensively utilize the roots of *R. rosea* for infections, colds, fatigue, and toothache and is one of the important medicinal plants of Labrador (Cuerrier and Hermanutz 2012).

2.2.3   RHODIOLA IN THE MODERN DAY

According to folkloric tradition, the Vikings used roseroot for its strengthening action while doing hard work (Brown, Gerbarg, and Ramazanov 2002; Panossian, Wikman, and Sarris 2010), enhancing effects on endurance (Magnusson 1992), and before going into battle (MacKinnon et al. 2009). Although the veracity of this is believed to be somewhat questionable by some (Panossian, Wikman, and Sarris 2010), there appears, nonetheless, to be an enormous amount of consensus as to the use of this plant in traditional and folkloric medicines for increasing energy.

At one point or another, *R. rosea* has appeared, and still appears in some cases, in the official Pharmacopoeia of multiple European countries, notably, Sweden (Sparschuch 1775), France, Estonia, and Russia (Panossian, Wikman, and Sarris 2010). Naturally, many reports and studies exist on the phytochemistry and pharmacology of *R. rosea* in Scandinavian literature (Khanum, Bawa, and Singh 2005), and particularly Russian literature (Galambosi 2006).

Although Russian literature has dominated the research on *R. rosea* as an adaptogen since the Cold War era, there has been a considerable increase in interest by the general scientific community since the turn of the millennium (for an extensive summary on the phytochemistry and pharmacology of *R. rosea*, see Brown, Gerbarg, and Ramazanov [2002]; Khanum, Nawa, and Singh [2005]; and Panossian, Wikman, and Sarris [2010]). The plant is now popular in Asia and Eastern Europe with a reputation of stimulating the nervous system, decreasing depression, enhancing work performance, eliminating fatigue, and preventing high-altitude sickness (Khanum, Bawa, and Singh 2005).

The increased attention on the medicinal potential of this plant, along with the popular *R. crenulata* in traditional Chinese medicine, has led to concerns about overharvesting. Indeed, the Evenkis sometimes sell dried roseroot rhizomes as a means of supplementary revenue, yet losses in populations have been reported near Mt. Dovyren in Siberia (Davydov 2011). In the search for substitutes to *R. rosea*, many species of *Rhodiola* from Kyrgyzstan have been the subject of phytochemical research, such as *R. linearifolia* Boriss. *R. coccinea*, *R. pamiroalaica*, *R. kaschgarica* Boriss. *R. gelida*, and *R. litwinowii* Boriss. (Chaldanbaeva, Nuralieva, and Kalykeeva 2012). In central Asia, salidroside (Panossian, Wikman, and Sarris 2010), one of the main active ingredients of *R. rosea*, has been isolated in multiple *Rhodiola* species of ethnobotanical importance, such as *R. kirilowii* (Wiedenfeld et al. 2007), *R. algida* (Li et al. 2009), *R. sacra*, and *R. sachalinensis* (Mook-Jung et al. 2002; Li et al. 2007).

R. sachalinensis is becoming one of the most popular traditional Chinese medicines from *Rhodiola* species (Li et al. 2007), and others, including *R. kirilowii* (Zuo et al. 2007). *R. quadrifida* (Wójcik et al. 2008), and *R. sacra* are now the focus of substantial interest in terms of medical and pharmacological properties (Alm 2004). These have been studied not only for their adaptogenic potential but also for their immunomodulatory (Li et al. 2009) and their neuroprotective effects (Mook-Jung et al. 2002). Needless to say, the increased interest in the medicinal potential of *Rhodiola* species and the pressing need to find a substitute for *R. rosea* and *R. crenulata* in the market place raises the question about the conservation of the genus.

2.3　CONSERVATION OF *RHODIOLA*: AN UPCOMING CHALLENGE

With such a great number of species presenting medicinal qualities, one can almost predict that this taxon will suffer from overharvesting (Kylin 2010). Indeed, like many other medicinal plants, nearly all of its supply comes from the wild (Galambosi 2005). Worldwide, over 50,000 medicinal plants are used and two-thirds come from natural populations (Edwards 2004). The World Health Organisation has warned, in a number of publications (see OMS 2003; Bodeker et al. 2005), against overharvesting of wild habitats. Nonetheless, relentless harvesting is still happening in many countries. (Bulgaria is one example and samples bought on the Internet are shipped from this country, although the source of origin cited on the website mentioned Russia.) *Rhodiola*'s are renown for their numerous health benefices, and with great "powers" comes an even greater reputation. The crown on *Rhodiola*'s reputation comes from the discoveries of powerful adaptogens in their root extract, explaining the biggest part of its success (Brekhman and Dardymov 1968; Germano and Ramazanov 1999). It also explains its possible demise in many countries. Since the root is the coveted part for medicinal use, harvesting *Rhodiola* has a direct negative impact on its survival. This, coupled with a distribution restricted to northern latitude and high mountains (Small and Catling 1999), the severe climate condition of such a distribution, the very low seed germination (in the field) and low vegetative propagation rate of wild specimen (Tasheva and Kosturkova 2010, 2012), leads to a slow establishment and development of the plant (Kylin 2010). This considerable sensitivity to harvest combined with its great reputation gained through several centuries and the growing world demand for medicinal plants (Cavalière et al. 2010) are the major reasons why many species of the genus are endangered in numerous parts of the world. Recently, concerns have been raised pertaining to the additional impact of climate change on the survival of alpine, arctic, and subarctic plants (Cavalière 2009; Downing and Cuerrier 2011) and *R. rosea* could be among the victims.

2.3.1 Conservation in the Old World

2.3.1.1 *Species Status and Protection*

Roseroot has been harvested for more than 1000 years in different parts of the Old World (Kylin 2010) and did not go extinct, so why would it be endangered now? This long-term use happened before commercial exploitation, at a time where only healers or some family members knew where to harvest the wild roseroots and the secret of its preparation (Galambosi 2006; Kylin 2010). In southwestern China, for example, Lei et al. (2006) report an accelerated and uncontrolled use of *Rhodiola* species during the 1980s, to the extent that species such as *R. sachalinensis* (Yan et al. 2003) are now listed as endangered throughout China.

Today, the whole taxon has not yet been assessed for the IUCN Red list (Bilz et al. 2011). Nevertheless, local assessments for many countries, using the same classification categories of the IUCN Red list, paint a bleak picture. A number of *Rhodiola* species are listed in categories ranging from lower risk to edge of extinction. (Table 2.3 presents *Rhodiola* conservation status in different countries.) Furthermore, the intensive geological exploration and extraction of minerals during the Socialist period of the twentieth century led to the exhaustion of natural roseroot populations in the Baikal region and northern and subarctic Urals. Roseroot has consequently been listed in Red Books of the former Soviet Union (Borodin 1985) and some of its constituting Republics, such as the Republic of Buryatia AR, the Republic of Yakut ASSR, the "Mongolian Red Book," the "Rare and Extinct Plant Species in Tuva A Republic," and the "Rare and Extinct Plant Species in Siberia" (Tasheva and Kosturkova 2012), as well as the Komi Republic (Taskaev 1999), the Central Urals, Arkhangelsk, Nenetz and Khanty Mansiysk Autonomous Area of the Russian Federation, and the Republic of Karelia (Kotiranta et al. 1998).

TABLE 2.3
Rhodiola Conservation Status in Different Countries of the Old World

Species	Countries	Species Status	Protection	Comments	References
R. fastigiata	China	Edge of extinction	–		Liu et al. (2006)
R. rosea L.	Bosnia-Herzegovina	Endangered	–		Galambosi (2006)
R. rosea L.	Bulgaria	Critically Endangered	Yes	Protected by law. Collection forbidden.	Petrova and Vladimirov (2009), Bulgarian Academy of Sciences (2011)
R. rosea L.	Czech Republic	Critically Endangered	–		Holub and Prochazka (2000)
R. rosea L.	Finland	Least concerned	–	Cultivated for introduction in Russia	Tasheva and Kosturkova (2012)
R. rosea L.	Norway	Lower risk (IUCN categories)	Yes	In the Svalbard archipelago 30 populations are protected	Engelskjøn, Lund, and Alsos (2003)
R. rosea L.	Russia	Seriously threatened	Yes	Collection is strongly regulated; Some population are included in reservation.	Galambosi (2006), Tasheva and Kosturkova (2012)
R. rosea L.	Russia: Murmansk	Vulnerable	Yes	Collection is strongly regulated	Kotiranta et al. (1998)
R. rosea L.	Russia: Republic of Karelia	Vulnerable	Yes	Collection is strongly regulated.	Kotiranta et al. (1998)
R. rosea L.	Russia: Komi Republic	–	–	–	Taskaev (1999)
R. rosea L.	Scandinavia	Little concern	–	*R. rosea* strains have been established in Norway, Sweden, Finland, and Iceland by a project of the Nordic Gene Bank	Galambosi (2005)

(Continued)

TABLE 2.3 (Continued)
Rhodiola Conservation Status in Different Countries of the Old World

Species	Countries	Species Status	Protection	Comments	References
R. rosea L.	Slovakia	Vulnerable	N3		Galambosi (2006)
R. rosea L.	Sweden	Rare	– Harvest is prohibited in Västra Götaland	In the Västra Götaland, Göteborg and Bohuslän Counties	Mossberg and Rydberg (1995), Länsstyrelsen Västra Götalands län (2010)
R. rosea L.	Switzerland	Least concerned	N5		NRL (2012)
R. rosea L.	Ukraine	Rare	– Yes	Collection of wild medicinal plant is strictly monitored	Didukh (2009), Minarchenko (2011)
R. rosea L.	Ukraine	Endangered	N2 Yes	Collection of wild medicinal plant is strictly monitored	Kricsfalusy and Budnikov (2007), Minarchenko (2011)
R. rosea L.	Great Britain	Least concerned	N5		Cheffings and Farrell (2005)
R. sachalinensis	China	Endangered	N1		Yan et al. (2003)
R. saxifragoides (Frod.) H. Ohba	Pakistan	Vulnerable (IUCN categories)	N3		Ali and Alam (2006), Alam and Ali (2010)
Rhodiola sp., *R. crenulata*	China	Class One Endangered species (China National checklist)	N1		Lei et al. (2006)

Conservation status meaning: 1, critically imperilled; 2, imperilled; 3, vulnerable; 4, apparently secure; 5, secure; –, the information was not specified in the reference and/or could not be found.

Compilation of the different threat status presented in Table 2.3 shows it is threatened, vulnerable, or rare in 15 countries out of the 20 entries, and is only protected in eight of those countries. Also, its lesser economic importance and much higher cost of collection in some European countries such as the Alps, Carpathian, and the Scandinavian countries (Galambosi 2005) make it less susceptible to exploitation in those regions, as can be seen in Table 2.3 from the least concerned status. However, in some parts of Norway, there is a well-developed network of harvesting, which includes transport of the root for commercial purposes.

Needless to say that, on a global scale, the rapidly growing demand and high price for *Rhodiola* products (Galambosi 2005; Minarchenko 2011) are increasing pressure on populations already highly exploited (Kylin 2010). To properly illustrate this, one only needs to consider Russia's annual exports, which are estimated at approximately 20–30 t of dry *Rhodiola* roots (Galambosi 2005), even though it is red listed in the country. However, the "Atlas of areas and resources of medicinal plants of Soviet Union" estimates the resources of South Siberia, the key-producing region, at 637 t (Tsikov 1980). Since the regeneration of wild populations requires 15–20 years, annual fresh root quantities are estimated at 30–40 t.

Currently, 46 companies worldwide are commercializing roseroot products, of which 30 are listed as ingredients suppliers (Ampong-Nyarko et al. 2005). And yet, there is good reason to believe that the market will continue to rise since many scientists are said to claim roseroot as a better supplement than ginkgo (*Ginkgo biloba* L.) and ginseng (*Panax* spp.) (Kylin 2010). With such an increased market pressure, high extraction rate, and increased habitat loss in different countries (Xia et al. 2005, 2007), a conservation plan is greatly needed to protect this important resource. Promising solutions for *Rhodiola*'s conservation include a greater elucidation of its genetic diversity, biotechnologies, and cultivation.

2.3.1.2 *Genetic Diversity*

Only a handful of studies have investigated *R. rosea's* genetic diversity (e.g., Elameen et al. 2008; Kozyrencko, Gontcharova, and Gontcharov 2011). Nevertheless, the genetic knowledge about the genus *Rhodiola* keeps increasing with several studies published on the subject in the past decade (e.g., Yan et al. 2003; Xia et al. 2005, 2007; Elameen et al. 2008).

In a study on 11 populations of *R. rosea* ranging from Poland to Kamtchatka in the Russian far east, Kozyrenko, Gontcharova, and Gontcharov (2011) showed a high level of diversity at the species level and a low level of diversity at the population level. The major part of the variation was found within the population, when all subpopulations were compared. These results correspond to what was found by Elameen et al. (2008) using populations from 10 counties in Norway. In the latter country, the variation was almost exclusively within the counties. Elameen et al. (2008) have also found a low level of genetic differentiation between populations, indicating a high level of gene flow that could be the result of pollen dissemination and/or seed dispersal. Furthermore, Kozyrenko, Gontcharova, and Gontcharov (2011) also demonstrate that when looking at the macro-population level, the genetic differentiation is greater between populations and the gene flow is low. Hence, when considering the Norwegian populations studied by Elameen et al. (2008) as one macro-population, it could explain the high gene flow and the higher within population variation for the range of their study. Thus, the global trend presented here for roseroot points toward regionally and genetically isolated macro-populations. It is also worth mentioning that Kozyrenko, Gontcharova, and Gontcharov (2011) found some differentiation for the Polish population that underlines two evolutionary lines of *R. rosea*.

Overall, since *R. rosea* presents a high genetic differentiation among macro-populations, a low gene flow among those populations, and at least two evolutionary lines (Kozyrenko et al. 2001), conservation strategies should consider protecting as many populations as possible *in situ* while collecting samples from more population for *ex situ* conservation. Moreover, Elameen et al. (2008), as well as Archambault (2009; see also Chapter 1), also suggested that more worldwide studies should be conducted on the genetic diversity of *R. rosea* in order to decipher the relationships among populations. In parallel, many other species of *Rhodiola* (e.g., *R. chrysanthemifolia* [Xia et al. 2007], *R. alsia* [Xia et al. 2005], and *R. sachalinensis* [Yan et al. 2003]) have also revealed the same genetic structure, with variation located mainly amongst populations, underlining the genus characteristically wide but patchy distribution. These species would also benefit from *in situ* and *ex situ* conservation policies. It is needless to say that more studies are warranted on the genetic diversity at the genus level before strict conservation measures are taken.

2.3.1.3 Biotechnology

Although genetics give a good idea of how conservation actions should be aimed, the fact remains that in order to keep exploiting a scarce and diminishing resource for medicinal purposes and providing the world growing demand in roseroot material, something must be done to start some new production, expand existing ones, and protect what is left of wild populations. Biotechnologies can bring some solutions to this conundrum. Among other things, using *in vitro* culture offers possibilities for germplasm conservation (Liu et al. 2004) and modification of the genotype of the plants (Verpoorte et al. 2002). It also provides the mean for faster cloning, mass-multiplication, production of plants with better properties, and production of plants with higher quantities of the desired metabolites (Tripathi and Tripathi, 2003; Liu et al. 2004; Khan et al. 2009).

There are many interesting studies on the *in vitro* culture of *Rhodiola* species (e.g., *R. rosea* [Tasheva and Kosturkova 2010, 2011, 2012; Ghiorghiţă et al. 2011] or *R. fastigiata* [Liu et al. 2006]). Of these, Liu et al. (2006) have found an efficient system for mass-propagation of *R. fastigiata*, a species on the edge of extinction, and mass production is already on the way. This might be exactly what was needed to save this species. Similarly, Tasheva and Kosturkova (2010) have also developed an efficient system for *in vitro* seedling micropropagation of *R. rosea*. Using this process, they can produce 100 regenerated plantlets in a 3-month period and using these for further clonal propagation, 1000 new plants could be produced in another 3 months. Those are very encouraging results for a species for which it is very hard to initiate *in vitro* culture (Ghiorghiţă et al. 2011). Even more exciting is the very high survival rate obtained after roseroot explant adaptation from production to greenhouse (85%) and from greenhouse to field conditions (wild), where 70% of the transplant survived over the winter (Tasheva and Kosturkova 2010). This is very promising for the conservation of roseroot. Using this technique, it will be possible to repopulate some depleted natural habitats and it could also be used to produce plant materials to supply the agricultural industry's interest in roseroot production. In the end, it could relieve the pressure on wild populations.

2.3.2 Conservation in the New World

Compared to the amounts of studies in the Old World, much less information concerning *Rhodiola* is coming from the New World. Nevertheless, three species are growing in Canada, Alaska, and the northern mountains of the United States: *R. rosea*, *R. integrifolia* (four subspecies known in North America), and *R. rhodantha* (Small and Catling 1999). In most of their range, they have not yet been assessed and until we have information for those regions, the global picture looks similar to what is observed in the Old World. Table 2.4 presents the Canadian and American status for the diverse species and can be used in conjunction with Table 2.3 for comparison with the country conservation status. In addition, Table 2.5 compares the status between provinces or states in North

---

TABLE 2.4

Rhodiola National Conservation Status in Canada and the United States (Status for Comparison with Table 2.3)

Species	Countries	Conservation Status		References
R. integrifolia	Canada	-	NNR	NatureServe (2012)
	USA	ESA[a]: Extirpated	NNR	NatureServe (2012)
R. integrifolia ssp. *integrifolia*	Canada	-	NNR	NatureServe (2012)
	USA	-	NNR	NatureServe (2012)
R. integrifolia ssp. *leedyi*	Canada	Not present	-	NatureServe (2012)
	USA	ESA[a]: Listed Threatened	N1	NatureServe (2012)
R. integrifolia ssp. *neomexicana*	Canada	Not present	-	NatureServe (2012)
	USA	Critically imperilled	N1	NatureServe (2012)
R. integrifolia ssp. *procera*	Canada	Not present	-	NatureServe (2012)
	USA	-	NNR	NatureServe (2012)
R. rhodantha	Canada	Not present	-	NatureServe (2012)
	USA	-	NNR	NatureServe (2012)
R. rosea	Canada	-	NNR	NatureServe (2012)
	USA	-	NNR	NatureServe (2012)

Conservation status meaning: 1, critically imperilled; 2, imperilled; 3, vulnerable; 4, apparently secure; 5, secure; NNR, Nnt yet assessed (Unranked); -, the information was not specified in the reference and/or could not be found elsewhere.

[a] U.S. ESA: Endangered Species Act.

TABLE 2.5

***Rhodiola* Conservation Status in States and Provinces of Canada and the United States**

Species	State/Province	Conservation Status		Protection	Comments	References
R. integrifolia	Can: Yukon Territories	Secure	S5	–		NatureServe (2012)
R. integrifolia	Can: Alberta	Vulnerable	S3	–		NatureServe (2012)
R. integrifolia	Can: British Columbia	Secure	S5	–		NatureServe (2012)
R. integrifolia	Can: Northwest Territories	Not assessed	SNR SU	–		NatureServe (2012)
R. integrifolia	USA: Washington	Not assessed	SNR SU	–		NatureServe (2012)
R. integrifolia	USA: Montana	Not assessed	SNR SU	–		NatureServe (2012)
R. integrifolia	USA: Oregon	Not assessed	SNR SU	–		NatureServe (2012)
R. integrifolia	USA: California	Not assessed	SNR SU	–		NatureServe (2012)
R. integrifolia	USA: Nevada	Not assessed	SNR SU	–		NatureServe (2012)
R. integrifolia	USA: Idaho	Not assessed	SNR SU	–		NatureServe (2012)
R. integrifolia	USA: Utah	Critically Imperilled	S1	–		NatureServe (2012)
R. integrifolia	USA: Colorado	Not assessed	SNR SU	–		NatureServe (2012)
R. integrifolia	USA: New Mexico	Not assessed	SNR SU	–		NatureServe (2012)
R. integrifolia	USA: Wyoming	Vulnerable	S3			NatureServe (2012)
R. integrifolia	USA: Minesota	Critically Imperilled	S1	Yes		NatureServe (2012), DFW (2008)
R. integrifolia	USA: New York	Critically Imperilled	S1	Yes	Protected under the Environmental Conservation Law	NatureServe (2012), NYSDEC (2012)
R. rhodantha	USA: Montana	Not assessed	SNR SU	–		NatureServe (2012)
R. rhodantha	USA: Wyoming	Apparently Secure	S4	–		NatureServe (2012)
R. rhodantha	USA: Utah	Imperilled	S2	–		NatureServe (2012)
R. rhodantha	USA: Colorado	Not assessed	SNR SU	–		NatureServe (2012)

Species	Location	Salvage restricted	SR			
R. rhodantha	USA: Arizona			Yes	Collection only with permit	USDA (2012), AZGFD (2012)
R. rhodantha	USA: New Mexico	Not assessed	SNR SU	–		NatureServe (2012)
R. rosea	Can: Nunavut	Not assessed	SNR SU	–		NatureServe (2012)
R. rosea	Can: Labrador	Apparently Secure	S4	–		NatureServe (2012)
R. rosea	Can: Newfoundland	Apparently Secure	S4	–		NatureServe (2012)
R. rosea	Can: Québec	Vulnerable	S3	–		NatureServe (2012)
R. rosea	Can: New Brunswick	Vulnerable	S3	–		NatureServe (2012)
R. rosea	Can: Nova Scotia	Apparently Secure	S4	–		NatureServe (2012)
R. rosea	USA: New York	Critically Imperilled	S1	Yes	Protected under the Environmental Conservation Law	NatureServe (2012), NYSDEC (2012)
R. rosea	USA: Pennsylvania	Critically Imperilled	S1	–		NatureServe (2012)
R. rosea	USA: Vermont	Critically Imperilled	S1	Yes	Protected under Vermont's Endangered Species Law	NatureServe (2012), VFWD (2011)
R. rosea	USA: Connecticut	Not assessed	SNR SU	–		NatureServe (2012)
R. rosea	USA: New Jersey	Not assessed	SNR SU	–		NatureServe (2012)
R. rosea	USA: North Carolina	Presumed Extirpated	SX	Yes		NatureServe (2012), NCDACS (2010)
R. rosea	USA: Utah	Not assessed	SNR SU	–		NatureServe (2012)
R. rosea	USA: Tennessee	Extirpated	SX	–		USDA (2012)

Conservation status meaning: 1, critically imperilled; 2, imperilled; 3, vulnerable; 4, apparently secure; 5, secure; SNR/SU, not yet assessed (Unranked); -, the information was not specified in the reference and/or could not be found elsewhere.

America. Of the 36 states or provinces listed in Table 2.5, only 20 were investigated for the species. Of the remaining 20, for which an assessment has been conducted, 14 listed *Rhodiola* as vulnerable, imperilled, critically imperilled, salvage restricted (collection only with permission), presume extirpated, and extirpated. In comparison, they are listed as secure or apparently secure for only six states or provinces (Table 2.5). Also, one can speculate from Table 2.5 that the species status is apparently better in Canada. Indeed of 10 entries pertaining to Canadian provinces, 5 are secure or apparently secure, 3 are vulnerable, and 2 are not assessed (Table 2.5). Thus, all imperilled and critically imperilled status are pertaining solely to regions of the United States. This great discrepancy between Canada and the United States probably arises more from the species preferred northern habitat than from a difference in conservation laws or conservation programs. Effectively, there must be fewer habitats for the species in the lower latitudes of the United States compared to the Canadian climate. Also, populations in Québec are less accessible as they are mostly located in Nunavik.

Suitable habitats for *R. rosea* are also found in the Québec Mingan Archipelago National Park Reserve. While the species is listed as vulnerable in the province (NatureServe 2012), it is not considered endangered and thus, does not benefit *ipso facto* from a legal protection. Regardless, its presence in the Canadian park system provides a protection against exploitation, since the only collection permitted in the park is for traditional purposes or research, which requires a permit (Parcs Canada 2005, 2011). This protection is not without importance. Small and Catling (1999) stated that roseroot did not, at the time, enjoy the same recognition as a valued medicinal plant in North America when compared to Europe and Asia. Nevertheless, considering that along with the American ginseng (*Panax quinquefolius* L.) it is the only other Canadian native plant known to be an adaptogen, and its reputation and value could develop rapidly. As suggested by these authors, *R. rosea* could be cultivated in the cold regions of Canada where few other plants could (Small and Catling 1999). The idea did not take long to be set in motion. Five years later, field experiments were initiated in Alberta to describe the best agronomic technique for *R. rosea* cultivation (Ampong-Nyarko et al. 2005, 2006; Ampong-Nyarko and Zhang 2007). In 2007 and 2008, small experimental plots were secured in Nunavik, although the research has temporarily been abandoned, while in 2012, new plots were added in Nunatsiavut (Labrador). This rather new research project involving Inuit communities based on the coast of Labrador is still ongoing.

2.3.3 CULTIVATION: THE FUTURE OF *RHODIOLA ROSEA*?

Successful cultivation of medicinal plants is not only bound to decrease pressure on wild populations, but, in the case of roseroot, it is believed to be the only hope in order to meet the required supply necessary for the industry (Galambosi 2006). In the Old World, roseroot is cultivated in various parts of Russia and other places in Eurasia (Small and Catling 1999). More specifically, experiments have been conducted on its cultivation in Russia (Elsakov and Gorelova 1999, cited by Galambosi 2006), Sweden, Findland (Galambosi 2006), Poland (Furmanowa et al. 1999), Germany (see Galambosi 2006), and Italy (Aiello, Scartezzini, and Vender 2010). Meanwhile, across the ocean, experiments have also been conducted in the province of Alberta in Canada (Ampong-Nyarko et al. 2005) and, to a smaller scale, in Québec and Labrador. Nonetheless, roseroot cultivation, as reported by Galambosi (2005) and Platikanov and Evstatieva (2008), is not without problems. The species is sensible to temperature, light, and moisture variation (Galambosi 2005; Ampong-Nyarko et al. 2006; Platikanov and Evstatieva 2008). Therefore, it can be grown in only a small number of places. It is also recognized that seeds are hard to germinate (Ampong-Nyarko et al. 2006) and that seedling has a slow growth rate (Galambosi 2005, 2006; Platikanov and Evstatieva 2008; Ghiorghiţă et al. 2011). Germination rate can be very high (90%) if gibberellic acid is used, but farmers may lose their organic stamps (A. Cuerrier, unpublished results). In general, seeds from Nunavik seem to have a higher rate of germination than the Russian accession used in Alberta, even without the use of hormones. Further studies are therefore needed to fully understand germination rates of seeds originating from different populations, countries, and continents. Galambosi (2006) has indicated that growth is faster under cultivation than in nature; however, the establishment of *Rhodiola* fields necessitates transplantation of seedlings (Galambosi 2006). Once properly established, it takes from 3 to 5 years before the first harvest (Galambosi 2005, 2006; Ampong-Nyarko and Zhang 2007; Platikanov and Evstatieva 2008), and harvesting is labor intensive (Galambosi 2006). Cost of cultivation can thus be very expensive. But once these barriers are overcome, cultivation can be efficient and could provide enough material to supply international markets. Yet again, the producer must be able to guarantee a certain stock supply, guarantee the quality, freshness, and traceability of the product, and most importantly, the production must be cost-effective (Ampong-Nyarko and Zhang 2007).

Overall, the outcome of these experiments seems to point toward salvation through cultivation. Ampong-Nyarko et al. (2006) have estimated a possible yield of 4 t/ha by the third year of their experiment and are aiming toward 160 ha of production by 2009. This is a successful introduction in the province of Alberta. In Bulgaria, roseroot has also been successfully adapted to the Rhodope Mountains and is now an economic crop of the country (Platikanov and Evstatieva 2008). Elsakov and Gorelova (1999), cited in Galambosi (2006), report a production of 5.5 kg (55 t/ha) of fresh root per square meter in the Russian cultivation experiment. Results from yet another research conducted in Italy are showing a yield of 120 g of dried root per plant in an experiment with a density of 8 plants/m^2 (Aiello, Scartezzini, and Vender 2010).

Once all the obstacles previously mentioned are tackled, *Rhodiola* plants may attain 100% seedling survival through their first winter (Ampong-Nyarko et al. 2006). Similarly, Platikanov and Evstatieva (2008) have reported a survival rate to be between 90% and 95%. These encouraging results are met with few major disease or pest problems during cultivation experiments (Galambosi 2005; Platikanov and Evstatieva 2008). For a successful method of propagation, Galambosi (2006) suggested that, for large-scale cultivation, mass production of seedlings should be promoted. On the other hand, Platikanov and Evstatieva (2008) proposed a mixed method using both clonal propagation and mass seedling production. In Nunavik, field experiments were started from seedlings raised at the Montreal Botanical Garden, whereas in Nunatsiavut rhizome division (clonal propagation) was done directly in the field from wild plants. Both experiments gave a good to high survival rate (A. Cuerrier, unpublished results).

In general, for *Rhodiola* conservation to be successful, future research needs to address genetic studies in a more comprehensive way. This is an important first step. Then, using biotechnologies will provide a mean to reintroduce the species with specific genotypes in natural habitats from where it was extirpated or in populations that have a low gene pool. Biotechnologies can also provide a high number of seedlings and plantlets for agricultural production, which in turn will hopefully be able to develop enough raw materials to supply international trade. Furthermore, *in situ* and *ex situ* protection and conservation measures must be complementary with the previous steps in order to insure lasting wild populations. Finally, the species status should be assessed for countries or regions where it has not been done and its global status should be assessed for the IUCN Red list.

REFERENCES

Ager, T. A., and L. P. Ager. 1980. Ethnobotany of the Eskimos of Nelson Island, Alaska. *Arctic Anthropology* 17 (1): 26–48. Available at: http://www .jstor.org/sTable/40315966.

Aiello, D. N., F. Scartezzini, and C. Vender. 2010. *Rhodiola rosea*: Dalla raccolta spontanea alla coltivazione. *Erboristeria Domani* 7 (8): 43–9.

Alam, J. A. N., and S. I. Ali. 2010. Contribution to the Red List of the Plants of Pakistan. *Pakistan Journal of Botany* 42 (5): 2967–71.

Ali, S. I., and J. Alam. 2006. *Contribution to the Red List of the Plants of Pakistan: Endemic Phanerogams of Gilgit and Baltistan*. Karachi: University of Karachi.

Allen, D. E., and G. Hatfield. 2004. *Medicinal Plants in Folk Tradition: An Ethnobotany of Britain & Ireland*. Portland, OR: Timber Press, Inc.

Alm, T. 1996. Bruk av rosenrot (*Rhodiola rosea*) mot skjørbuk. *Polarflokken* 20 (1): 29–32.

Alm, T. 2004. Ethnobotany of *Rhodiola rosea* (Crassulaceae) in Norway. *SIDA* 21 (1): 324–44.

Ampong-Nyarko, K., Brown, J., De Mulder, J., Chaudhary, N., Zhang, Z., and A. Jiao. 2006. Specialty Crop Report: *Rhodiola rosea* commercialization in Alberta, 41 pp., Alberta Agriculture, Food and Rural development, Industry Development Sector. Available at: http://www1.agric.gov.ab.ca/$department/deptdocs.nsf/all/sdd11117 [Accessed November 14, 2012].

Ampong-Nyarko, K., J. Brown, J. De Mulder, Z. Zhang, and B. Henrique. 2005. Specialty Crop Report: New Crop Development: *Rhodiola* potential commercialization in Alberta, 45 pp., Alberta Agriculture, Food and Rural development, Industry Development Sector. Available at: http://www1.agric.gov.ab.ca/$department/deptdocs.nsf/all/sdd11117 [Accessed November 14, 2012].

Ampong-Nyarko, K., and Z. Zhang. 2007. New crop development—Crop Diversification Centre North: Development of basic agronomic recommendations for economically growing *Rhodiola rosea* in Alberta, 35 pp., Alberta Agriculture, Food and Rural development, Industry Development Sector. Available at: http://www1.agric.gov.ab.ca/$department /deptdocs.nsf/all/sdd11117 [Accessed November 14, 2012].

Anderson, J. P. 1939. Plants used by the Eskimo of the northern Bering Sea and Arctic regions of Alaska. *American Journal of Botany* 26 (9): 714–6.

Angmo, K., B. S. Adhikari, and G. S. Rawat. 2012. Changing aspects of traditional healthcare system in western Ladakh, India. *Journal of Ethnopharmacology* 143 (2): 621–30.

Archambault, M. 2009. Étude moléculaire des populations de *Rhodiola rosea* L. du Nunavik (Québec, Canada). M.Sc. thesis. Department of Biological Sciences, University of Montreal.

Arizona Game and Fish Department (AZGFD) 2012. Available at: http://www.azgfd.com /w_c/edits/hdms_status_definitions.shtml [Accessed November 13, 2012].

Bilz, M., S.P. Kell,N. Maxted, and R.V. Lansdown. 2011. *European Red List of Vascular Plants*. Luxembourg: Publications Office of the European Union.

Birket-Smith, K. 1928. The Greenlanders of the present day. In *Greenland, Vol. 2. The Past and Present Population of Greenland*, edited by M. Vahl, G. C Amdup, L. Bobé, and A. S. Jensen, 1–207. London: C. A. Reitzel, Copenhagen: Humphrey Milford/Oxford University Press.

Blondeau, M. 2009. *La Flore Vasculaire des Environs de Wemindji*, Baie James, Québec et Nunavut. Québec.

Blondeau, M., C. Roy, and A. Cuerrier. 2010. *Plants of the Villages and the Parks of Nunavik*. Québec: Éditions MultiMondes.

Bodeker, G., C. K. Ong, C. Grundy, G. Burford, and K. Shein. 2005. *WHO Global Atlas of Traditional, Complementary and Alternative Medicine*. Kobe, Japan: WHO, Centre for Health Development.

Borodin, A. M. 1985. *The Red Book of the USSR*. Moscow: Lesnaya Promyshlennost.

Brekhman, I. I., and I. V. Dardymov. 1968. New substances of plant origin which increase non-specific resistance. *Annual Review of Pharmacology* 8: 419–30.

Brown, R. P., P. L. Gerbarg, and Z. Ramazanov. 2002. *Rhodiola rosea*: A phytomedicinal overview. *HerbalGram* 56: 40–52.

Bulgarian Academy of Sciences. (Eds.) 2011. *Red Data Book of the Republic of Bulgaria: Vol. 1, Plants and Fungi*. Sofia, Bulgaria: Ministry of Environment and Waters of Bulgaria (MEWB). Available at: http://www.e-ecodb.bas.bg/rdb/en/vol1/ [Accessed November 13, 2012].

Byg, A., J. Salick, and W. Law. 2010. Medicinal plant knowledge among lay people in five eastern tibet villages. *Human Ecology* 38 (2): 177–91.

Cavalière, C. 2009. The effects of climate change on medicinal and aromatic plants. *HerbalGram* 81: 44–57.

Cavalière, C., R. P. Lynch, M. E., and M. Blumenthal. 2010. Herbal supplement sales rise in all channels in 2009. *HerbalGram* 86: 62–5.

Chaldanbaeva, A. K., J. C. Nuralieva, and A. A. Kalykeeva. 2012. First International Biology Congress in Kyrgyzstan. In *Phytochemical Study of Kyrgyzstan Rhodiola*, edited by S. Canbulat, 59. Bishkek: Kyrgyzstan-Turkey Manas University. Available at: http://www.biocong.manas.edu.kg.

Cuerrier, A., and Elders of Kangiqsualujjuaq. 2011. *The Botanical Knowledge of the Inuit of Kangiqsualujjuaq, Nunavik*. Inukjuak, QC: Avataq Cultural Institute.

Cuerrier, A., and L. Hermanutz. 2012. *Our Plants... Our Land. Plants of Nain and Torngat Mountains Basecamp and Research Station (Nunatsiavut)*. Montréal & St. John's: Institut de Recherche en Biologie Végétale & Memorial University of Newfoundland.

Davydov, V. 2011. Public healthhand folk medicine among North Baikal Evenkis. In *The Healing Landscapes of Central and Southeastern Siberia*, edited by D. G. Anderson, 129–46. Edmonton: CCI Press.

Didukh, Y.P. (Ed.) 2009. *Red Data Book of Ukraine: Flora*. Kiev, Ukraine: Ukrainian Scientific Publishers, 900 pp.

Dines, T.D., R.A. Jones, S.J. Leach, D.R. McKean, D.A. Pearman, C.D. Preston, F.J. Rumsey, and I. Taylor. 2005. The vascular plant red data list for Great Britain. Species Status 7: 1-116. Peterborough: Joint Nature Conservation Committee. Available at: http://www.jncc.defra.gov.uk/pdf/pub05_speciesstatusvpredlist3_web.pdf [Accessed November 10, 2012].

Division of Fish and Wildlife (DFW) 2008. Minesota Administrative Rules: Chapter 6134, Endangered, Threatened, Special Concern Species, 6134.0300 Vascular plants. Minnesota Department of Natural Resources, Minnesota. Available at: https://www.revisor.mn.gov/rules/?id=6134 [Accessed November 13, 2012].

Downing, A., and A. Cuerrier. 2011. A synthesis of the impacts of climate change on the first nations and inuit of Canada. *International Journal of Traditional Knowledge* 10 (1): 57–70.

Dragland, S. 2001. *Rosenrot, Botanikk, Innholdsstoff, Dyrking Og Bruk*. Planteforsk, Grønn Forskning.

Edwards, R. 2004. No remedy in sight for herbal ransack. *New Scientist* 181 (2429): 10–1.
eFloras. 2008. Flora of China. Available at: http://www.efloras.org/flora_page.aspx?flora_id = 2.
Egede, H. 1818. *A Description of Greenland*. 2nd ed. London: T. and J. Allman.
Elameen, A., S. S. Klemsdal, S. Dragland, S. Fjellheim, and O. A. Rognli. 2008. genetic diversity in a germplasm collection of roseroot (*Rhodiola rosea*) in Norway Studied by AFLP. *Biochemical Systematics and Ecology* 36 (9): 706–15.
Elsakov, G.V. and A. P. Gorelova. 1999. Fertilizer effects on the yield and biochemical composition of rose-root stonecrop in North Kola region. *Agrokhimiya* 10: 58–61.
Engelskjøn, T., L. Lund, and I. G. Alsos. 2003. Twenty of the most thermophilous vascular plant species in svalbard and their conservation state. *Polar Research* 22 (2): 317–39.
Fuchs, L. 1964. *New Kreüterbüch in Welchem Nit Allein Die Gantz*. Reprint. München, Germany: Verlag.
Furmanowa, M., B. Kedzia, M. Hartwich et al. 1999. Phytochemical and pharmacological properties of *Rhodiola rosea* L. *Herba Polonica* 45 (2): 108–13.
Galambosi, B. 2005. *Rhodiola rosea* L. from wild collection to field production. *Medicinal Plant Conservation* 11: 31–5.
Galambosi, B. 2006. Demand and availability of *Rhodiola rosea* L. raw material. In *Medicinal and Aromatic Plants*, edited by R. J. Bogers, L. E. Craker, and D. Lange, 223–36. The Netherlands: Springer.
Germano, C., and Z. Ramazanov. 1999. *Arctic Root (Rhodiola rosea): The Powerful New Ginseng Alternative*. New York: Kensington Press.
Ghimire, S. K., and Y. Aumeeruddy-Thomas. 2009. Ethnobotanical classification and plant nomenclature system of high altitude agro-pastoralists in Dolpo, Nepal. *Botanica Orientalis: Journal of Plant Science* 6: 56–68.
Ghiorghiţă, G., M. Hârţan, D. L. Maftei, and D. Nicuţă. 2011. Some considerations regarding the *in vitro* culture of *Rhodiola rosea* L. *Romanian Biotechnological Letters* 16 (1): 5902–8.
Gray, B. 2011. *The Boreal Herbal: Wild Food and Medicine Plants of the North*. Whitehorse and Edmonton: Aroma Borealis Press and CCI Press.
Griffin, D. 2001. Contributions to the ethnobotany of the Cup'it Eskimo, Nunivak Island, Alaska. *Journal of Ethnobiology* 21 (2): 91–132.
Gupta, S., M. S. Bhoyar, J. Kumar, et al. 2012. Genetic diversity among natural populations of *Rhodiola imbricata* Edgew. from trans-himalayan cold arid desert using random amplified polymorphic DNA (RAPD) and inter simple sequence repeat (ISSR) markers. *Journal of Medicinal Plants Research* 6 (3): 405–15.
Hanelt, P. 2001. *Mansfeld's Encyclopedia of Agricultural and Horticultural Crops*. Vol. 1. Berlin: Springer.
Hedman, S. 2000. *Rosenrot: Nordens Mirakelört*. Mikas Förlag: Ölandstryckarna.
Heller, C. A. 1953. *Edible and Poisonous Plants of Alaska*. Alaska: University of Alaska and United States Department of Agriculture.
Hermsmeier, U., J. Grann, and A. Plescher. 2012. *Rhodiola integrifolia*: hybrid origin and asian relatives. *Botany* 90 (11): 1186–90.
Hjaltalin, O. J. 1830. *Isländs Botanik*. Köpenhamn, Denmark: Hins islezka bokmenntafelags.
Holub, J., and F. Procházka. 2000. Red list of the flora of the Czech Republic (State in the Year 2000). *Preslia* 72: 187–230.
Høygaard, A. 1941. The nutrition of the Angmagssalik Eskimo. In *Studies on the Nutrition and Physio-Pathology of Eskimos. Undertaken at Angmagssalik East-Greenland 1936–1937*. Oslo, Norway: I Kommisjon Hos Jacob Dybwad.
Hughes, C. C. 1960. *An Eskimo Village in the Modern World*. Ithaca, NY: Cornell Univeristy Press.
Hui, Y., M. Shuangxi, P. Liyan, L. Zhongwen, and S. Handong. 2002. A new clucoside from *Rhodiola fastigiata* (Crassulaceae). *Acta Botanica Sinica* 44 (2): 224–6. Available at: http://www.europepmc.org/abstract/CBA/384067.

Humagain, K, and K. K. Shrestha. 2010. Medicinal plants in Rasuwa district, Central Nepal: Trade and livelihood. *Botanica Orientalis: Journal of Plant Science* 6 (5): 39–46.

Iljina, I. 1997. *Komi Folk Medicine*. Syktyvkar, Russia: Komi khisnoje isdatelstvo.

Ismailov, A. É., Z. A. Kuliev, A. D. Vdovin, N. D. Abdullaev, and B. M. Murzubraimov. 1998. Oligomeric proanthocyanidin glycosides of *Rhodiola Pamiroalaica*. *Chemistry of Natural Compounds* 34 (4): 450–5.

Ivanitskiy, N. A. 1890. Materialy Po Etnografii Vologodskoy Gubernii. Izvestiya Imperatorskogo Obshchestva Lyubiteley Yestestvoznaniya, Antropologii i Etnografii. T.LXIX. *Trudy Etnograficheskogo Otdela* 9: 1–2.

Ji, H., P. Shengji, and L. Chunlin. 2004. An ethnobotanical study of medicinal plants used by the Lisu people in Nujiang, Northwest Yunnan, China. *Economic Botany* 58 (sp1): S253–64.

Jones, A. 2010. *Plants That We Eat: Nauriat Niġiñaqtuat*. 2nd ed. Fairbanks, AK: University of Alaska Press.

Källman, S. 1997. *Vilda Växter Som Mat & Medicin*. Västerås, Sweden: Ica Bokförlag.

Kassam, K., M. Karamkhudoeva, M. Ruelle, and M. Baumflek. 2010. Medicinal plant use and health sovereignty: Findings from the Tajik and Afghan Pamirs. *Human Ecology: An Interdisciplinary Journal* 38 (6): 817–29.

Kelly, G. S. 2001. *Rhodiola rosea*: A possible plant adaptogen. *Alternative Medicine Review* 6 (3): 293–302.

Kershaw, L. 2000. *Edible and Medicinal Plants of the Rockies*. Edmonton: Lone Pine Publishing.

Khan, M. Y., S. Aliabbas, V. Kumar, and S. Rajkumar. 2009. Recent advances in medicinal plant biotechnology. *Indian Journal of Biotechnology* 8: 9–22.

Khanum, F., A. S. Bawa, and B. Singh. 2005. *Rhodiola rosea:* A versatile adaptogen. *Comprehensive Reviews in Food Science and Food Safety* 4 (3): 55–62.

Kotiranta, H., P. Uotila, S. Sulkava, and S.-L. Peltonen. (Eds.) 1998. *Red Data Book of East Fennoscandia*. 351 pp., Ministry of the Environment, Finnish Environment Institute & Botanical Museum, Finnish Museum of Natural History. Helsinki.

Kozyrenko, M. M., S.B. Gontcharova, and A. A. Gontcharov. 2011. Analysis of the genetic structure of *Rhodiola rosea* (Crassulaceae) using inter-simple sequence repeat (ISSR) polymorphisms. *Flora Morphology, Distribution, Functional Ecology of Plants* 206 (8): 691–6.

Kricsfalusy, V., and G. Budnikov. 2007. Threatened vascular plants in the Ukrainian Carpathians: Current status, distribution and conservation. *Thaiszia Journal of Botany* 17: 11–32.

Kumar, G. P., R. Kumar, O. P. Chaurasia, and S. B. Singh. 2011. Current status and potential prospects of medicinal plant sector in trans-Himalayan Ladakh. *Journal of Medicinal Plants Research* 5 (14): 2929–40.

Kumar, M., Y. Paul, and V. K. Anand. 2009. An ethnobotanical study of medicinal plants used by the locals in Kishtwar, Jammu and Kashmir, India. *Ethnobotanical Leaflets* 2009 (10): 5.

Kylin, M. 2010. *Genetic Diversity of Roseroot (Rhodiola rosea L.) from Sweden, Greenland and Faroe Islands*. M.Sc. thesis. Alnarp: Department of Biology, Swedish University of Agricultural Sciences, Uppsala.

Lagerberg, T., J. Holmboe, and R. Nordhagen. 1955. Rosenrot. In *Våre ville planter. Bind 3*, edited by T. Lagerberg, J. Holmboe, and R. Nordhagen, 231–7. Oslo, Norway: Tanum.

Länsstyrelsen Västra Götalands län. 2010. *Fridlysta arter i Västra Götalands län*. Available at: http://www5.o.lst.se/projekt/frida/artdetaljSv.asp?id=Rosenrot [Accessed November 07, 2012]. (In Swedish).

Lantis, M. 1946. The social culture of the Nunivak Eskimo. *Transactions of the American Philosophical Society* 35 (3): 153–323.

Lantis, M. 1959. Folk medicine and hygiene. *Anthropological Papers of the University of Alaska* 8: 1–75.

Lei, Y., H. Gao, T. Tsering, S. Shi, and Y. Zhong. 2006. Determination of genetic variation in *Rhodiola crenulata* from the Hengduan Mountains region, China using inter-simple sequence repeats. *Genetics and Molecular Biology* 29 (2): 339–44.

Lei, Y., P. Nan, T. Tsering, Z. Bai, C. Tian, and Y. Zhong. 2003. Chemical composition of the essential oils of two *Rhodiola* species from Tibet. *Zeitschrift Fur Naturforschung* 58 (3–4): 161–4.

Li, H. X., S. C. W. Sze, Y. Tong, and T. B. Ng. 2009. Production of Th1- and Th2-dependent cytokines induced by the Chinese medicine herb, *Rhodiola algida*, on human peripheral blood monocytes. *Journal of Ethnopharmacology* 123 (2): 257–66.

Li, T. 2002. *Chinese and Related North American Herbs. Phytopharmacology and Therapeutic Values*. Boca Raton, FL: CRC Press.

Li, T., G. Xu, L. Wu, and C. Sun. 2007. 'Pharmacological studies on the sedative and hypnotic effect of salidroside from the Chinese medicinal plant *Rhodiola sachalinensis*.' *Phytomedicine: International Journal of Phytotherapy and Phytopharmacology* 14 (9): 601–4.

Li, T., and H. Zhang. 2008a. Identification and comparative determination of rhodionin in traditional tibetan medicinal plants of fourteen *Rhodiola* species by high-performance liquid chromatography-photodiode array detection and electrospray ionization-mass spectrometry. *Chemical and Pharmaceutical Bulletin* 56 (6): 807–14.

Li, T., and H. Zhang. 2008b. Application of microscopy in authentication of traditional Tibetan medicinal plants of five *Rhodiola* (Crassulaceae) Alpine species by comparative anatomy and micromorphology. *Microscopy Research and Technique* 71 (6): 448–58.

Liu, C., S. Murch, J. Jain, and P. Saxena. 2004. Goldenseal (*Hydrastis canadensis* L.): *In vitro* regeneration for germplasm conservation and elimination of heavy metal contamination. *In Vitro Cellular & Developmental Biology* 40: 75–9.

Liu, H., Y. Xu, Y. Liu, and C. Liu. 2006. Plant regeneration from leaf explants of *Rhodiola fastigiata*. *In Vitro Cellular & Developmental Biology—Plant* 42 (4): 345–7.

Long, C., S. Li, B. Long, Y. Shi, and B. Liu. 2009. Medicinal plants used by the Yi ethnic group: a case study in Central Yunnan." *Journal of Ethnobiology and Ethnomedicine* 5: 13.

MacKinnon, A., L. Kershaw, J. T. Arnason, P. Owen, A. Karst, and F. Hamersley-Chambers. 2009. *Edible and Medicinal Plants of Canada*. Edmonton: Lone Pine Publishing.

Magnusson, B. 1992. *Fägringar: Växter Som Berör Oss*. Östersung, Sweden: Berndtssons.

Mamedov, N. 2005. Adaptogenic, geriatric, stimulant and antidepressant plants of Russian far east. *Journal of Cell Molecular Biology* 4: 71–5.

Mamedov, N., Z. Gardner, and L. E. Craker. 2005. Medicinal plants used in Russia and Central Asia for the treatment of selected skin conditions. *Journal of Herbs, Spices & Medicinal Plants* 11 (1–2): 191–222.

Manandhar, N. P. 2002. *Plants and People of Nepal*. Portland, OR: Timber Press.

Mayuzumi, S., and H. Ohba. 2004. The phylogenetic position of Eastern Asian Sedoideae (Crassulaceae) inferred from chloroplast and nuclear DNA sequences. *Systematic Botany* 29 (3): 587–98.

Metzo, K. 2011. Medical pluralism and expert knowledge in Buriatiia. In *The Healing Landscapes of Central and Southeastern Siberia*, edited by D. G. Anderson, 29–44. Edmonton: CCI Press.

Minarchenko, V. 2011. Medicinal plants of Ukraine: Diversity, resource, legislation. *Medicinal Plant Conservation* 14: 7–13.

Mook-Jung, I., H. Kim, W. Fan, Y. Tezuka, S. Kadota, H. Nishijo, and M. W. Jung. 2002. Neuroprotective effects of constituents of the oriental crude drugs, *Rhodiola sacra*, *R. sachalinensis* and Tokaku-Joki-to, against beta-amyloid toxicity, oxidative stress and apoptosis. *Biological and Pharmaceutical Bulletin* 25 (8): 1101–4.

Mossberg, B., and H. Rydberg. 1995. *Alla Sveriges Fridlysta Växter*. Växsjö: Wahlströms & Wid-strand, Naturskyddsföreningen.

NatureServe. 2012. NatureServe Explorer: An online encyclopedia of life [web application]. Version 7.1. NatureServe, Arlington, Virginia. Available at: http://www.natureserve.org /explorer [Accessed November 13, 2012].

Nedelcheva, A. 2012. Traditional knowledge and modern trends for Asian medicinal plants in Bulgaria from an ethnobotanical view. *EurAsian Journal of BioSciences* 6: 60–9.

New York State Department of Environmental Conservation (NYSDEC), 2012. Environmental Conservation Law, § 3-0301, 9-0105, 9-1503, Available at: http://www.dec.ny.gov /regs/15522.html [Accessed November 13, 2012].

Nordenskiöld, A. E. 1882. *The Voyage of the Vega Round Asia and Europe: With a Historic Review of Previous Journeys Along the North Coast of the Old World*. New York: MacMillan and Co.

North Carolina Department of Agriculture and Consumer Services (NCDACS). 2010. Section.0300, Endangered plant species list: Threatened plant species list: list of species of special concern. Available at: http://www.ncagr.gov/plantindustry/plant/plantcon-serve/plist.htm [Accessed November 13, 2012].

NRL 2012. National Red List. Available at: http://www.nationalredlist.org/species-search [Accessed November 07, 2012].

Ohsugi, M., W. Fan, K. Hase, et al. 1999. Active-oxygen scavenging activity of traditional nourishing-tonic herbal medicines and active constituents of *Rhodiola sacra*. *Journal of Ethnopharmacology* 67 (1): 111–9.

OMS. 2003. *Directives OMS sur les bonnes pratiques agricoles et les bonnes pratiques de récolte (BPAR) relatives aux plantes médicinales*. Genève, Switzerland: OMS.

Panossian, A, G. Wikman, and J. Sarris. 2010. Rosenroot (*Rhodiola rosea*): Traditional use, chemical composition, pharmacology and clinical efficacy. *Phytomedicine* 17 (7): 481–93.

Parcs Canada. 2005. Plan directeur: Réserve de parc national du Canada de l'Archipel-de-Mingan.121 pp. Available at: http://www.pc.gc.ca/fra/pn-np/qc/mingan/plan/plan4.aspx [Accessed November 24, 2012].

Parcs Canada. 2011. Plan directeur: Document de consultation publique. Réserve de parc national du Canada de l'Archipel-de-Mingan. 20 pp. Available at: http://www.pc.gc.ca/ fra/pn-np/qc/mingan/plan/plan5.aspx [Accessed November 24, 2012].

Petrova, A., and V. Vladimirov. 2009. Red list of Bulgarian vascular plants. *Phytologia Balcanica* 15 (1): 63–94.

The Plant List. 2010. Version 1. Available at: http://www.theplantlist.org/.

Platikanov, S., and L. Evstatieva. 2008. Introduction of wild golden goot (*Rhodiola rosea* L.) as a potential economic crop in Bulgaria. *Economic Botany* 62 (4): 621–7.

Popov, K. 1974. "Zyryane i Zyryanskiy Kray." *Izvestiya Imperatorskogo Obshchestva Lyubiteley Yestestvoznaniya, Antropologii i Etnografii* 13 (2): 18.

Porsild, A. E. 1953. Edible plants of the Arctic. *Arctic* 6 (1): 15–34.

Rana, J. C., K. Pradheep, O. P. Chaurasia et al. 2011. Genetic resources of wild edible plants and their uses among tribal communities of cold arid region of India. *Genetic Resources and Crop Evolution* 59 (1): 135–49.

Rink, H. 1974. *Danish Greenland: Its People and Products*. London: C. Hurst & Co.

Rokaya, M. B., Z. Münzbergová, M. R. Shrestha, and B. Timsina. 2012. Distribution patterns of medicinal plants along an elevational gradient in Central Himalaya, Nepal. *Journal of Mountain Science* 9 (2): 201–13.

Ryvarden, L. 1993. Bergknappfamilien. In *Norges Planter. Bind 1*, edited by L. Ryvarden, 175–82. Oslo, Norway: J.W. Cappelens Forlag.

Scoggan, H. J. 1978. *The Flora of Canada. Part 1–4*. Ottawa, ON: National Museums of Canada.

Scott, P. J. 2010. *Edible Plants of Newfounland and Labrador*. Portugal Cove – St. Philip's, NL: Boulder Publications.

Shah, A., G. Abass, and M. P. Sharma. 2012. Ethnobotanical study of some medicinal plants from Tehsil BudhaL, district Rajouri (Jammu and Kashmir). *International Multidisciplinary Research Journal* 2 (6): 05-06.

Shang, X., C. Tao, X. Miao et al. 2012. Ethno-veterinary survey of medicinal plants in Ruoergai Region, Sichuan Province, China. *Journal of Ethnopharmacology* 142 (2): 390–400.

Sharma, P. K., S. K. Thakur, S. Manuja et al. 2011. Observations on traditional phytotheraphy among the inhabitants of Lahaul Valley through Amchi System of medicine—A cold desert Area of Himachal Pradesh in North Western Himalayas, India. *Chinese Medicine* 02 (03): 93–102.

Shatar, S,, R. P. Adams, and W. Koenig. 2007. Comparative study of the essential oil of *Rhodiola rosea* L. from Mongolia. *Journal of Essential Oil Research* 19 (3): 215–7.

Small, E., and P. M. Catling. 1999. *Canadian Medicinal Crops*. Ottawa, ON: NRC Research.

Smith, W. G. 1973. Artic pharmacognosia. *Arctic* 26 (4): 324–33. Available at: http://www .jstor.org/sTable/40509174.

Sparschuch, H. 1775. *Pharmacopoea Svecica*. Holmia: H. Fougt.

Stevens, P. F. 2001. Angiosperm Phylogeny Website. Version 12, July 2012 [more or less continuously updated since]. Available at: http://www.mobot.org/MOBOT/research /APweb/.

Tasheva, K., and G. Kosturkova. 2010. Bulgarian golden root *in vitro* cultures for micropropagation and reintroduction. *Central European Journal of Biology* 5 (6): 853–63.

Tasheva, K., and G. Kosturkova. 2011. *Rhodiola rosea* L. *in vitro* plants morphophysiological and cytological characteristics. *Romanian Biotechnological Letters* 16 (6): 79–85.

Tasheva, K., and G. Kosturkova. 2012. The role of biotechnology for conservation and biologically active substances production of *Rhodiola rosea*: Endangered medicinal species. *The Scientific World Journal* 2012: 1–21.

Taskaev, A. 1999. *Red Book of Komi Republic. Rare and Endangered Species of Plants and Animals*. Moscow-Syktyvkar, Russia: Design and Cartography Publisher.

Tripathi, L., and J. N. Tripathi. 2003. Role of biotechnology in medicinal plants. *Tropical Journal of Pharmaceutical Research* 2 (2): 243–53.

Tsikov, P. S. 1980. *Atlas of Areas and Resources of Medicinal Plants of Soviet Union*. Moscow: Principal Office of Geodesy and Cartography.

Uprety, Y., H. Asselin, E. K. Boon, S. Yadav, and K. K. Shrestha. 2010. Indigenous use and bio-efficacy of medicinal plants in the Rasuwa district, Central Nepal. *Journal of Ethnobiology and Ethnomedicine* 6 (1): 1–10.

USDA, NRCS. 2012. The PLANTS Database Available at: http://www.plants.usda.gov [Accessed November 24, 2012]. National Plant Data Team, Greensboro, NC.

Vermont Fish and Wildlife Department (VFWD). 2011. Regulation of the Secretary of Agency of Natural Resources. 10 V.S.A. App § 10. Vermont endangered and threatened species rule. Available at: http://www.leg.state.vt.us/statutes/fullsection.cfm?Title=10APPEND IX&Chapter=001&Section=00010 [Accessed Novembre 13, 2012].

Verpoorte, R., Contin, A., and J. Memelink. 2002. Biotechnology for the production of plant secondary metabolites. *Phytochemistry Reviews* 1 (1): 13–25.

Virey. 1811. *Traité de Pharmacie Théorique et Pratique. Tome 1*. Paris, France: Rémont et Ferra.

Wang, S., Z. Zheng, Y. Weng, et al. 2004. Angiogenesis and anti-angiogenesis activity of Chinese medicinal herbal extracts. *Life Sciences* 74 (20): 2467–78.

Wiedenfeld, H., M. Zych, W. Buchwald, and M. Furmanowa. 2007. New compounds from *Rhodiola kirilowii*. *Scientia Pharmaceutica* 75 (1): 29–34.

Wójcik, R., A. K. Siwicki, E. Skopińska-Różewska et al. 2008. The *in vitro* influence of *Rhodiola quadrifida* extracts on non-specific cellular immunity in pigs. *Central European Journal of Immunology* 33: 193–6.

Wójcik, R., A. K. Siwicki, E. Skopińska-Różewska, A. Wasiutyński, E. Sommer, and M. Furmanowa. 2009. The effect of chinese Medicinal Herb *Rhodiola kirilowii* extracts on cellular immunity in mice and rats. *Polish Journal of Veterinary Sciences* 12 (3): 399–405.

Wong, C., H. Li, K. Cheng, and F. Chen. 2006. A systematic survey of antioxidant activity of 30 Chinese medicinal plants using the ferric reducing antioxidant power assay. *Food Chemistry* 97 (4): 705–11.

World Health Organization. 2013. Medicinal plants in Mongolia. *Flora of North America*. Geneva. Available at: http://www.wpro.who.int/publications/Medicinal_Plants_in _Mongolia_VF.pdf#page = 179.

Xia, T., S. Chen, S. Chen, and X. Ge. 2005. Genetic variation within and among populations of *Rhodiola alsia* (Crassulaceae) native to the Tibetan Plateau as detected by ISSR markers. *Biochemical Genetics* 43 (3–4): 87–101.

Xia, T., S. Chen, S. Chen, et al. 2007. ISSR analysis of genetic diversity of the Qinghai-Tibet plateau endemic *Rhodiola chrysanthemifolia* (Crassulaceae). *Biochemical Systematics and Ecology* 35 (4): 209–214.

Yan, T., y. Zu, X. Yan, and F. Zhou. 2003. Genetic structure of endangered *Rhodiola sachalinensis*. *Conservation Genetics* 4: 213–8.

Yoshikawa, M., H. Shimada, H. Shimoda, et al. 1995. Rhodiocyanosides a and b, new antiallergic cyanoglycosides from Chinese natural medicine Si Lie Hong Jing Tian, the underground part of *Rhodiola quadrifida* (Pall.) Fisch. et Mey. *Chemical & Pharmaceutical Bulletin* 43 (7): 1245–7.

Yousef, G. G., M. H. Grace, D. M. Cheng, I. V. Belolipov, I. Raskin, and M. A. Lila. 2006. Comparative phytochemical characterization of three *Rhodiola species*. *Phytochemistry* 67 (21): 2380–91.

Yu, H., L. Ma, J. Zhang, G. Shi, Y. Hu, and Y. Wang. 2011. Characterization of glycosyltransferases responsible for salidroside biosynthesis in *Rhodiola sachalinensis*. *Phytochemistry*.

Zaurov, D. E., I. V. Belolipov, A. G. Kurmukov, I. S. Sodombekov, A. A. Akimaliev, and S. W. Eisenman. 2013. The medicinal plants of Uzbekistan and Kyrgyzstan. In *Medicinal Plants of Central Asia: Uzbekistan and Kyrgyzstan*, edited by S. W. Eisenman, D. E. Zaurov, and L. Struwe, 15–273. New York: Springer.

Zhang, M., Y. Ma, and Z. Yuan. 2010. Chemical constituents from the roots of *Rhodiola algida* var. Tangutica. *Asian Journal of Traditional Medicines* 5 (4): 138–44.

Zhou, J., G. Xie, and X. Yan. 2011. *Encyclopedia of Traditional Chinese Medicines: Molecular Structures, Pharmacological Activities, Natural Sources and Applications, Vol. 5*. Berlin, Heidelberg: Springer Berlin Heidelberg.

Zuo, G., Z. Li, L. Chen, and X. Xu. 2007. Activity of compounds from Chinese herbal medicine *Rhodiola kirilowii* (Regel) maxim against HCV NS3 serine protease. *Antiviral Research* 76 (1): 86–92.

3 Phytochemistry of *Rhodiola rosea*

Fida Ahmed, Vicky Filion,
Ammar Saleem, and John T. Arnason

CONTENTS

3.1 Introduction 65
3.2 Phytochemical Constituents 66
    3.2.1 Salidroside 70
    3.2.2 Rosavins 72
    3.2.3 Phenolic Compounds 72
    3.2.4 Terpenes 72
    3.2.5 Essential Oils 73
3.3 Canadian (Nunavik) Populations of *R. rosea* 73
3.4 Metabolic Profiling of *R. rosea* 74
3.5 Biosynthesis of Phenylethanol and Phenylpropanoid Derivatives 74
    3.5.1 The Biosynthesis of Salidroside 74
    3.5.2 The Biosynthesis of Rosavins 76
3.6 Bioactivity of Phytochemicals of *R. rosea* 76
    3.6.1 Salidroside 76
    3.6.2 Tyrosol 78
    3.6.3 Rosavin 78
    3.6.4 Other Bioactive Compounds 78
3.7 Discussion and Conclusions 78
References 80

3.1 INTRODUCTION

Rhodiola rosea L. (Crassulaceae), commonly known as roseroot, golden root, or Arctic root, is a highly valued medicinal plant in the traditional pharmacopeia of certain regions of Europe, particularly Russia, the Scandinavian countries, and central Asia, including Northern China and Mongolia (Brown et al. 2002). *R. rosea* has been used for centuries as a remedy for a variety of ailments including nervous system disorders, depression, headaches, fatigue, high altitude sickness, anemia, impotence, gastrointestinal disorders, infections, cold, and flu symptoms (Brown et al. 2002). The efficacy of *R. rosea* as well as its safety for use in humans had begun to be assessed in preclinical and clinical trials primarily by researchers in Eastern Europe as early as the 1960s, with a dramatic rise in interest among the global research community since the last decade (Panossian et al. 2010). *R. rosea* has been reported to have antioxidant (Calcabrini et al. 2010), anti-inflammatory (Bawa and Khanum 2009), neuroprotective (Palumbo et al. 2012), anti-cancer (Liu et al. 2012), antidepressant (Perfumi and Mattioli 2007), antistress (Mattioli et al. 2009), and antidiabetic (Kwon et al. 2006) activities among many others.

The long-standing traditional use of *R. rosea* and its multiple biological activities have necessitated a closer examination of its phytochemical constituents, especially from its roots and rhizomes, which are used most frequently in medicinal preparations. The phytochemical composition of the plant determines its biological activity; variation in the concentration and proportion of bioactive compounds may contribute to the varied pharmacological properties of *R. rosea* observed in the literature. Phytochemicals in *R. rosea* may fluctuate based on several factors, including genotype, geographical location, gender, biotic and abiotic factors, cultivation conditions, method and season of harvest, and extraction and storage methods (Galambosi 2006; Elameen et al. 2008). In addition, the increased availability and popularity of commercial *Rhodiola* products advertising beneficial health claims make it even more important to accurately identify and profile the phytochemical constituents. Identity markers are crucial for reasons of both safety and efficacy in order to verify that the correct species of *Rhodiola* was used and to prevent substitution by other morphologically similar species. Identification of active principles is needed if standardization of a product with reliable efficacy is to be achieved. In Sections 3.2 through 3.6, the phytochemicals present in Eurasian *R. rosea* and recently discovered Canadian populations of *R. rosea* are described. The importance of profiling marker compounds via metabolic fingerprinting techniques is also discussed. Additionally, the pharmacological activities of key phytochemicals are briefly mentioned.

3.2   PHYTOCHEMICAL CONSTITUENTS

Studies on the chemical composition of *R. rosea* were initiated in the 1960s by researchers in Eastern Europe and published mainly in Slavic and Scandinavian languages (Khnykina and Zotova 1966; Saratikov et al. 1967; Revina et al. 1976; Komar et al. 1980; Kurkin et al. 1986; Dubichev et al. 1991; Furmanowa et al. 1999; cited in Rohloff 2002). Recent phytochemical work using high-performance liquid chromatography (HPLC) and gas chromatography (GC) coupled with mass spectrometry (MS) techniques have achieved more efficient and rapid separation and identification of known as well as novel minor compounds (Ganzera et al. 2001; Rohloff 2002; Tolonen et al. 2003a, 2003b; Ma et al. 2006; Petsalo et al. 2006; Yousef et al. 2006; Ali et al. 2008; Avula et al. 2009; Ma et al. 2013; Mudge et al. 2013).

Over 140 phytochemicals belonging to several distinct biosynthetic classes have been isolated from *R. rosea* plants, mainly from its roots and (or) rhizomes as well as from the aerial parts (Panossian et al. 2010). These include phenylethanol derivatives, salidroside and *p*-tyrosol, phenylalkanoids, particularly the phenylpropanoid glycosides, rosavin, rosarin, and rosin (collectively known as the "rosavins"), terpenes, essential oils, simple phenolics, flavonoids (flavonols, flavonolignans), proanthocyanidins, gallic acid esters, cyanogenic glucosides, and tannins. Table 3.1 shows a detailed list of *R. rosea* phytochemical compounds grouped by biosynthetic class and Figure 3.1 shows the chemical structures of selected compounds. Of these, salidroside and the rosavins are the most intensively studied compounds.

TABLE 3.1
Phytochemical Compounds of *Rhodiola rosea*

Biosynthetic Class	Phytochemicals	References
Phenylmethanoids	Benzyl alcohol O-α-L-arabinopyranosyl-(1→6)-O-β-D-glucopyranoside	Avula et al. (2009)
	Phenyl methyl O-α-L-arabinofuranosyl-(1→6)-O-β-D-glucopyranoside	Avula et al. (2009)
	Benzyl-O-β-D-glucopyranoside	Mudge et al. (2013)
Phenylethanoids	2-Phenylethyl O-α-L-arabinopyranosyl-(1→6)-O-β-D-glucopyranoside	Avula et al. (2009)
	Mongrhoside	Avula et al. (2009)
	Salidroside: 2-(4-Hydroxyphenyl) ethyl-O-β-D-glucopyranoside	Troshchenko and Kutikova (1967)
	Tyrosol: 4-(2-Hydroxyethyl)phenol	Troshchenko and Kutikova (1967)
	Viridoside	Avula et al. (2009)
Phenylpropanoids	Cinnamyl alcohol	Zapesochnaya and Kurkin (1982)
	Rosarin (trans-cinnamyl O-(6′-O-α-L-arabinofuranosyl)-β-D-glucopyranoside)	Zapesochnaya and Kurkin (1982)
	Rosavin (trans-cinnamyl O-(6′-O-α-L-arabinopyranosyl)-β-D-glucopyranoside)	Zapesochnaya and Kurkin (1982)
	Rosin (trans-cinnamyl O-β-D-glucopyranoside)	Zapesochnaya and Kurkin (1982)
	Cinnamyl-(6′-O-β-D-xylopyranosyl)-O-β-glucopyranoside	Tolonen et al. (2003a)
	4-methoxy-cinnamyl-(6'-O-α-arabinopyranosyl)-O-β-glucopyranoside	Tolonen et al. (2003a)
	Triandrin, Sachaliside 1 (4-hydroxy-cinnamyl-O-β-D-glucopyranoside)	Kurkin et al. (1991)
	Vimalin (4-methoxy-cinnamyl-O-β-D-glucopyranoside)	Kurkin et al. (1991)
	(-)-Lariciresinol (lignans)	Kurkin et al. (1991)
	(-)-Lariciresinol 4-O-β-D-glucopyranoside	Kurkin et al. (1991)

(Continued)

TABLE 3.1 (*Continued*)
Phytochemical Compounds of *Rhodiola rosea*

Biosynthetic Class	Phytochemicals	References
Phenolic acids	Caffeic acid, Caffeic acid 3-O-β-D-glucopyranoside	Kurkin et al. (1991)
	Chlorogenic acid	Kurkin et al. (1991)
	p-Coumaric acid 4-O-β-D-glucopyranoside	Kurkin et al. (1991)
	p-Coumaric acid 1-O-β-D-glucopyranoside (melilotoside)	Kurkin et al. (1991)
	Gallic acid esters, methyl gallate	Kurkin et al. (1984b)
Flavonoids	Rhodionin (herbacetin 7-O-α-rhamnopyranoside)	Zapesochnaya and Kurkin (1983)
	Rhodiosin (herbacetin 7-O-(3"-O- β-D-glucopyranosyl-α-L-rhamnopyranoside)	Zapesochnaya and Kurkin (1983)
	Rhodiolinin	Zapesochnaya and Kurkin (1983)
	Tricin	Brown et al. (2002)
	Kaempferol 3-O-β-D-xylopyranosyl-(1→2)-β-D-glucopyranoside	Avula et al. (2009)
	Kaempferol	Dubichev et al. (1991)
	8-Methylherbacetin	Kurkin et al. (1984b)
	Acetylrhodalgin	Kurkin et al. (1984b)
	Kaempferol 7-O-α-L-rhamnopyranoside	Kurkin et al. (1984b)
	Herbacetin	Jeong et al. (2009)
	Rhodiolinin	Jeong et al. (2009)
	Rhodionidin (herbacetin-7-O-α-L-rhamnopyranose-8-O-β-D-glucopyranoside)	Kurkin et al. (1984a)
	Rhodiolgin (gossypetin-7-O-α-L-rhamnopyranoside)	Kurkin et al. (1984a)
	Rhodiolgidin (gossypetin-7-O-α-L-rhamnopyranose-8-O-β-D-glucopyranoside)	Kurkin et al. (1984a)
	Rhodalin (herbacetin-8-O-β-D-xylopyranoside)	Kurkin et al. (1984a)
	Rhodalidin (herbacetin-8-O-β-D-xylopyranose-3-O-β-D-glucopyranoside)	Kurkin et al. (1984a)
	Gossypetin-di-O-glucoside	Petsalo et al. (2006)
	OH-gossypetin-7-O-rhamnose-8-O-glucose	Petsalo et al. (2006)
	Herbacetin-di-O-glucoside	Petsalo et al. (2006)
	Kaempferol-3-O-glucose-7-O-glucose	Petsalo et al. (2006)
	Quercetin-3-O-rhamnose-7-O-glucose	Petsalo et al. (2006)
	Gossypetin-di-O-glucoside or O-diglucoside	Petsalo et al. (2006)
	Gossypetin-3-O-glucose-7-O-xylose/arabinose	Petsalo et al. (2006)

(*Continued*)

TABLE 3.1 (*Continued*)
Phytochemical Compounds of *Rhodiola rosea*

Biosynthetic Class	Phytochemicals	References
	Kaempferol-3-O-rhamnose-7-O-glucose	Petsalo et al. (2006)
	Herbacetin-3-O-glucose-7-O-xylose/arabinose	Petsalo et al. (2006)
	Quercetin-3'/4'-rhamnose	Petsalo et al. (2006)
Oligomeric/ polymeric proanthocyanidins	Prodelphinidin gallates/esters (epigallocatechin gallate dimers)	Yousef et al. (2006)
Monoterpenes/ glycosides	Rosiridol (3,7-dimethylocta-2,6-diene-1,4-diol)	Kurkin et al. (1985)
	Rosiridin(e) (3,7-dimethylocta-2,6-diene-1,4-diol 1-O-β-D-glucopyranoside)	Kurkin et al. (1985)
	Sachalinol A	Avula et al. (2009)
	Rhodioloside A ((2E,6E,4R)-4,8-dihydroxy-3,7-dimethyl-2,6-octadienyl β-D-glucopyranoside)	Ma et al. (2006)
	Rhodioloside B ((2E,4R)-4-hydroxy-3,7-dimethyl-2,6-octadienyl α-D-glucopyranosyl(1→6)-β-D-glucopyranoside)	Ma et al. (2006)
	Rhodioloside C ((2E,4R)-4-hydroxy-3,7-dimethyl-2,6-octadienyl β-D-glucopyranosyl(1→3)-β-D-glucopyranoside)	Ma et al. (2006)
	Rhodioloside D ((2E,4R)-4,7-dihydroxy-3,7-dimethyl-2-octenyl β-D-glucopyranoside)	Ma et al. (2006)
	Rhodioloside E ((2E)-7-hydroxy-3,7-dimethyl-2-octenyl α-L-arabinopyranosyl(1→6)-β-D-glucopyranoside)	Ma et al. (2006)
	Rhodioloside F ((2E, 4R)-4-hydroxy-3,7-dimethyl-2,6-octa-dienyl α-L-arabinopyranosyl(1→6)-β-D-glucopyranoside)	Ma et al. (2006)
	Geraniol	Ali et al. (2008)
	Myrtenol	Evstatieva et al. (2010)
	Geranyl	Evstatieva et al. (2010)
	1-O-α-L-arabinopyranosyl(1→6)-β-D-glucopyranoside	Mudge et al. (2013)
Triterpenes	Daucosterol	Kurkin et al. (1985)
	β-Sitosterol	Dubichev et al. (1991)
Cyanogenic glucosides	Lotaustralin	Akgul et al. (2004)
	Rhodiocyanoside A	van Diermen et al. (2009)

3.2.1 SALIDROSIDE

Salidroside (**1**), perhaps the most widely tested compound in pharmacological bioassays, was isolated from *R. rosea* roots along with its aglycone precursor, tyrosol
(**2**) and termed "rhodioloside" (Troshchenko and Kutikova 1967). Rhodioloside was
later reidentified as salidroside (Thieme 1969) based on its previous isolation from
Salix triandra (Brigel and Beguin 1926; in Gÿorgy 2006).

Salidroside was initially the main phytochemical marker compound for the identification and standardization of *R. rosea* extracts (Brown et al. 2002; Wang et al. 2012).
However, salidroside has since been isolated from many other *Rhodiola* spp., including *R. sachalinensis* A. Bor (Bi et al. 2009), *R. sexifolia* S. H. Fu, *R. chrysanthemifolia* (Le'vl.) S. H. Fu, *R. alsia* (Fröd.) S. H. Fu, *R. bupleuroides* (Wall. ex Hk. f.
et Thoms.) S. H. Fu, *R. macrocarpa* (Praeg.) S. H. Fu, *R. sacra* (Prain ex Hamet)

Phenylethanol derivatives

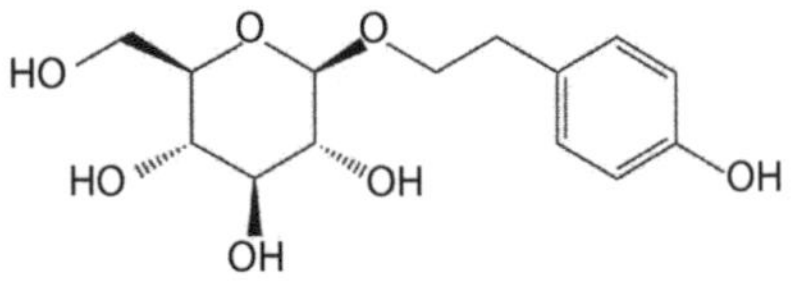

(1) 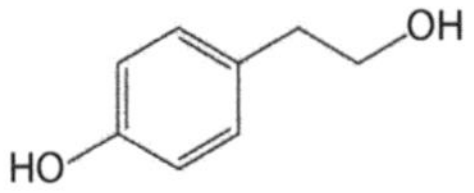   (2)

Phenylpropanoids

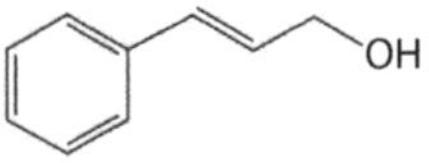

(3) 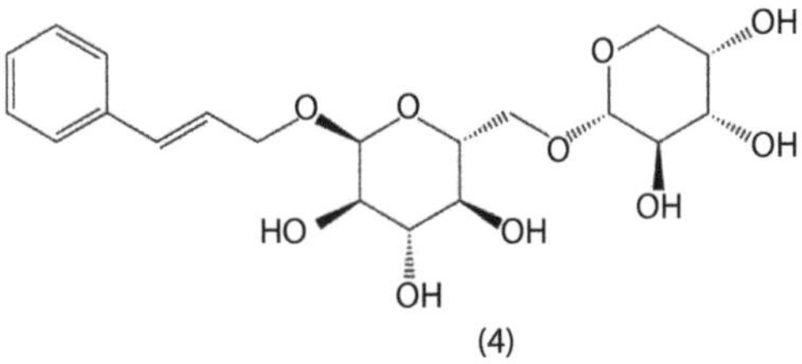   (4)

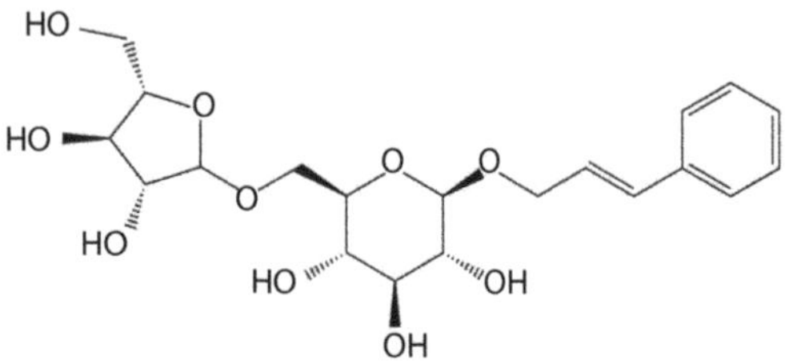

(5) 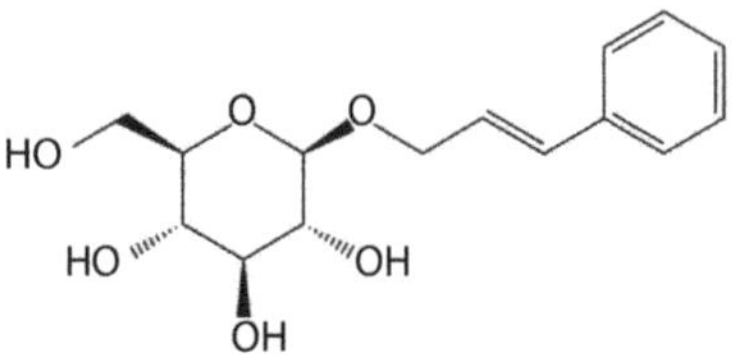   (6)

FIGURE 3.1   (*Continued*)

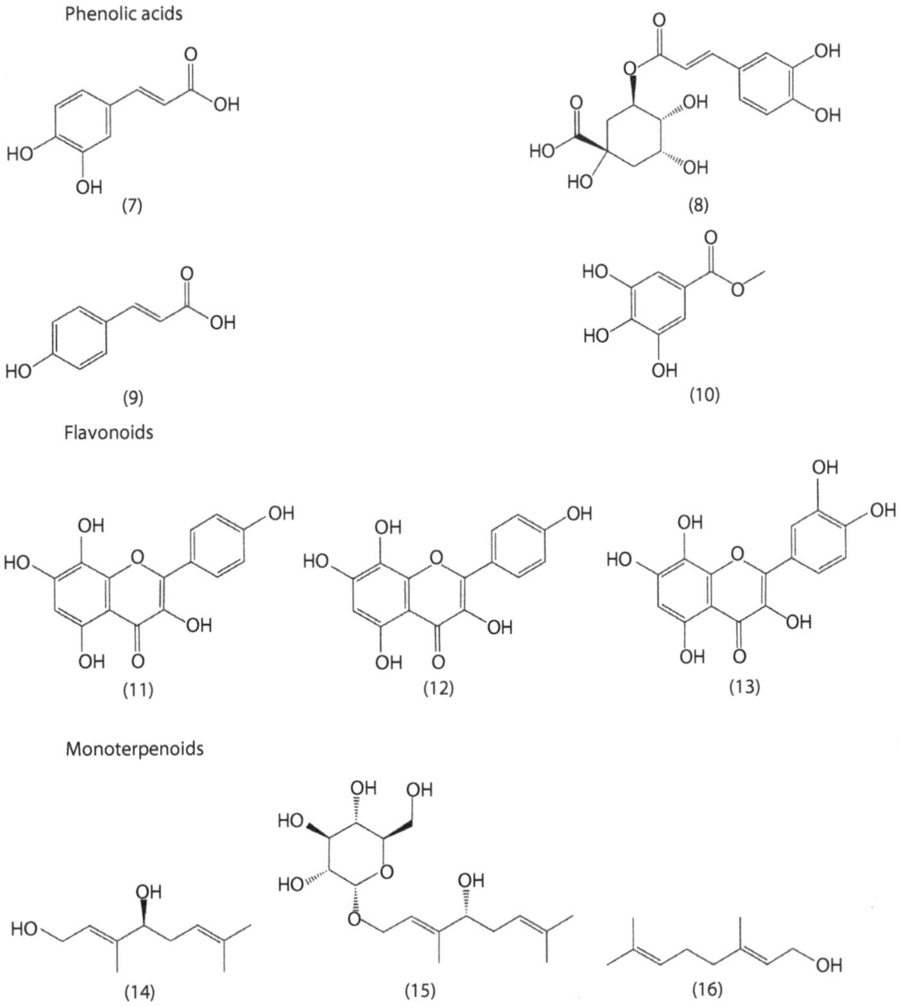

Phenolic acids

(7)

(8)

(9)

(10)

Flavonoids

(11)

(12)

(13)

Monoterpenoids

(14)

(15)

(16)

FIGURE 3.1 Chemical structures of selected *Rhodiola rosea* phytochemicals. (1) Salidroside, (2) *p*-tyrosol, (3) Cinnamyl alcohol, (4) Rosavin, (5) Rosarin, (6) Rosin, (7) Caffeic acid, (8) Chlorogenic acid, (9) *p*-Coumaric acid, (10) Methyl gallate, (11) Kaempferol, (12) Herbacetin, (13) Gossypetin, (14) (-)-Rosiridol, (15) Rosiridin, and (16) Geraniol.

S. H. Fu, *R. kirilowii* (Regel) Maxim, *R. sinuata* (Royle ex Edgew.) S. H. Fu, *R. himalensis* (D. Don) S. H. Fu, *R. coccinea* (Royle) Borrisova, *R. crenulata* (Hk. f. et Thoms.) H. Ohba, *R. tieghemii* (Hamet) S. H. Fu, *R. yunnanensis* (Franch.) S. H. Fu, *R. fastigiata* (Hk. f. et Thoms) S. H. Fu (Chen et al. 2008a), *R. heterodonta* (Hk. f. et Thoms.) Boriss. (Yousef et al. 2006), and *R. quadrifida* (Pall.) Fisch. et Mey. (Troshchenko and Kutikova 1967; Wiedenfeld et al. 2007) in varying concentrations. In addition, the presence of salidroside is not restricted to the Crassulaceae family; it has been identified in *Vaccinium vitis-idaea* (Ericaceae) (Thieme and Winkler 1966),

Olea europaea (Oleaceae) (Ryan and Robards 1988), and *Betula platyphylla* (Betulaceae) (Shen et al. 1999) to name a few. Thus, the presence of salidroside alone is not an adequate phytochemical marker to distinguish *R. rosea* from other species.

3.2.2 Rosavins

Rosavin (**4**), rosarin (**5**), and rosin (**6**), collectively known as the "rosavins," belong to the group of phenylpropanoids, and are glycosides of cinnamyl alcohol (**3**) (Zapesochnaya and Kurkin 1982). The rosavins are generally used in conjunction with salidroside as diagnostic marker compounds for *R. rosea*. Commercial extracts of *R. rosea* are now standardized to both salidroside (0.8 %–1 %) and the rosavins (minimum 3 %) in a 3 : 1 ratio reflecting their approximate concentrations in plant extracts (Brown et al. 2002). Other phenylpropanoids have been identified from *R. rosea* rhizomes and tissue cultures: cinnamyl-(6'-O-β-D-xylopyranosyl)-O-β-glucopyranoside (Tolonen et al. 2003a), sachaliside 1, triandrin (4-hydroxy-cinnamyl-O-β-D-glucopyranoside) (Kurkin et al. 1991), vimalin (4-methoxy-cinnamyl-O-β-D-glucopyranoside) (Kurkin et al. 1991), and 4-methoxy-cinnamyl-(6'-O-α-arabinopyranosyl)-O-β-glucopyranoside (Tolonen et al. 2003a).

3.2.3 Phenolic Compounds

R. rosea contains a large number of simple phenolics, including hydroxycinnamic acids (Brown et al. 2002), caffeic acid (**7**) (Kurkin et al. 1991), chlorogenic acid (**8**) and *p*-coumaric acid (**9**) (Kurkin et al. 1991), gallic acid derivatives (**10**) (Kurkin et al. 1984b), flavonoids, and tannins (Pooja et al. 2006). Flavonoids from *R. rosea* are often found as glycosides of kaempferol (**11**), herbacetin (**12**), and gossypetin (**13**). The flavonoids include flavonols such as rhodiolinin (Jeong et al. 2009), rhodionin, rhodiosin (Zapesochnaya and Kurkin 1983), flavolignans such as rhodiolin (Zapesochnaya and Kurkin 1983), and proanthocyanidins (Yousef et al. 2006).

The aerial portions of *R. rosea* also contain flavonoids, but not phenylpropanoids. These include glycosides of herbacetin and gossypetin, including rhodionin, rhodionidin, rhodiolgin, rhodiolgidin, rhodalin, and rhodalidin (Kurkin et al. 1984a). Recently, 10 new flavonoids were identified from the leaves and flowers of *R. rosea* by Petsalo et al. (2006) (Table 3.1).

3.2.4 Terpenes

R. rosea roots contain monoterpenes, including rosiridol (**14**) and its glycoside rosiridin (**15**) (Kurkin et al. 1985). In a study by Ma et al. (2006), five new monoterpene glycosides, rhodiolosides A–E, were isolated and their structures elucidated. Recently, another new monoterpene glycoside, rhodioloside F, was identified (Ali et al. 2008). Triterpenes daucosterol and β-sitosterol were also identified from *R. rosea* roots (Kurkin et al. 1985; Dubichev et al. 1991).

3.2.5 ESSENTIAL OILS

The essential oil of *R. rosea* roots contains different mixtures of compounds depending on the source of the plant material as well as the extraction method. *R. rosea* roots from Norway were found to contain around 86 different compounds in the essential oil, consisting mainly of monoterpene hydrocarbons, oxygenated monoterpenes, and aliphatic alcohols (Rohloff 2002). In these samples, *n*-decanol (30.38%), geraniol (12.49%), and 1,4-*p*-menthadien-7-ol (5.10%) were the most abundant volatiles detected. Geraniol (**16**) is primarily responsible for the characteristic rose-like fragrance of *R. rosea* roots; other compounds including geranyl formate, geranyl acetate, benzyl alcohol, and phenylethyl alcohol play minor roles (Rohloff 2002). A comparative study of *R. rosea* root essential oils from Bulgaria, China, and India showed geraniol to be the primary compound from the former, similar to that observed in roots from Mongolia (Shatar et al. 2007), while phenethylalcohol was the most important compound in the Indian sample (Evstatieva et al. 2010). Finnish populations of *R. rosea* contained myrtenol, *trans*-pinocarveol, and geraniol as the primary volatiles in the essential oil (Héthelyi et al. 2005).

3.3 CANADIAN (NUNAVIK) POPULATIONS OF *R. ROSEA*

Most of our knowledge about *R. rosea* phytochemistry stems from Eurasian populations. Recently, *R. rosea* populations were discovered in Nunavik, Northern Québec, Canada. *R. rosea* is used by the indigenous Inuit people as a tonic to maintain mental and physical health or to prevent illness (Alm 2004). Comparative phytochemical analyses within local populations as well as with Eurasian *R. rosea* samples have shown interesting trends in phytochemical profiles (Filion 2008; Filion et al. 2008; Avula et al. 2009).

HPLC and LC-ESI-TOF analyses (liquid chromatography-electrospray ionization-time of flight) show that salidroside, rosavin, rosarin, and rosin, the key marker compounds of *R. rosea*, are also present in the Nunavik populations (Filion et al. 2008; Avula et al. 2009). A comparative study of Eurasian and Nunavik *R. rosea* populations by Avula et al. (2009) reported that concentrations of key phytochemicals, including salidroside, the rosavins, and rosiridin were generally lower in the Nunavik populations. Rhodioloside F, a monoterpene glycoside, was detected in the Eurasian population, but only in low amounts in one of the four Nunavik populations analyzed. Rhodioloside D and mongrhoside were detected only in the Eurasian population, but not in the Nunavik ones, suggesting the potential for these to be distinguishing marker compounds, although more samples are needed to validate these findings. Benzyl alcohol *O*-α-l-arabinopyranosyl-(1→6)-*O*-β-D-glucopyranoside, viridoside, and kaempferol 3-*O*-β-D-xylopyranosyl-(1→2)-β-D-glucopyranoside were also reported in both the Nunavik and the Eurasian populations (Avula et al. 2009).

Within different local Nunavik *R. rosea* populations, there was a lot of variation in the levels of phytochemicals detected, possibly due to environmental factors (Filion 2008). The geographical location also impacts the concentration of phytochemicals, particularly the rosavins (Kucinskaite et al. 2007). The phytochemicals in the Nunavik populations did not vary significantly with soil nutrient conditions; however, Graglia et al. (2001) showed that salidroside content in *Betula nana* varies with soil acidity. In addition, Galambosi (2006) reported that with increased soil nutrient availability, concentrations of active metabolites increased in cultivation experiments. Salidroside levels decreased in Nunavik plants impacted by herbivory by bud mites, indicating that salidroside may be a precursor for plant defence compounds or too expensive to produce by plants already stressed by other environmental stressors to maintain normal state concentrations (Filion 2008). Gender also seemed to impact salidroside production in Nunavik *R. rosea* populations; males contained higher levels than females (Filion 2008). Overall, the levels of salidroside seemed to vary with environmental stresses while levels of rosavins were relatively stable in Nunavik *R. rosea* populations (Filion 2008).

3.4 METABOLIC PROFILING OF *R. ROSEA*

Advances in chromatographic techniques have enabled the rapid detection of phytochemical marker compounds from *R. rosea*. However, since phytochemicals in *R. rosea* populations are subject to high genotypic (Elameen et al. 2008) as well as phenotypic variation, often, more than a few marker compounds are necessary to establish diagnostic phytochemical profiles. To this end, nontarget fingerprinting methods, including ^{1}H NMR spectroscopy have recently been applied to separate *R. rosea* populations from different geographical locations harvested at different times based on phytochemical profiles (Ioset et al. 2011). Interestingly, using principal component analyses models based on only the aromatic portion of NMR spectra, the authors reported that the geographical populations clustered separately based on changes in salidroside and rosavin content, while using the entire spectra of all chemical constituents did not separate the populations as well, thus providing validation for the use of these compounds as markers (Ioset et al. 2011). The authors also demonstrated that the time of harvest had an impact on phytochemical content; salidroside and rosavin increased from May to August and steadily declined at the end of summer.

Wang et al. (2012) used a fast, sensitive HPLC-UV-based technique to assess the quality of 10 different batches of *R. rosea* extracts from different manufacturers in an effort to test whether they could be clustered based on the origin of their raw materials. Indeed, using multivariate analyses, they showed that the extracts separated well based on their province of origin and correlated strongly with the content of rosavin, but not salidroside or tyrosol (Wang et al. 2012), thus demonstrating the importance of using rosavin as a marker for confirming the presence of *R. rosea* in herbal extracts.

3.5 BIOSYNTHESIS OF PHENYLETHANOL AND PHENYLPROPANOID DERIVATIVES

3.5.1 THE BIOSYNTHESIS OF SALIDROSIDE

Salidroside is synthesized from its aglycone precursor, tyrosol, by the enzymatic addition of glucose via UDP-glucosyltransferases (Ma et al. 2007) or by β-D-glucosidases (Shi et al. 2007). Figure 3.2 shows the biosynthesis of salidroside and the rosavins.

The biosynthesis of tyrosol occurs through the shikimic acid pathway, which produces phenylalanine and tyrosine. At this point, there is an ongoing debate in the literature about which amino acid is the precursor of tyrosol (Figure 3.2 paths 1 or 2) (Ma et al. 2008). There are two possible routes to tyrosol production; phenylalanine is deaminated by phenylalanine ammonium lyase (PAL), eventually forming para-coumaric acid and then tyrosol (Xu and Su 1997; Li et al. 2005; Ma et al. 2008). This can happen via direct decarboxylation by para-coumaric acid decarboxylase (Liang and Zheng 1981) or by the conversion of para-coumaric acid into para-coumaryl alcohol by cinnamyl alcohol dehydrogenase (CAD) by a series of enzymatic reactions and then into tyrosol through at least two, as yet uncharacterized steps (Wang et al. 2007).

The other reigning hypothesis on tyrosol formation is that tyrosine is converted into tyramine by tyrosine decarboxylase (TyrDC). Tyramine is oxidized by tyramine-oxidase to 4-hydroxy-phenylacetaldehyde (4-HPAA), which is then reduced to 4-hydroxy-phenylacetalcohol, also known as tyrosol (Ellis 1983; Landtag et al. 2002; Ma et al. 2008). There is evidence from biotransformation studies in *Rhodiola* plant callus cultures showing that tyrosine, and not phenylalanine, may be the limiting step for salidroside formation (Ma et al. 2008).

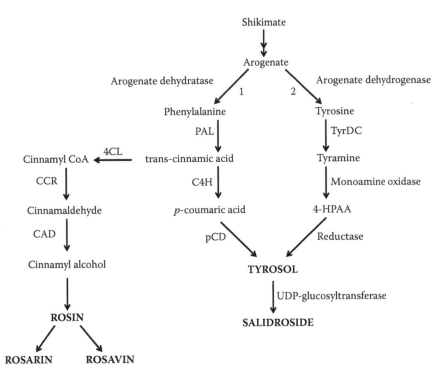

FIGURE 3.2 Schematic of the biosynthesis of key marker phytochemicals salidroside, tyrosol and the rosavins. See text for pathway description. Salidroside biosynthesis: PAL: Phenylalanine ammonia lyase; TyrDC: tyrosine decarboxylase; C4H: Cinnamate 4-hydroxylase; 4-HPAA: 4-hydroxy-phenylacetaldehyde; pCD: *p*-coumaric acid decarboxylase. Rosavin biosynthesis: 4CL hydroxycinnamate: CoA ligase; CCR: cinnamyl-CoA reductase; CAD: cinnamyl alcohol dehydrogenase.

3.5.2 THE BIOSYNTHESIS OF ROSAVINS

The rosavins are cinnamyl alcohol glycosides and therefore are products of the shikimic acid pathway derived from the deamination of phenylalanine by PAL, which forms cinnamic acid. Figure 3.2 shows the key steps in the synthesis of the rosavins. The next step is the formation of cinnamyl-CoA ester catalyzed by hydroxycinnamate: CoA ligase (4CL), which is reduced to cinnamaldehyde by cinnamyl-CoA reductase (CCR). The cinnamaldehyde is further reduced by CAD to cinnamyl alcohol. The enzymes that take part in the formation of the glycosides of cinnamyl alcohol have not yet been elucidated. Rosin is formed by the transfer of one glucose unit to cinnamyl alcohol, and rosarin and rosavin are synthesized by further addition of sugar molecules to rosin (Gÿorgy 2006).

3.6 BIOACTIVITY OF PHYTOCHEMICALS OF *R. ROSEA*

Previous phytochemical investigations of *R. rosea* have led to the isolation and identification of over 140 compounds belonging to different classes of secondary metabolites. However, the biological activity of the *R. rosea* extract has been primarily attributed to a few of these phytochemicals, including salidroside, tyrosol, and rosavin. It is important to keep in mind that other bioactive compounds including phenolics, flavonoids, monoterpenes, and triterpenes are also present in *R. rosea* and that the combination of these compounds in the extract acting synergistically is probably responsible for the sum total of its pharmacological effects.

3.6.1 SALIDROSIDE

Of all the compounds present in *R. rosea*, salidroside is the best studied for its pharmacological effects both *in vitro* and *in vivo*. Salidroside has been reported to have multiple beneficial biological activities, including antioxidant (Chen et al. 2009), neuroprotective (Chen et al. 2008b), anticancer (Hu et al. 2010a), hepatoprotective (Wu et al. 2009), antibacterial (Cybulska et al. 2011), antiviral (Wang et al. 2009), antihypoxic (Tan et al. 2009), anti-inflammatory (Guan et al. 2011a), and antihypoglycemic (Yu et al. 2008) activities.

Cellular oxidative stress occurs when there is an imbalance between the production of reactive oxygen species (ROS) and the antioxidant capacity of the cell. Oxidative stress underlies the pathophysiology of multiple disorders, including neurodegenerative diseases, cardiovascular dysfunction, metabolic syndrome, as well as cancer. Salidroside has been shown to be protective against oxidative stress in many cellular and animal models, commonly induced by hydrogen peroxide (Zhang et al. 2007; Cai et al. 2008; Chen et al. 2009; Mao et al. 2010; Yu et al. 2010; Guan et al. 2011b; Zhu et al. 2011; Qian et al. 2012; Shi et al. 2012). Some of the underlying mechanisms of action of salidroside include inhibiting ROS accumulation, attenuating lipid peroxidation, and DNA damage, stabilizing intracellular Ca^{2+} ion levels, restoring the balance of pro- and antiapoptotic proteins and inhibiting the activity of certain caspases.

In addition to the antioxidant activities of salidroside, it is also a potent neuroprotective agent. Salidroside has been shown to be protective against amyloid-β *in vitro*, a key peptide involved in the etiology of Alzheimer disease, by the induction of antioxidant enzymes, inhibition of ROS accumulation, and reduction of apoptosis (Jang et al. 2003; Zhang et al. 2010). Amyloid-β oligomers also lead to Ca^{2+} ion dysregulation by overactivation of glutamate receptors, a phenomenon known as glutamate excitotoxicity. Salidroside protects against glutamate excitotoxicity by buffering the excess influx of Ca^{2+} ions and inhibiting the activity of caspase-3 (Cao et al. 2006; Chen et al. 2008b). Salidroside is also protective against *in vitro* and *in vivo* models of Parkinson's disease by attenuating damage to dopaminergic cells by blocking the nitric oxide pathway, inhibiting ROS production (Chen et al. 2007; Li et al. 2011), and elevating levels of glial-derived neurotrophic factor (Zhang et al. 2006a). Salidroside can also reverse apoptosis in neural stem cells isolated from rat hippocampi subjected to streptozotocin insult via its antioxidant effects (Qu et al. 2012).

Salidroside protects from hypoxia/ischemic/reperfusion injury in the brain by reducing cerebral edema, reducing markers of lipid peroxidation, increasing levels of antioxidant enzymes, and reducing apoptosis or markers of inflammation including tumor necrosis factor-α (TNF-α), interleukin-1β (IL-1β), and interleukin (IL-6) (Song et al. 2006; Zou et al. 2009; Liang et al. 2010; Zhong et al. 2010; Shi et al. 2012). A detailed mechanistic analysis revealed that salidroside exerts protective effects under hypoxic conditions by increasing expression of hypoxia-inducible factor-1α (HIF-1α) and vascular endothelial growth factor (VEGF) (Zhang et al. 2009).

Recently, salidroside has been demonstrated to possess anticancer properties and, therefore, could be developed as a potential chemopreventive agent, although more research is required to validate these effects. Salidroside has shown cytotoxic activity against multiple cancer cell lines, including hormone-sensitive and hormone-resistant breast cancer cells (Hu et al. 2010a, 2010b), gastric cancer, lung cancer (Hu et al. 2010a), and bladder cancer cells (Liu et al. 2012). This compound acts by inducing cell cycle arrest and the subsequent induction of apoptosis. Salidroside also inhibits metastasis or the ability of the cancerous mass to migrate to other sites from the primary site by suppressing matrix metalloproteinases (Sun et al. 2012). Salidroside also reduces tumor angiogenesis (Skopińska-Rózewska et al. 2008).

Salidroside exhibits numerous other pharmacological activities. It is a potent anti-inflammatory agent; it reduces the levels of several cytokines produced upon lipopolysaccharide stimulation, including TNF-α, IL-1β, and IL-6 (Guan et al. 2011a). It has extensive hepatoprotective (Jiang et al. 2002; Zeng et al. 2005; Wu et al. 2008, 2009; Ouyang et al. 2010), antistress (Cifani et al. 2010), antifatigue (Ji et al. 2007), immune-modulating (Lin et al. 2011), hematopoietic (Zhang et al. 2005, 2006b; Zheng et al. 2012), antibacterial (Cybulska et al. 2011; Coenye et al. 2012), and antiviral (Wang et al. 2009) properties. Attempts have been made at synthesizing analogs of salidroside and testing their biological activity under conditions of hypoglycemia and serum limitation in PC12 cells (a model of ischemic stroke). Salidroside analogs exhibited equal or greater protective activity by modulating apoptotic gene expression and restoring mitochondrial membrane potential (Guo et al. 2010, 2011; Meng et al. 2011). Clearly, salidroside is a highly active compound with pleiotropic effects on multiple pharmacological targets.

3.6.2 TYROSOL

In the context of *R. rosea*, tyrosol has been shown to be protective against adrenaline and $CaCl_2$-induced arrythmia (Maimeskulova and Maslov 1998). *p*-Tyrosol protects bone marrow cells from subacute lead toxicity by attenuating lipid peroxidation (Pashkevich et al. 2003). It is a better antioxidant/neuroprotective agent against cerebral ischemia compared to salidroside both *in vitro* (rat cortical neuron cultures) and *in vivo* (rats-MCAO); it acts by restoring the balance of pro- and antiapoptotic proteins (Shi et al. 2012).

3.6.3 ROSAVIN

Rosavin shows antibacterial inhibitory activity against resistant strains of *Neisseria gonorrhoeae* when tested in disk diffusion assays (Cybulska et al. 2011). Daily oral administration of rosavin reduces angiogenesis in L-1 sarcoma cells grafted onto the skin of Balb/c mice, providing a possible mechanism of action for the observed anticancer effects of *R. rosea* (Skopińska-Różewska et al. 2008). Rosavin exhibits antidepressant activity *in vivo* in rats in the forced swim test by significantly decreasing freezing time and increasing swim duration compared to untreated controls (Kurkin et al. 2006). Rosavin was found to be the most potent among several synthesized phenylpropanoid glycosides in terms of inhibiting acetylcholinesterase activity, an enzyme which degrades the neurotransmitter acetylcholine, a key drug target in Alzheimer disease (Li et al. 2011).

3.6.4 OTHER BIOACTIVE COMPOUNDS

Apart from salidroside, tyrosol, and rosavin, other compounds isolated from *R. rosea* roots also have potent biological activity. Flavonols isolated from the roots of *R. rosea*, particularly kaempferol, showed strong inhibitory activity against two influenza viruses (Jeong et al. 2009). Bioassay-guided fractionation of *R. rosea* roots using monoamine oxidase A and B inhibitory activity led to the identification of rosiridin, rhodioloside B and C isomers, cinnamyl alcohol, triandrin, and EGCG dimers as the most active compounds; in contrast, salidroside or the rosavins possessed no significant activity (van Diermen et al. 2009). Gossypetin-7-*O*-L-rhamnopyranoside and rhodioflavonoside from the stems of *R. rosea* showed strong inhibitory activity against *Staphylococcus aureus* (Ming et al. 2005). These two compounds were cytotoxic against prostate cancer cells (Ming et al. 2005) and inhibited acetylcholinesterase activity (Hillhouse et al. 2004).

3.7 DISCUSSION AND CONCLUSIONS

The phytochemistry of *R. rosea*, particularly several European and Chinese populations, has been assessed extensively, leading to the isolation and identification of over a hundred different compounds. Salidroside, despite its widespread prevalence among many *Rhodiola* species and several other plant families, continues to be used as a marker compound for the standardization of *R. rosea* extracts (Brown et al. 2002). In light of recent metabolic profiling studies that demonstrate that rosavin, and not salidroside, is a better factor for distinguishing different *R. rosea* populations, these practices might be reconsidered (Ioset et al. 2011; Wang et al. 2012). The biological activity of *R. rosea* is attributed primarily to the presence of salidroside in the published literature, even though the diverse activities of its other constituents have been well established (Kurkin 2003; van Diermen et al. 2009). In addition, numerous studies report the biological activity of high concentrations of chemically synthesized, commercially available forms of salidroside; these may not reflect the actual levels of salidroside present in the *R. rosea* extract nor the synergistic effect of salidroside with the other phytochemical constituents present in the *R. rosea* extract.

Rapid, sensitive and cost-effective HPLC-based techniques to profile herbal extracts are crucial and are beginning to be utilized for *R. rosea* not only for the accurate distinction of morphologically similar *Rhodiola* species and (or) different geographical populations of the same species (Wang et al. 2012) but also for the quantification of key phytochemicals that fluctuate depending on genetic (Elameen et al. 2008) and environmental factors (Galambosi 2006). An important consideration in the case of phytochemical analyses of *R. rosea* crude extracts is the high concentration of tannins; Yousef et al. (2006) reported that oligomeric and polymeric proanthocyanidins account for approximately 30% of the dry weight of the acetone extract of *R. rosea*. These tannins can hinder the detection of other compounds, destabilize extracts in solution, and interfere with protein-based bioassays (Wang et al. 2012). However, very few studies perform a tannin removal step, usually by polyamide filtration (Wang et al. 2012).

The beneficial health benefits of *R. rosea* and its phytochemicals have led to the overexploitation of wild populations; in fact, *R. rosea* is classified as an endangered or threatened species in several regions (Galambosi 2006). Apart from cultivation, biotechnological tools such as plant callus aggregate cultures are used to overexpress enzymes in the biosynthetic pathway to improve salidroside or rosavin production, although these are dependent on culture conditions, such as pH, media, and sugar molecules provided as substrate (Xu et al. 1998). Tissue culture techniques have led to the synthesis of entirely new sets of compounds not present in the original roots/rhizomes of the plant, or low amounts of known bioactive compounds (Kurkin 2003).

R. rosea phytochemicals have been combined with active ingredients from other adaptogenic plants. One such herbal formulation is ADAPT-232®, manufactured by the Swedish Herbal Institute, which consists of a fixed combination of dried extracts from roots of *R. rosea*, berries of *Schizandra chinensis*, and roots of *Eleutherococcus senticosus*, and is standardized with respect to salidroside (0.32%), rosavin, (0.5%), tyrosol (0.05%), schizandrin (0.37%), γ-schizandrin (0.24%), and eleutherosides B and E (0.15%) (Aslanyan et al. 2010). These mixtures of adaptogenic plants have been shown to improve cognitive function via synergistic effects, particularly under stressful situations (Aslanyan et al. 2010).

Thus, the study of the phytochemistry of *R. rosea* provides insight into its biological activity, as well as underlines the importance of establishing appropriate marker compounds. Phytochemical content of *R. rosea* varies from population to population. Geographical location, biotic and abiotic stress factors, and gender are the important factors responsible for this variation in phytochemistry, and consequently, biological activity. Fingerprinting and nontarget profiling using HPLC or NMR-based techniques will gain prominence in terms of extract standardization and quality control in the recent future.

REFERENCES

Akgul, Y., D. Ferreira, E. A. Abourashed, and I. A. Khan. 2004. Lotaustralin from *Rhodiola rosea* roots. *Fitoterapia* 75 (6): 612–4.

Ali, Z., F. R. Fronczek, and I. A. Khan. 2008. Phenylalkanoids and monoterpene analogues from the roots of *Rhodiola rosea*. *Planta Medica* 74 (2): 178–81.

Alm, T. 2004. Ethnobotany of *Rhodiola rosea* (crassulaceae) in Norway. *SIDA, Contributions to Botany* 21 (1): 321–44.

Aslanyan, G., E. Amroyan, E. Gabrielyan, M. Nylander, G. Wikman, and A. Panossian. 2010. Double-blind, placebo-controlled, randomised study of single dose effects of ADAPT-232 on cognitive functions. *Phytomedicine* 17 (7): 494–9.

Avula, B., Y. H. Wang, Z. Ali, T. J. Smillie, V. Filion, A. Cuerrier, J. T. Arnason, et al. 2009. RP-HPLC determination of phenylalkanoids and monoterpenoids in *Rhodiola rosea* and identification by LC-ESI-TOF. *Biomedical Chromatography* 23: 865–72.

Bawa, P. A. S., and F. Khanum. 2009. Anti-inflammatory activity of *Rhodiola rosea*—"A second-generation adaptogen." *Phytotherapy Research* 23 (8): 1099–102.

Bi, H. M, S. Q. Zhang, C. J Liu, and C. Z Wang. 2009. High hydrostatic pressure extraction of salidroside from *Rhodiola sachalinensis*. *Journal of Food Process Engineering* 32 (1): 53–63.

Brigel, M. and C. Beguin. 1926. Isolation of rutoside, asparagines and a new glycoside, hydrolysable by emulsion, salidroside from *Salix triandra* L. *Comptes rendus hebdomadaires des seances del'Academie des sciences* 183: 321–3.

Brown, R., P. Gerbarg, and Z. Ramazanov. 2002. *Rhodiola rosea*: A phytomedicinal overview. *HerbalGram* 56: 40–52.

Cai, L., H. Wang, Q. Li, Y. Qian, and W. Yao. 2008. Salidroside inhibits H_2O_2-induced apoptosis in PC 12 cells by preventing cytochrome c release and inactivating of caspase cascade. *Acta Biochimica Et Biophysica Sinica* 40 (9): 796–802.

Calcabrini, C., R. De Bellis, U. Mancini, L. Cucchiarini, L. Potenza, R. De Sanctis, V. Patrone, et al. 2010. *Rhodiola rosea* ability to enrich cellular antioxidant defences of cultured human keratinocytes. *Archives of Dermatological Research* 302 (3): 191–200.

Cao, L. L., G. H. Du, and M. W. Wang. 2006. The effect of salidroside on cell damage induced by glutamate and intracellular free calcium in PC12 cells. *Journal of Asian Natural Products Research* 8 (1–2): 159–65.

Chen, H., J. Z. Chen, R. R. Hou, A. L. Yang, X. G. Kang, Y. Zhang, Y. L. Li, and H. L. Li. 2007. Inhibition of salidroside on apoptosis of PC12 cells induced by paraquat and its related mechanisms. *Chinese Journal of Pharmacology and Toxicology* 21 (3): 190–6.

Chen, S., D. Zhang, S. Chen, T. Xia, Q. Gao, Y. Duan, and F. Zhang. 2008a. Determination of salidroside in medicinal plants belonging to the *Rhodiola* L. genus originating from the Qinghai-Tibet plateau. *Chromatographia* 68 (3–4): 299–302.

Chen, X., Q. Zhang, Q. Cheng, and F. Ding. 2009. Protective effect of salidroside against H_2O_2-induced cell apoptosis in primary culture of rat hippocampal neurons. *Molecular and Cellular Biochemistry* 332 (1–2): 85–93.

Chen, X., S. L. Zhou, X. S. Gu, and F. Ding. 2008b. Protective effects of salidroside on glutamate induced neurotoxicity in cultured hippocampal neurons. *Acta Anatomica Sinica* 39 (3): 355–9.

Cifani, C., M. V. Micioni Di B., G. Vitale, V. Ruggieri, R. Ciccocioppo, and M. Massi. 2010. Effect of salidroside, active principle of *Rhodiola rosea* extract, on binge eating. *Physiology and Behavior* 101 (5): 555–62.

Coenye, T., G. Brackman, P. Rigole, E. De Witte, K. Honraet, B. Rossel, and H. J. Nelis. 2012. Eradication of *Propionibacterium acnes* biofilms by plant extracts and putative identification of icariin, resveratrol and salidroside as active compounds. *Phytomedicine* 19 (5): 409–12.

Cybulska, P., S. D. Thakur, B. C. Foster, I. M. Scott, R. I. Leduc, J. T. Arnason, and J. -A R. Dillon. 2011. Extracts of Canadian first nations medicinal plants, used as natural products, inhibit *Neisseria gonorrhoeae* isolates with different antibiotic resistance profiles. *Sexually Transmitted Diseases* 38 (7): 667–71.

Dubichev, A. G., V. A. Kurkin, G. G. Zapesochnaya, and E. D. Vorontsov. 1991. Chemical composition of the rhizomes of the *Rhodiola rosea* by the HPLC method. *Chemistry of Natural Compounds* 27 (2): 161–4.

Elameen, A., S. S. Klemsdal, S. Dragland, S. Fjellheim, and O. A. Rognli. 2008. Genetic diversity in a germplasm collection of roseroot (*Rhodiola rosea*) in Norway studied by AFLP. *Biochemical Systematics and Ecology* 36 (9): 706–15.

Ellis, B. E. 1983. Production of hydroxyphenylethanol glycosides in suspension cultures of *Syringa vulgaris. Phytochemistry* 22: 1941–3.

Evstatieva, L., M. Todorova, D. Antonova, and J. Staneva. 2010. Chemical composition of the essential oils of *Rhodiola rosea* L. of three different origins. *Pharmacognosy Magazine* 6 (24): 256–8.

Filion, V. J. 2008. A novel phytochemical and ecological study of the Nunavik medicinal plant *Rhodiola rosea* L. (M.Sc. Thesis, University of Ottawa).

Filion, V. J., A. Saleem, G. Rochefort, M. Allard, A. Cuerrier, and J. T. Arnason. 2008. Phytochemical analysis of Nunavik *Rhodiola rosea* L. *Natural Product Communications* 3 (5): 721–6.

Furmanowa, M., Kedzia, B., Hartwich, M., Kozlowski, J., Krajewska-Patan, A., Mscisz, A., Jankowiak, J. 1999. Phytochemical and pharmacological properties of *Rhodiola rosea* L. *Herba Polonica* 45 (2): 108–13.

Galambosi, B. 2006. Demand and availability of *Rhodiola rosea* L. raw material In *Medicinal and Aromatic Plants*. Eds. R. J. Bogers, L. E. Craker, and D. Lange (Netherlands: Springer, 2006), 223–36.

Ganzera, M., Y. Yayla, and I. A. Khan. 2001. Analysis of the marker compounds of *Rhodiola rosea* l. (golden root) by reversed phase high performance liquid chromatography. *Chemical and Pharmaceutical Bulletin* 49 (4): 465–7.

Graglia, E., R. Julkunen-Tiitto, G. R. Shaver, I. K. Schmidt, S. Jonasson, and A. Michelsen. 2001. Environmental control and intersite variations of phenolics in betula nana in tundra ecosystems. *New Phytologist* 151 (1): 227–36.

Guan, S., H. Feng, B. Song, W. Guo, Y. Xiong, G. Huang, W. Zhong, et al. 2011a. Salidroside attenuates LPS-induced pro-inflammatory cytokine responses and improves survival in murine endotoxemia. *International Immunopharmacology* 11 (12): 2194–9.

Guan, S., W. Wang, J. Lu, W. Qian, G. Huang, X. Deng, and X. Wang. 2011b. Salidroside attenuates hydrogen peroxide-induced cell damage through a cAMP-dependent pathway. *Molecules* 16 (4): 3371–9.

Guo, Y., X. Li, Y. Zhao, Y. Si, H. Zhu, and Y. Yang. 2011. Synthesis and biological evaluation of two salidroside analogues in the PC12 cell model exposed to hypoglycemia and serum limitation. *Chemical and Pharmaceutical Bulletin* 59 (8): 1045–7.

Guo, Y., Y. Zhao, C. Zheng, Y. Meng, and Y. Yang. 2010. Synthesis, biological activity of salidroside and its analogues. *Chemical and Pharmaceutical Bulletin* 58 (12): 1627–9.

Gÿorgy, Z. 2006. Glycoside production by *in vitro Rhodiola rosea* cultures. (Ph.D. diss., University of Oulu).

Héthelyi, E. B., K. Korány, B. Galambosi, J. Domokos, and J. Pálinkás. 2005. Chemical composition of the essential oil from rhizomes of *Rhodiola rosea* L. grown in Finland. *Journal of Essential Oil Research* 17 (6): 628–9.

Hillhouse, B. J., D. S. Ming, C. J. French, and G. H. N. Towers. 2004. Acetylcholine esterase inhibitors in *Rhodiola rosea. Pharmaceutical Biology* 42 (1): 68–72.

Hu, X., S. Lin, D. Yu, S. Qiu, X. Zhang, and R. Mei. 2010a. A preliminary study: The anti-proliferation effect of salidroside on different human cancer cell lines. *Cell Biology and Toxicology* 26 (6): 499–507.

Hu, X., X. Zhang, S. Qiu, D. Yu, and S. Lin. 2010b. Salidroside induces cell-cycle arrest and apoptosis in human breast cancer cells. *Biochemical and Biophysical Research Communications* 398 (1): 62–7.

Ioset, K. N., N. T. Nyberg, D. Van Diermen, P. Malnoe, K. Hostettmann, A. N. Shikov, and J. W. Jaroszewski. 2011. Metabolic profiling of *Rhodiola rosea* rhizomes by ^{1}H NMR spectroscopy. *Phytochemical Analysis* 22 (2): 158–65.

Jang, S. I., H. O. Pae, B. M. Choi, G. S. Oh, S. Jeong, H. J. Lee, H. Y. Kim, et al. 2003. Salidroside from *Rhodiola sachalinensis* protects neuronal PC12 cells against cytotoxicity induced by amyloid-β. *Immunopharmacology and Immunotoxicology* 25 (3): 295–304.

Jeong, H. J., Y. B. Ryu, S. J. Park, J. H. Kim, H. J. Kwon, J. H. Kim, K. H. Park, et al. 2009. Neuraminidase inhibitory activities of flavonols isolated from *Rhodiola rosea* roots and their *in vitro* anti-influenza viral activities. *Bioorganic and Medicinal Chemistry* 17 (19): 6816–23.

Ji, C. F., X. Geng, and Y. B. Ji. 2007. Effect of salidroside on energy metabolism in exhausted rats. *Journal of Clinical Rehabilitative Tissue Engineering Research* 11 (45): 9149–51.

Jiang, M. D., X. Y. Gan, F. W. Xie, W. Z. Zeng, and X. L. Wu. 2002. Effect of salidroside on the proliferation and collagen mRNA transcription in rat hepatic stellate cells stimulated by acetaldehyde. *Yaoxue Xuebao* 37 (11): 841–4.

Khnykina, L. A., and M. I. Zotova. 1966. To the pharmacognostic study of *Rhodiola rosea* *Aptechnoe Delo* 15: 34–38.

Komar, V. V., Z. V. Karpliuk, S. M. Kit, L. V. Komar, and V. O. Smolins'ka. 1980. Macro- and microelement composition of root extracts of *Rhodiola rosea* (golden root). *Farmatsevtychnyĭ Zhurnal* 3: 58–60.

Kucinskaite, A., L. Pobłocka-Olech, M. Krauze-Baranowska, M. Sznitowska, A. Savickas, and V. Briedis. 2007. Evaluation of biologically active compounds in roots and rhizomes of *Rhodiola rosea* L. cultivated in Lithuania. *Medicina (Kaunas, Lithuania)* 43 (6): 487–94.

Kurkin, V. A. 2003. Phenylpropanoids from medicinal plants: distribution, classification, structural analysis, and biological activity. *Chemistry of Natural Compounds* 39 (2): 123–53.

Kurkin, V. A., A. V. Dubishchev, G. G. Zapesochnaya, I. N. Titova, V. B. Braslavskii, O. E. Pravdivtseva, V. N. Ezhkov, et al. 2006. Effect of phytopreparations containing phenylpropanoids on the physical activity of animals. *Pharmaceutical Chemistry Journal* 40 (3): 149–50.

Kurkin, V. A., G. G. Zapesochnaya, Y. N. Gorbunov, E. L. Nukhimovskii, A. I. Shreter, and A. N. Shchavlinskii.1986. Chemical investigation of some species of the genera *Rhodiola* L. and *Sedum* L. and problems of their chemotaxonomy. *Rastitel'nye Resursy* 22: 310–9.

Kurkin, V. A., G. G. Zapesochnaya, and A. N. Shchavlinskii. 1984a. Flavonoids of the epigeal part of *Rhodiola rosea*. I. *Chemistry of Natural Compounds* 20 (5): 623–4.

Kurkin, V. A., G. G. Zapesochnaya, and A. N. Shchavlinskii 1984b. Flavonoids of the rhizomes of *Rhodiola rosea*. III. *Chemistry of Natural Compounds* 20 (3): 367–8.

Kurkin, V. A., G. G. Zapesochnaya, A. G. Dubichev, E. D. Vorontsov, I. V. Aleksandrova, and R. V. Panova. 1991. Phenylpropanoids of callus culture of *Rhodiola rosea*. *Khimiya-Prirodnykh_Soedinenii* 4: 481–90.

Kurkin, V. A., G. G. Zapesochnaya, and A. N. Shchavlinskii. 1985. Terpenoids of the rhizomes of *Rhodiola rosea*. *Chemistry of Natural Compounds* 21 (5): 593–7.

Kwon, Y. I, H. D. Jang, and K. Shetty. 2006. Evaluation of *Rhodiola crenulata* and *Rhodiola rosea* for management of type II diabetes and hypertension. *Asia Pacific Journal of Clinical Nutrition* 15 (3): 425–32.

Landtag, J., A. Baumert, T. Degenkolb, J. Schmidt, V. Wray, D. Scheel, D. Strack, et al. 2002. Accumulation of tyrosol glucoside in transgenic potato plants expressing a parsley tyrosine decarboxylase. *Phytochemistry* 60: 683–9.

Li, W., G. S. Du, and Q. N. Huang. 2005. Salidroside contents and related enzymatic activities in *Rhodiola kirilowii* callus. *Acta Botanica Boreali-Occidentalia Sinica* 25: 1645–8.

Li, X., X. Ye, X. Li, X. Sun, Q. Liang, L. Tao, X. Kang, et al. 2011. Salidroside protects against MPP +-induced apoptosis in PC12 cells by inhibiting the NO pathway. *Brain Research* 1382: 9–18.

Li, X. D., S. T. Kang, G. Y. Li, X. Li, and J. H. Wang. 2011. Synthesis of some phenylpropanoid glycosides (PPGs) and their acetylcholinesterase/xanthine oxidase inhibitory activities. *Molecules* 16 (5): 3580–96.

Liang, X. Q., P. Xie, Y. Zhang, T. Shi, T. H. Yan, and Q. J. Wang. 2010. Effects of salidroside on myocardial ischemia/reperfusion injury in rats. *Chinese Journal of Natural Medicines* 8 (2): 127–31.

Liang, Z. H., and G. Z. Zheng. 1981. Secondary metabolism of high plants. *Plant Physiology Communications* 17 (1): 14–21.

Lin, S. S. C., L. W. Chin, P. C. Chao, Y. Y. Lai, L. Y. Lin, M. Y. Chou, M. C. Chou, et al. 2011. *In vivo* Th1 and Th2 cytokine modulation effects of *Rhodiola rosea* standardised solution and its major constituent, salidroside. *Phytotherapy Research* 25 (11): 1604–11.

Liu, Z., X. Li, A. R. Simoneau, M. Jafari, and X. Zi. 2012. *Rhodiola rosea* extracts and salidroside decrease the growth of bladder cancer cell lines via inhibition of the mTOR pathway and induction of autophagy. *Molecular Carcinogenesis* 51 (3): 257–67.

Ma, C., L. Hu, Z. Lou, H. Wang, and X. Gu. 2013. Preparative separation and purification of four phenylpropanoid glycosides from *Rhodiola rosea* by high-speed counter-current chromatography. *Journal of Liquid Chromatography and Related Technologies* 36 (1): 116–26.

Ma, G., W. Li, D. Dou, X. Chang, H. Bai, T. Satou, J. Li, et al. 2006. Rhodiolosides A–E, monoterpene glycosides from *Rhodiola rosea*. *Chemical and Pharmaceutical Bulletin* 54 (8): 1229–33.

Ma, L. Q., B. Y. Liu, D. Y. Gao, X. B. Pang, S. Y. Lü, H. S. Yu, H. Wang, et al. 2007. Molecular cloning and overexpression of a novel UDP-glucosyltransferase elevating salidroside levels in *Rhodiola sachalinensis*. *Plant Cell Reports* 26 (7): 989–99.

Ma, L. Q., D. Y. Gao, Y. N. Wang, H. H. Wang, J. X. Zhang, X. B. Pang, T. S. Hu, et al. 2008. Effects of overexpression of endogenous phenylalanine ammonia-lyase (PALrs1) on accumulation of salidroside in *Rhodiola sachalinensis*. *Plant Biology* 10 (3): 323–33.

Maimeskulova, L. A., and L. N. Maslov. 1998. The anti-arrhythmia action of an extract of *Rhodiola rosea* and of n-tyrosol in models of experimental arrhythmias. *Eksperimental'Naia i Klinicheskaia Farmakologiia* 61 (2): 37–40.

Mao, G. X., Y. Wang, Q. Qiu, H. B. Deng, L. G. Yuan, R. G. Li, D. Q. Song, et al. 2010. Salidroside protects human fibroblast cells from premature senescence induced by H_2O_2 partly through modulating oxidative status. *Mechanisms of Ageing and Development* 131 (11–12): 723–31.

Mattioli, L., C. Funari, and M. Perfumi. 2009. Effects of *Rhodiola rosea* L. extract on behavioural and physiological alterations induced by chronic mild stress in female rats. *Journal of Psychopharmacology* 23 (2): 130–42.

Meng, Y., Y. Guo, Y. Ling, Y. Zhao, Q. Zhang, X. Zhou, F. Ding, et al. 2011. Synthesis and protective effects of aralkyl alcoholic 2-acetamido-2-deoxy-β-D-pyranosides on hypoglycemia and serum limitation induced apoptosis in PC12 cells. *Bioorganic and Medicinal Chemistry* 19 (18): 5577–84.

Ming, D. S., B. J. Hillhouse, E. S. Guns, A. Eberding, S. Xie, S. Vimalanathan, and G. H. N. Towers. 2005. Bioactive compounds from *Rhodiola rosea* (crassulaceae). *Phytotherapy Research* 19 (9): 740–3.

Mudge, E., D. Lopes-Lutz, P. N. Brown, and A. Schieber. 2013. Purification of phenylalkanoids and monoterpene glycosides from *Rhodiola rosea* L. roots by high-speed counter-current chromatography. *Phytochemical Analysis* 24 (2): 129–34.

Ouyang, J. F., J. Lou, C. Yan, Z. H Ren, H. X. Qiao, and D. S Hong. 2010. *In vitro* promoted differentiation of mesenchymal stem cells towards hepatocytes induced by salidroside. *Journal of Pharmacy and Pharmacology* 62 (4): 530–8.

Palumbo, D. R., F. Occhiuto, F. Spadaro, and C. Circosta. 2012. *Rhodiola rosea* extract pro-
     tects human cortical neurons against glutamate and hydrogen peroxide-induced cell
     death through reduction in the accumulation of intracellular calcium. *Phytotherapy
     Research* 26 (6): 878–83.
Panossian, A., G. Wikman, and J. Sarris. 2010. Rosenroot (*Rhodiola rosea*): traditional use,
     chemical composition, pharmacology and clinical efficacy. *Phytomedicine* 17 (7):
     481–93.
Pashkevich, I. A., I. A. Uspenskaya, V. V. Nefedova, and A. B. Egorova. 2003. A comparative
     in vivo study of the effects of p-tyrosol and *Rhodiola rosea* extract on bone marrow cells.
     Eksperimental'Naya i Klinicheskaya Farmakologiya 66 (4): 50–2.
Perfumi, M., and L. Mattioli. 2007. Adaptogenic and central nervous system effects of single
     doses of 3% rosavin and 1% salidroside *Rhodiola rosea* L. extract in mice. *Phytotherapy
     Research* 21 (1): 37–43.
Petsalo, A., J. Jalonen, and A. Tolonen. 2006. Identification of flavonoids of *Rhodiola rosea* by
     liquid chromatography-tandem mass spectrometry. *Journal of Chromatography A* 1112
     (1–2): 224–31.
Pooja, K. R. A., F. Khanum, and A. S. Bawa. 2006. Phytoconstituents and antioxidant potency
     of *Rhodiola rosea*—A versatile adaptogen. *Journal of Food Biochemistry* 30 (2): 203–14.
Qian, E. W., D. T. Ge, and S. K. Kong. 2012. Salidroside protects human erythrocytes against
     hydrogen peroxide-induced apoptosis. *Journal of Natural Products* 75 (4): 531–7.
Qu, Z. Q., Y. Zhou, Y. S. Zeng, Y. K. Lin, Y. Li, Z. Q. Zhong, and W. Y. Chan. 2012. Protective
     effects of a *Rhodiola crenulata* extract and salidroside on hippocampal neurogenesis
     against streptozotocin-induced neural injury in the rat. *PloS ONE* 7 (1): e29641.
Revina, T. A., E. A. Krasnov, T. P. Sviridova, G. Y. Stepanyuk, and Y. P. Surov. 1976. Biological
     characteristics and chemical composition of *Rhodiola rosea* L. grown in Tomsk.
     Rastitel'nyne Resursy 12, 355–60.
Rohloff, J. 2002. Volatiles from rhizomes of *Rhodiola rosea* L. *Phytochemistry* 59 (6): 655–61.
Ryan, D. and K. Robards. 1998. Critical review: Phenolic compounds in olives. *The Analyst*
     123 (5): 31R–44R.
Saratikov, A. S., E. A. Krasnov, L. A. Khnykina, and L. M. Duvidzon. 1967. Separation and
     study of individual biologically active agents from *Rhodiola rosea* and *Rhodiola quadri-
     fida Izv. Sib. Otd. Akad. Nauk SSSR, Ser. Biol.-Med. Nauk* 1: 54–60.
Shatar, S., R. P. Adams, and W. Koenig. 2007. Comparative study of the essential oil of
     Rhodiola rosea L. from Mongolia. *Journal of Essential Oil Research* 19 (3): 215–7.
Shen, Y., Y. Kojima, and M. Terazawa. 1999. Four glucosides of p-hydroxyphenyl derivatives
     from birch leaves. *Journal of Wood Science* 45(4): 332–6.
Shi, L. L., L. Wang, Y. X. Zhang, and Y. J. Liu. 2007. Approaches to biosynthesis of salidroside
     and its key metabolic enzymes. *Forestry Studies in China* 9 (4): 295–9.
Shi, T. Y., S. F. Feng, J. H. Xing, Y. M. Wu, X. Q. Li, N. Zhang, Z. Tian, et al. 2012.
     Neuroprotective effects of salidroside and its analogue tyrosol galactoside against
     focal cerebral ischemia *in vivo* and H_2O_2-induced neurotoxicity *in vitro*. *Neurotoxicity
     Research* 21 (4): 358–67.
Skopińska-Rózewska, E., M. Malinowski, A. Wasiutyński, E. Sommer, M. Furmanowa,
     M. Mazurkiewicz, and A. K. Siwicki. 2008. The influence of *Rhodiola quadrifida* 50%
     hydro-alcoholic extract and salidroside on tumor-induced angiogenesis in mice. *Polish
     Journal of Veterinary Sciences* 11 (2): 97–104.
Song, Y. Y., G. Qi, J. T. Han, and Y. P. Li. 2006. Effect of salidroside on changes of interleu-
     kin beta of rats with global cerebral ischemia-reperfusion injury. *Chinese Journal of
     Clinical Rehabilitation* 10 (11): 30–2.
Sun, C., Z. Wang, Q. Zheng, and H. Zhang. 2012. Salidroside inhibits migration and invasion
     of human fibrosarcoma HT1080 cells. *Phytomedicine* 19 (3–4): 355–63.

Tan, C. B., M. Gao, W. R. Xu, X. Y. Yang, X. M. Zhu, and G. H. Du. 2009. Protective effects of salidroside on endothelial cell apoptosis induced by cobalt chloride. *Biological and Pharmaceutical Bulletin* 32 (8): 1359–63.

Thieme, H. 1969. On the identity of glucoside rhodioloside and salidroside. *Pharmazie* 24 (2): 118–9.

Thieme, H. and H. Winkler. 1966. On the occurrence of salidroside in the leaves of the red whortleberry. *Pharmazie* 21(3): 182.

Tolonen, A., A. Hohtola, and J. Jalonen. 2003b. Liquid chromatographic analysis of phenyl-propanoids from *Rhodiola rosea* extracts. *Chromatographia* 57 (9–10): 577–9.

Tolonen, A., M. Pakonen, A. Hohtola, and J. Jalonen. 2003a. Phenylpropanoid glycosides from *Rhodiola rosea*. *Chemical and Pharmaceutical Bulletin* 51 (4): 467–70.

Troshchenko, A. T., and G. A. Kutikova. 1967. Rhodioloside from *Rhodiola rosea* and *Rh. quadrifida*. I. *Chemistry of Natural Compounds* 3 (4): 204–7.

van Diermen, D., A. Marston, J. Bravo, M. Reist, P. A. Carrupt, and K. Hostettmann. 2009. Monoamine oxidase inhibition by *Rhodiola rosea* L. roots. *Journal of Ethnopharmacology* 122 (2): 397–401.

Wang, H., Y. Ding, J. Zhou, X. Sun, and S. Wang. 2009. The *in vitro* and *in vivo* antiviral effects of salidroside from *Rhodiola rosea* L. against coxsackievirus B3. *Phytomedicine* 16 (2–3): 146–55.

Wang, L., L. L. Shi, and Y. J. Liu. 2007. Effects of different light treatments on growth and PAL activity of the suspension-cultured cells of *Rhodiola fastigiata*. *Scientia Silvae Sinicae* 43 (6): 49–53.

Wang, Z., H. Hu, F. Chen, L. Zou, M. Yang, A. Wang, J. E. Foulsham, et al. 2012. Metabolic profiling assisted quality assessment of *Rhodiola rosea* extracts by high-performance liquid chromatography. *Planta Medica* 78 (7): 740–6.

Wiedenfeld, H., M. Dumaa, M. Malinowski, M. Furmanowa, and S. Narantuya. 2007. Phytochemical and analytical studies of extracts from *Rhodiola rosea*. *Rhodiola quadrifida*. *pharmazie* 62 (4): 308–11. Erratum in *Pharmazie* 62 (5): 400.

Wu, Y. L., D. M. Piao, X. H. Han, and J. X. Nan. 2008. Protective effects of salidroside against acetaminophen-induced toxicity in mice. *Biological and Pharmaceutical Bulletin* 31 (8): 1523–9.

Wu, Y. L., L. H. Lian, Y. Z. Jiang, and J. X. Nan. 2009. Hepatoprotective effects of salidro-side on fulminant hepatic failure induced by D-galactosamine and lipopolysaccharide in mice. *Journal of Pharmacy and Pharmacology* 61 (10): 1375–82.

Xu, J., Z. Su, and P. Feng. 1998. Suspension culture of compact callus aggregate of *Rhodiola sachalinensis* for improved salidroside production. *Enzyme and Microbial Technology* 23 (1–2): 20–7.

Xu, J. F., and Z. G. Su.1997. Regulation of metabolism for improved salidroside production in cell suspension culture of *Rhodiola sachalinensis* A. Bor: The effect of precursors. *Natural Product Research and Development* 10: 8–14.

Yousef, G. G., M. H. Grace, D. M. Cheng, I. V. Belolipov, I. Raskin, and M. A. Lila. 2006. Comparative phytochemical characterization of three *Rhodiola* species. *Phytochemistry* 67 (21): 2380–91.

Yu, S., M. Liu, X. Gu, and F. Ding. 2008. Neuroprotective effects of salidroside in the PC12 cell model exposed to hypoglycemia and serum limitation. *Cellular and Molecular Neurobiology* 28 (8): 1067–78.

Yu, S., Y. Shen, J. Liu, and F. Ding. 2010. Involvement of ERK1/2 pathway in neuroprotec-tion by salidroside against hydrogen peroxide-induced apoptotic cell death. *Journal of Molecular Neuroscience* 40 (3): 321–31.

Zapesochnaya, G. G., and V. A. Kurkin. 1982. Glycosides of cinnamyl alcohol from the rhizomes of *Rhodiola rosea*. *Chemistry of Natural Compounds* 18 (6): 685–8.

Zapesochnaya, G. G., and V. A. Kurkin. 1983. The flavonoids of the rhizomes of *Rhodiola rosea*. II. A flavonolignan and glycosides of herbacetin. *Chemistry of Natural Compounds* 19 (1): 21–9.

Zeng, W. Z., X. L. Wu, M. D. Jiang, P. L. Wang, and G. Z. Chu. 2005. Effect of salidroside on gene expression of CBP and smad in CCl 4-induced liver fibrosis in rats. *World Chinese Journal of Digestology* 13 (3): 341–5.

Zhang, J., A. Liu, R. Hou, J. Zhang, X. Jia, W. Jiang, and J. Chen. 2009. Salidroside protects cardiomyocyte against hypoxia-induced death: A HIF-1α-activated and VEGF-mediated pathway. *European Journal of Pharmacology* 607 (1–3): 6–14.

Zhang, L., H. Yu, X. Zhao, X. Lin, C. Tan, G. Cao, and Z. Wang. 2010. Neuroprotective effects of salidroside against beta-amyloid-induced oxidative stress in SH-SY5Y human neuroblastoma cells. *Neurochemistry International* 57 (5): 547–55.

Zhang, L., H. Yu, Y. Sun, X. Lin, B. Chen, C. Tan, G. Cao, et al. 2007. Protective effects of salidroside on hydrogen peroxide-induced apoptosis in SH-SY5Y human neuroblastoma cells. *European Journal of Pharmacology* 564 (1–3): 18–25.

Zhang, X., B. Zhu, S. Jin, S. Yan, and Z. Chen. 2006b. Effects of salidroside on bone marrow matrix metalloproteinases of bone marrow depressed anemic mice. *Sheng Wu Yi Xue Gong Cheng Xue Za Zhi = Journal of Biomedical Engineering* 23 (6): 1314–9.

Zhang, X. S., B. D. Zhu, X. Q. Huang, and Y. Chen. 2005. Effect of salidroside on bone marrow cell cycle and expression of apoptosis-related proteins in bone marrow cells of bone marrow depressed anemia mice. *Journal of Sichuan University (Medical Science Edition)* 36 (6): 820, 823–46.

Zhang, Y. H., S. D. Chen, J. L. Li, H. Q. Yang, R. Zheng, H. Y. Zhou, G. Wang, et al. 2006a. Salidroside promotes the expression of GDNF in the MPTP model of Parkinson's disease. *Chinese Journal of Neurology* 39 (8): 540–3.

Zheng, K. Y. Z., Z. X. Zhang, A. J. Y. Guo, C. W. C. Bi, K. Y. Zhu, S. L. Xu, J. Y. X. Zhan, et al. 2012. Salidroside stimulates the accumulation of HIF-1α protein resulted in the induction of EPO expression: a signaling via blocking the degradation pathway in kidney and liver cells. *European Journal of Pharmacology* 679 (1–3): 34–9.

Zhong, H., H. Xin, L. X. Wu, and Y. N. Zhu. 2010. Salidroside attenuates apoptosis in ischemic cardiomyocytes: A mechanism through a mitochondria-dependent pathway. *Journal of Pharmacological Sciences* 114 (4): 399–408.

Zhu, Y., Y. P. Shi, D. Wu, Y. J. Ji, X. Wang, H. L. Chen, S. S. Wu, et al. 2011. Salidroside protects against hydrogen peroxide-induced injury in cardiac H9c2 cells via PI3K-akt dependent pathway. *DNA and Cell Biology* 30 (10): 809–19.

Zou, Y. Q., Z. Y. Cai, Y. F. Mao, J. B. Li, and X. M. Deng. 2009. Effects of salidroside-pretreatment on neuroethology of rats after global cerebral ischemia-reperfusion. *Journal of Chinese Integrative Medicine* 7 (2): 130–4.

4 Cultivation of *Rhodiola rosea* in Europe

Bertalan Galambosi

CONTENTS

4.1   Introduction 88
4.2   Distribution of *R. rosea* 89
4.3   Traditional Knowledge of *R. rosea*...89
    4.3.1   Northern Scandinavia 89
    4.3.2   Northern Urals, Russia 90
    4.3.3   Other Parts of Europe..90
    4.3.4   Information from Russia after 1990 90
4.4   *R. rosea* Supply...91
4.5   Research on *R. rosea* 92
    4.5.1   Russia...92
        4.5.1.1   Plant Size and Weight 92
    4.5.2   Scandinavia..94
        4.5.2.1   Sweden..94
        4.5.2.2   Finland..94
        4.5.2.3   Norway..95
    4.5.3   Alpine Countries..96
        4.5.3.1   Austria...96
        4.5.3.2   Italy 96
        4.5.3.3   Germany 97
        4.5.3.4   Switzerland 97
    4.5.4   Carpathian Countries...98
        4.5.4.1   Poland 98
        4.5.4.2   Romania..98
        4.5.4.3   Bulgaria...98
        4.5.4.4   Moldova 99
        4.5.4.5   Czech Republic 99
        4.5.4.6   Slovak Republic 99
    4.5.5   Other Countries 99
        4.5.5.1   Hungary 99
        4.5.5.2   Estonia..100
        4.5.5.3   Great Britain 100
        4.5.5.4   Mongolia 100
    4.5.6   Present Status of *R. rosea* in Europe 100

4.6   Life Cycle and Cultivation of *R. rosea* 101
    4.6.1   Life Cycle.. 101
        4.6.1.1   Seed Phase 101
        4.6.1.2   First Year of Growth.. 102
        4.6.1.3   Second Year of Growth.. 102
        4.6.1.4   Plant Growth from 3 to 40 Years......................... 102
        4.6.1.5   Summer Growth Rhythm 102
    4.6.2   General Environmental Requirements 103
    4.6.3   Field Choice... 103
    4.6.4   Fertilization 103
    4.6.5   Variety 104
    4.6.6   Propagation Methods.. 105
        4.6.6.1   Root Division 105
        4.6.6.2   Seed Propagation 105
        4.6.6.3   Seedling Age... 105
        4.6.6.4   Transplanting Seedlings....................................... 106
    4.6.7   Weed Control.. 106
    4.6.8   Pests and Diseases 107
    4.6.9   Root Harvest... 107
        4.6.9.1   Harvesting Process 107
        4.6.9.2   Timing the Harvest 108
        4.6.9.3   Male–Female Plants 109
        4.6.9.4   Predrying Treatments 109
        4.6.9.5   Drying Temperature.. 112
        4.6.9.6   Postdrying Process.. 112
    4.6.10 Yield Potential 113
    4.6.11 Seed Harvest.. 114
4.7   Summary and Future Trends... 114
References.. 115

4.1 INTRODUCTION

This chapter reviews our knowledge of the use of *Rhodiola rosea* in traditional folk medicine and more recently as a pharmaceutical product with adaptogenic properties and mechanisms by which its increasing popularity contributes to precarity of supply. To ensure an adequate supply in the future, cultivation of this medicinal herb is of particular interest, and presents extensive research findings on effective agronomic techniques. Research on this species has been published around the world. In nearly all European countries where *Rhodiola* is indigenous, a key feature of research on the species is that interest is recent, dating back only two decades. This chapter is structured according to the geographical distribution of *R. rosea* across Europe. After a brief historical overview, information from those regions where *Rhodiola* grows naturally, Russia, Scandinavia, European Alpine, and Carpathian Mountains, and the main techniques for cultivating *Rhodiola* are reviewed.

4.2   DISTRIBUTION OF *R. ROSEA*

R. rosea is an extremely variable circumpolar species found in cool temperate and subarctic areas of the northern hemisphere, including North America, Greenland, Iceland, and Eurasia. In the Asian continent, its distribution pattern includes the Arctic, the Altai Mountains, eastern Siberia, Tien-Shan, and the Far-East, and extends south to the Himalayas.

The European distribution includes northern and arctic areas in Europe, the northern parts of the Ural Mountains, most of the high-altitude and mountainous regions of Central Europe, south of the Pyrenees, Central Italy, and Bulgaria (Hegi 1963). The distribution in Europe extends from Iceland and the British Isles across Scandinavia as far south as the Pyrenees, the Alps, the Carpathian Mountains, and other mountainous Balkan regions. It grows in varying density in the following countries: Austria, Bulgaria, Czech Republic, Finland, France, Greenland, Iceland, Ireland, Italy, Norway, Poland, Romania, Russian Federation (European parts), Slovakia, Spain, Switzerland, Sweden, United Kingdom, and in some of the Balkan states (Brown et al. 2002).

Rhodiola, a relatively advanced group of the *Crassulaceae* family, has been interpreted (mostly in Eurasia) as a separate genus, or (particularly in North America) as a section of the genus *Sedum* L. Hegi (1963) identified more than 50 species of *Rhodiola* and reestablished it as a separate genus. In the former Soviet Union, 21 species of the *Rhodiola* have been specified (Komarov 1939; Pautova and Tkachenko 2004). In North America, the *Rhodiola* subgenus includes at least five North American taxa of the *Sarracenia rosea* complex (Small and Catling 1999).

4.3   TRADITIONAL KNOWLEDGE OF *R. ROSEA*

4.3.1   Northern Scandinavia

In the northern parts of Sweden and Norway, a law forbade access to areas where roseroot grew naturally, due to its magic power. When the Vikings sailed to Iceland, they found vast populations of roseroot on the island and used it to treat wounds. When Catholic monks settled in Iceland, some knowledge about the plant's use was written.

R. rosea grows naturally only in a few regions of the Finnish Lapland, for example, around municipalities and villages including Enontekiö, Kilpisjärvi, and Utsjoki, the municipality of Kittilä, in the Olostunturi and Pyhätunturi fields. Plant presence in proximity to dwellings suggests that it may also have a decorative use (Väre et al. 1998). Owing to its small area of distribution in Finnish Lapland, there are few written observations on its traditional use.

The last report on the use of roseroot against scurvy in Norway was made in 1766 by Gunnerus, who referred to the root as *Radix Scorbuticis salutaris* (Alm 1996, 2004). Roseroot is not a plant comparatively rich in vitamin C. The leaves have been found to contain 33 mg vitamin C per gram, and the rhizomes contain only 12 mg/g, but its importance lies in the fact that it is the first source of vitamin C available in early spring. The first written recommendation of the roseroot cultivation is dated back to 1762 from Norway. The plants were planted on the roofs of the wooden houses to prevent the risk of fire (Hedman 2000).

The plant has been used extensively in folk medicine, not only to treat burns but also internally, against scurvy and lung inflammation, and to facilitate urination. The priest Herman Ruge wrote from Valdres in Norway in 1762, following his observations of the plant's use among the Native people of Greenland: "I have myself eaten it, both fried and roasted as well as boiled. And I have either in taste or in effect found it disagreeable" (Hedman 2000). "The leaves were ground and mixed with the ingredients to make bread. The children eat the leaves raw," reveals a description reported by Hoeg (1976).

In 1985, *R. rosea* was recognized in Sweden as an herbal medicinal product, and it has been described as an antifatigue agent in the *Textbook of Phytomedicine for Pharmacists* (Sandberg and Bohlin 1993). In the Swedish textbook of pharmacology, *R. rosea* is mentioned as a plant with stimulant properties. The pharmaceutical book from 1997/98 mentions *R. rosea* as one of the most commonly used psychostimulants among the group of officially registered herbal products (Sandberg 1998).

Rhodiola is an endemic plant and the Icelandic flora is rich with roseroot. Botanical descriptions can be found in the relevant literatures (Arnajörg 1992; Hördur 1987; Hjaltalin 1830).

4.3.2 NORTHERN URALS, RUSSIA

The locations favored by these plants were kept secret, and special rituals, including absolute silence, accompanied by the collection of the plants. The Komi people also believed that the plants could not be seen by just any picker. According to their belief, there had to be a "similarity of blood" between the plant and the person for whom the magic plant was intended, otherwise the plant disappeared (Iljina 1997; Popov 1874).

Roseroot was a rare and valuable plant for the Komi people since it grows in very remote locations. Those who hunted in the Ural Mountains, along the upper Pechora River, and the Vishera and the Vym rivers, were the main pickers of this plant. The Komi did not have precise recipes, but instead adhered to the rule that "Every person needs his own handful." According to this belief, each person is different—big or small in stature, heavy or slim—so the remedy would be beneficial only when a person drinks his/her own limit, and this limit is defined by his/her own handful. Therefore, as a rule, remedies were prepared individually for each patient. A handful of the dried roots was steeped in half a liter of vodka or boiled water and kept in a warm place. Broth could also be evaporated on the stove in a vessel covered with dough. Collecting golden root in such remote places was extremely difficult in the past. So, the Komi people tried to cultivate plants near their dwellings, but these attempts were generally unsuccessful (Galambosi et al. 2010a).

4.3.3 OTHER PARTS OF EUROPE

German and French researchers described the benefits of roseroot for pain, headache, scurvy, hemorrhoids, as a stimulant, and as an anti-inflammatory (Brown et al. 2002; Virey 1819).

4.3.4 INFORMATION FROM RUSSIA AFTER 1990

In Russia, *R. rosea* has been an important remedy not only in traditional medicine (under the name of golden root) but also in modern medicine. Following extensive medical and pharmacological tests in the former Soviet Union, its alcohol-based extract (*Rhodiola Extract Liquid*) was officially registered in 1975 as a medicine and herbal tonic. This registration required was supported by a number of previous clinical and phytochemical investigations, but since they were published in Russian, the results remain largely unknown in the West. Numerous pharmacological effects of the roseroot extract (antifatigue, antistress, antioxidant, immune-system enhancing) have been described in this literature (Saratikov and Krasnov 1997). A new phase in the phytochemical investigation of *R. rosea* was carried out by a research group led by Professor A.V. Kurkin (Kurkin et al. 1985, 1986, 1989; Sokolov et al. 1985; Zapesochnaya et al. 1995).

After the political changes following the dissolution of the Soviet Union, information on the benefits of adaptogenic medicinal plants, including *R. rosea*, have been dispersed to Europe and the rest of the world. The first large phytomedicinal overview in English was published in 2002 by two American doctors and a Russian scientist (Brown et al. 2002). From 1930 to 1960, no scientific papers were apparently published on roseroot at all. Prior to 1985, less than four papers per year were published, but the publication of 12–18 papers per year between 2004 and 2009 demonstrates growing interest.

There have been two channels for dispersing this scientific information from Russia to the West. In eastern Europe, Russian-speaking researchers have had access to this information, but this scattered knowledge has had less impact. The other channel, through Swedish researchers, has reached a broader audience.

Swedish researchers were the first to conduct intensive research with adaptogens, the term for the medicinal effect of plants such as *R. rosea* that has only recently become more well known (Sandberg and Bohlin 1993). The scientists who spurred interest in this field were Dr. George Vikman and Dr. Alexander Panossian from the Swedish Herbal Institute. Beginning in 1976, their research team conducted numerous biochemical experiments on the mechanisms of adaptogenic action (Panossian 2003; Panossian et al. 1999, 2010; Panossian and Wagner 2005). On the basis of these experiments, the first commercial roseroot extract was launched in Sweden during the late 1980s. Numerous subsequent clinical trials have been carried out with their roseroot preparations, in collaboration with Russian scientists. The results of double-blind placebo-controlled clinical trials supported a more precise definition of adaptogens as a separate group of antifatigue and antistress metabolic regulators that increase mental and physical performance against a background of fatigue or stress. The results obtained showed that roseroot extract has potential for use in geriatrics to maintain the health status of the elderly and perhaps even prolong life span (Darbinyan et al. 2000; Panossian et al. 2000; Shevtsov et al. 2003; Spasov et al. 2000).

Inspired by the Swedish roseroot preparation, cultivation experiments with roseroot were begun in Finland during the early 1990s, and soon after in Norway (Dragland 2001). Interest in the utilization of native *Rhodiola* species then surfaced in several countries, including Italy (Aiello and Fusani 2004), Switzerland (Malnoe et al. 2009), Germany (Schittko 2004), Bulgaria, Romania, and other countries.

4.4 *R. ROSEA* SUPPLY

Up to now, most of the raw materials for the production of roseroot is taken from the natural populations, causing, in many countries, that the natural populations of roseroot are endangered (Kylin 2010). Lei et al. (2006) report an accelerated and uncontrolled use of *Rhodiola* species during the 1980s in southwestern China to the extent that they are now listed as endangered in China (Lei et al. 2006). One example of an endangered species in China is *R. sachalinensis*; heavy disturbances by human activities are the main reason for its status as in danger of extinction (Yan et al. 2003). Other factors reported to threaten *Rhodiola* is deforestation of natural habitat and excessive grazing (Xia et al. 2007). Galambosi (2006) confide the same pattern is shown in several countries, among them Russia, where roseroot is nowadays on the Russian Red List and the collection is restricted (Borodin 1995). Roseroot is the only *Rhodiola* species found in Scandinavia, including Greenland, Faroe Islands, and Svalbard, and to date it is not on the endangered species lists in this region (Kotiranta et al. 1998).

The endangered status of *R. rosea* has led to research along three themes: (a) domestication and cultivation of *R. rosea*, (b) intensive studies of other *Rhodiola* species, mainly by Chinese researchers, and (c) biochemical and biotransformation studies of the main compounds of roseroot.

4.5 RESEARCH ON *R. ROSEA*

Over the last 3–4 decades, numerous agronomical and biochemical studies in different countries have reported on the domestication of roseroot and described effective cultivation and postharvest processing methods (Galambosi 2006). We present the main findings of this research chronologically and by country.

The studies were published in the Soviet Union in the 1970s. In Sweden, the first studies on cultivation began in the 1980s, in Uppsala. Domestication of the plant in Finland dates from 1993, and in Norway from 1996. In Poland, the first phytochemical studies were published in 1990, and they were followed by several studies on domestication and agrotechnical methods in 2000. Research on roseroot cultivation in other European countries, in Italy and Switzerland, began in 2000. Recently, since 2010, reports on domestication experiments in Carpathian countries including Bulgaria have been published.

4.5.1 Russia

In the former Soviet Union, a large number of experiments were published covering a range of biological questions (Revina et al. 1976), phenology, seed production, and plant age in cultivation (Kim 1976; Polozhii and Reviakina 1976), ecological morphology (Nukhimovski 1976; Nukhimovski et al. 1987), biomorphogenesis (Nukhimovski et al. 1987), and the accumulation of bioactive compounds in relation to plant age (Revina et al. 1976). The potential natural resources of the country have been determinated (Tsikov 1980).

4.5.1.1 Plant Size and Weight

Polozhii and Reviakina (1976) carried out extensive botanical and biological studies of natural populations of the root in the Altai Mountains. Plant size ranged widely, depending on soil conditions. For example, at an altitude of 2700 m in stony tundra, average total fresh plant weight was 290 g (shoots: 20 g, roots: 220 g, rhizome: 50 g). At 1950 m, in more fertile valley soil, average fresh weight was 1265 g, with more developed rhizomes (shoot: 300 g, roots: 125 g, rhizome: 840 g).

Similarly, plant density varied greatly among plant populations in this region. The number of flowering shoots and the productivity of collectable seed yield were high in quite open moraine soils (430–503 shoot/100 m², 400–520 g seed/100 m²), while among higher species and more dense vegetation, the number of flowering shoots and seed yield were much lower (55–290 shoot/100 m², 31–270 g/100 m²).

Nukhimovski et al. (1987) reported that natural 50- to 60-year-old plants studied in the Altai Mountains had 20–40 new shoots. The largest individual *R. rosea* plant they found growing naturally in the region had 249 shoots and was estimated to be about 200 years old. Total fresh weight of the roots and rhizomes was 7.8 kg, of which live parts represented 3.5 kg. Cultivated plants accumulate mass in their underground parts more rapidly. Galambosi et al. (2009) reported that the total fresh underground weight of a 12-year-old cultivated plant was 7.98 kg, of which roots represented 1.355 kg and the rhizome represented 6.63 kg.

In introduction experiments, Kim (1976) compared the seedlings of propagated roseroot to those of the naturally grown seedlings (*n* = 50). In nature, the total fresh weight of the 1–2–3-year-old individual plants was 2.1–1.7–3.9 g/plant and rhizome fresh weight was 0.12–0.42–1.66 g, respectively. Total fresh weight under cultivated conditions was 2.8–30.3–100.7 g and rhizome weight was 1.19–11.01–52.41 g.

Other studies have reported important experimental results on postharvest processing technology of raw materials. Dubichev et al. (1991) demonstrated that the salidroside content of fresh roots decreases by 30% within 2 hours, standing at 40°C, due to the enzyme activity (autofermentation).

Agronomical results were published on field fertilization of roseroot (Elsakov and Gorelova 1999) and seed biology (Kovaleva et al. 1997; Tikhonova et al. 1997). The biomorphological potential of the species in the northern and subarctic Urals has been evaluated, its seed biology has been studied, and recommendations for its cultivation have been given (Frolov and Poletaeva 1998). Regulation of the biosynthesis of secondary metabolites in golden roots by means of water deficiency in order to obtain plant raw material with high contents of salidrosid and rosavins has been studied (Bacharov et al. 2009). According to the results of this experiment, the artificial water deficiency have increased the contents of the active metabolites in the roots. From a practical point of view, if we keep the whole dig plants on the soil surface for 1–2 weeks, we induce a water deficiency stress in the plants and by this simple means, it could be possible to increase the metabolite content in the root yield. Further studies are necessary to check this hypothesis.

In an experiment that could hold promise for rapidly increasing future yields, the biomass production of *R. rosea* in its natural habitat was compared with its growth under "photo culture," with controlled temperature (Kovaleva et al. 2003). Under laboratory conditions, plants grew without a dormancy period. The rhizome weight of plants grown for 135 days and 245 days under intensive photoculture was 2.3 and 10.5 g, respectively. The rhizome weight of a 3-year-old plant in natural habitat was 1.6 g, and in field stands of 2-year-old plants was 11 g/plant. These experimental results suggest that it may be possible to accelerate the plant's life cycle. Due to the international collaboration, phytochemical features of Russian and European origin *Rhodiola* accessions have been presented in some congresses (Makarov et al. 2003). In a newest publication, the polymorphism and genetic structure of several *Rhodiola* species were evaluated (Kozyrenko et al. 2011).

4.5.2 SCANDINAVIA

4.5.2.1 Sweden

In accordance with the phytochemical and pharmacological studies, the product development needed domestic raw material in Sweden, and its first field cultivations have been started in connection with the Örtmedicinska Institute from 1991. The cultivation activity was started by leading Professor Georg Wikman. Several farmers started the introduction of this natural plant into cultivation, and the farms are situated mainly in the southern parts of Sweden, near to Göteborg, first in Wallberga, later in Olofstorp (Anon 1992) and Hulukvars, near Jönköping (Frisk 2008).

4.5.2.2 Finland

The initiative for studying on *R. rosea* has been started by the adviser service of Lapland in 1996 (Mäkitalo and Jankkila 1994). Additionally, the University of Oulu has started a developing "Poherika" project, including *R. rosea* as well. At the same time, the analytical skill has been developed in the university (Tolonen 2003; Tolonen et al. 2003, 2004). In the "Sunare" research program, studies on long-term phytochemicals have been started (György 2006).

The domestication experiments have been started as a part of some larger research projects organized by Agrifood Research Finland at Mikkeli, in the southeastern part of the country (*Cultivation techniques of herbs suitable for the Finnish climate [1989–1995] and Introduction and acclimatization of new medicinal plants [1997–2001]*). Observation experiments explored basic questions of propagation biology, growth, biomass accumulation, and characteristics of root yield (Dragland and Galambosi 1996; Tuominen et al. 1999; Galambosi 1999, 2004, 2005; Galambosi et al. 2003). The chemical composition of the essential oil of the rhizome was studied as well (Hethelyi et al. 2005).

The research has been continued focusing on several new questions: comparison of wild and cultivated *R. rosea* strains (Galambosi et al. 2007) and comparison of the phytochemical features of male–female plants (Galambosi et al. 2009).

One important feature of the *R. rosea* research was the intensive international interaction between researchers, whose countries roseroot is an endemic species. The researchers exchanged genetic resources, ideas in elaboration with the growing techniques of this new species. Visits and research cooperation resulted several common papers between Russian, Norvegian, Italian, Switzerland, German, Hungarian, and Canadian researchers. Comparison of roseroot strains of different origin was carried out. In this cooperation, firstly rhizomes of 6-year-old Norwegian—origin roseroots, cultivated in south Finland, Mikkeli were compared to commercial Russian samples. The origins of the Russian samples were unknown, probably from the Altai Mountains, Siberia (Makarov et al. 2003). There were significant differences in the contents of the main compounds. The average salidroside contents of the three Scandinavian and the three Russian samples were 0.19% and 0.87%, while the rosavin contents were 0.51% and 2.67%, respectively.

In a second research study, dry root samples from Russia, Italy, Norway, and Finland were analyzed ($n = 2$–4 per origin) (Kosman et al. 2004). The salidroside contents ranged between 0.31% and 2.19%: from Altai, Russia (0.86%–1.63%), North-Ural, Russia (0.56%–1.82%), North-Italy (0.45%–2.19%), Central Norway (0.71%–1.62%), and North-Finland (0.31%–0.86%). The highest mean rosavin contents were measured in the samples from Altai, Russia (mean: 5.38%), followed by North Italy (*x*: 3.94%), Norway (*x*: 3.21%), North Ural, Russia (*x*: 2.39%), and North Finland (*x*: 1.37%). The results have strengthened the general belief of the high quality of the roseroot originated in the Altai Mountains, Russia, and emphasize the importance of the analytics of all local strains.

Roseroot was one of the target plants in a cooperative project "Barents herbs" of Russian, Norwegian, and Finnish researchers. The cooperation has resulted in several papers (Fjelldal et al. 2010; Thomsen et al. 2011; Martinussen et al. 2008; Galambosi et al. 2008).

International cooperation has resulted in the creation of a *Rhodiola* gene bank at MTT-Mikkeli, Finland, containing approximately 25 different accessions and a long-term conservation methodology (Asdal et al. 2006; Galambosi and Valo 2006). The collection will facilitate phytochemical comparison of different species (György ét al. 2011). Research results to date have been distributed through numerous lectures and demonstration field days, and in 2004, a seminar was organized on the adaptogenic properties of this plant (Galambosi et al. 2003).

4.5.2.3 Norway

Research work has been started by the analyses of the composition of essential oil obtained from a Norwegian roseroot accession (Rohloff 2002). With large cooperation of people from all parts of the country, a national germplasm collection was created, including 97 roseroot clones from all regions of the country.

The morphological and phytochemical features and the genetic diversity of different clones have been studied intensively. To initiate a selective breeding program, amplified fragment length polymorphism (AFLP) analysis was used to estimate genetic diversity within the Norwegian *R. rosea* germplasm collection, including 97 clones (Elameen et al. 2008). The genetic analysis showed that there was no close genetic similarity among clones related to their original growing county, and the results indicate high gene flow among *R. rosea* clones that might be a result of seed dispersal rather than cross-pollination.

Six clones of *R. rosea*, obtained from plants originating from widely different areas in Norway, were investigated for their *in vitro* inhibitory potential on CYP3A4-mediated metabolism and P-gp efflux transport activity (Hellum et al. 2010). Presumed active constituents in the ethanol extracts of the different clones were quantified. All clones showed potent inhibition of CYP3A4 and P-gp activities, with IC50 values ranging from 1.7 to 3.1 µg/mL and from 16.7 to 51.7 µg/mL, respectively, being below that reported for other herbs and some known classic drug inhibitors, such as St. John's wort and fluoxetine. The concentration of presumed biologically active constituents in the different clones varied considerably, but this variation was not related to the clones' inhibitory potential on CYP3A4 or P-gp activities. Other constituents might thus be responsible for the observed inhibitory properties. The place of origin seemed to be of minor importance for CYP3A4 or P-gp inhibition.

Three high rosavin-containing, seed-propagated *R. rosea* clones were cultivated for four growing seasons at different locations in Norway, between 60° and 64° Nordic latitudes (Thomsen 2012). The average total rosavine contents were slightly higher in the northern parts of the country (mean: 7.09%), than in the southern parts (4.17%–5.04%).

Plants from 10 geographic regions in the very northern part of Norway in Finnmark County were collected and evaluated (Fjelldal et al. 2010). The results show large geographical variations in the content of pharmacological metabolites. The total rosavin content ranged from 1.37% to 3.40% with a mean value of 2.67%, while the salidroside concentration ranged from 1.11% to 4.98% with mean value of 2.27%. A most recent result of a phytotron experiment revealed that the measured adaptogen components of *R. rosea* (salidroside, total rosavins) seemed to be in higher concentration in accession originated from the more Nordic latitudes, compared to the southern ones (Martinussen et al. 2011).

The product development in Norway has been carried out in the company of Rosenrot Norge AS, and the first roseroot preparation has been released in 2004 (www.rosenrot.no). The products are based on the first collection of the wild growing populations, but the company supported the development of the cultivation research as well. Field cultivation techniques, and postharvest processing experimental results have been forwarded to the growers (Dragland and Galambosi 1996; Thomsen et al. 2011). These experiments were led by Dr. S. Dragland in Bioforsk Ag, KISE, in Central Norway. KISE is the centrum of the roseroot growers' network founded during 2005.

4.5.3 ALPINE COUNTRIES

4.5.3.1 Austria

No special agronomical research exists in Austria. Preliminary geobotanical, biological observations and phytochemical considerations of natural and cultivated *R. rosea* were firstly presented by Kump (1991). The original plant material collected from the Obertauern Mountain by Dr. Kump later have been included into the gene collection of Finland, and some research studies have been carried out on it (Galambosi et al. 2009).

4.5.3.2 Italy

In Italy, intensive pharmacological research was carried out on the effects of the nervous system, on oxidative stress, antioxidative activity, and stress protective effects of *R. rosea* (Mattioli et al. 2008; Mattioli and Perfumi 2007). Battistelli et al. (2005) have investigated the effect of the *R. rosea* root aqueous extract on *in vitro* human erythrocytes exposed to hypochlorous acid (HOCl)–oxidative stress. The results obtained are consistent with a significant protection of the extract in the presence of the oxidative agent.

Perfumi and Mattioli (2007) have reported the results of a study in which they reinvestigated the effects produced by a single oral administration of an *R. rosea* hydroalcohol extract (containing 3% rosavin and 1% salidroside) on the central nervous system in mice. The extract was tested on antidepressants, adaptogenic, anxiolytic, nociceptive, and locomotor activities at doses of 10, 15, and 20 mg/kg, using predictive behavioral tests and animal models. The results show that this *R. rosea* extract significantly, but not dose-dependently, induced antidepressant-like, adaptogenic, anxiolytic-like, and stimulating effects in mice. This study thus provides evidence of the efficacy of *R. rosea* extracts after a single administration and confirms many preclinical and clinical studies indicating the adaptogenic and stimulating effects of such *R. rosea* extracts. Moreover, in this study, the antidepressant-like and anxiolytic-like activities of *R. rosea* were shown in mice for the first time.

In an interesting study of the same research group (Mattioli and Perfumi 2011), the effects of a *R. rosea* L. extract on the prevention of the development of nicotine dependence and for the reduction of abstinence suffering following nicotine cessation in mice were studied. Dependence was induced in mice by subcutaneous injections of nicotine (2 mg/kg, 4 times/day) for 8 days. The spontaneous abstinence syndrome was evaluated 20 hours after the last nicotine administration, by analysis of withdrawal signs, as affective (anxiety-like behavior) and physical (somatic signs and locomotor activity). *R. rosea* L. extract was administered orally during nicotine treatment (10, 15, and 20 mg/kg) or during nicotine withdrawal (20 mg/kg). Results show that both affective and somatic signs (head shaking, paw tremors, body tremors, ptosis, jumping, piloerection, and chewing) induced by nicotine withdrawal are abolished by administration of *R. rosea* L. extract in a dose-dependent fashion, during both nicotine exposure and nicotine cessation. In conclusion, the authors suggest for additional studies to define the use of *R. rosea* L. as a therapeutic approach in the treatment of smoking cessation.

In Northern Italy, the natural occurrence of the roseroot field experiments has been started from 2004 mainly in Trento province (Aiello et al. 2011) with local North Italy's Alps roseroot strains. Results have been reported on the genetic characterization (D'Ambrosio et al. 2008; Zini et al. 2007), seed germination (Aiello and Fusani 2004; Aiello et al. 2010), and the accumulation of biomass and active compounds in relation to different altitudes (Egger et al. 2007; Scartezzini et al. 2010). Commercial herb companies have started to cultivate roseroot in Trento province, and the area of this semi-large-scale experimental cultivation is estimated about 0.3 ha (Fusani, 2011, pers. comm.)

4.5.3.3 Germany

The interest of the medical industry has been arisen in Germany as well and field expedition was organized for the collection of original Siberian strains in Altai area. Acclimatization observation experiments by large gene collection have been started in Thuringia (Schittko 2004). Five years later, the experiences of the pilot cultivation have been presented in a seminar (Plescher et al. 2010). *R. rosea* has been successfully introduced into cultivation in south Germany. The essential precondition of the roseroot cultivation seems to be 500–600 mm/year precipitation. For the commercial cultivation, the present 3–3.5-year long life cycle should be shortened, aiming 2 t/ha dry root yield.

At the same time, a new *Rhodiola*-based preparation has been developed by a German medical company. The preparation has been tested clinically, and the product (Vitalgo) have launched to the market (Muehlhoff 2010).

4.5.3.4 Switzerland

The phytochemical and agronomical experiments on *R. rosea* have been started with local Alpine accessions in the Agroscope Changins-Wädenswill (ACW), Centre de Recherche Contheny (Malnoe et al. 2009) and as a result of the selection work, the first *R. rosea* variety "Mattmark" has been launched (Vouillamoz et al. 2011). At the same time, a chemical analytical service has been initiated (Slacanin 2011), and the first commercial cultivation resulted in commercial products (http://ilis.ch).

4.5.4 CARPATHIAN COUNTRIES

4.5.4.1 Poland

There is a long tradition for the phytochemical and agronomical research on different medicinal plants in Poland. The studies on the endemic roseroot strains started at the end of the 80th. The subjects of the studies are wide, including the phytochemical and pharmacological properties of the drug (Furmanowa et al. 1999; Hartwich 2010; Kedzia et al. 2006; Kucinskaite et al. 2007), the cell cultures of *R. rosea* and *R. kirillovii* (Furmanowa et al. 1995, 1999; Krajewska-Patana et al. 2007), technological methods for extraction (Zych et al. 2005), and the application of Rhodiola preparations for sportsmen (Skarpanska-Stejnborn et al. 2011).

The introduction research is concentrated at the University of Warsaw and several basic agronomical questions have been studied: seed biology and plant age (Przybyl et al. 2005), yield and accumulation of secondary metabolites (Altantsetseg et al. 2007; Kozlowski and Szczygliewska 2001; Weglarz et al. 2008), and the quality of the raw materials (Buchwald et al. 2006). The optimum life cycle of roseroot seems to be 5 years. During the sixth growing year, the symptoms of plant aging were already noticed (Przybyl et al. 2008). There is some field cultivation in Poland, but no correct data are available, the estimated area is less than half hectare.

4.5.4.2 Romania

In Romania, the research interest focused on the determination of the phytochemical characteristics of endemic *R. rosea* (Calugaru et al. 2007; Toma 2009). *In vitro* culture of *R. rosea* has been studied for propagating and replanting the threatened domestic strains (Ghiorghita et al. 2011; Krasimira et al. 2011). The essential oil content of the Romanian origin roseroot has been determined as well and some phytocoenological and biochemical aspects of the endemic population have been studied (Costica et al. 2007). There is no information on the cultivated area.

4.5.4.3 Bulgaria

Close to the south Carpathian Mountains, *R. rosea* is endemic species in the Bulgarian Rhodope Mountains, which is one of the most southern existence of roseroot in Europe. It is an endangered medicinal plant in Bulgaria. The physiological effects of *R. rosea* have been studied already during the 1980s (Petkov et al. 1986).

From the year 2000, the research focused on its phytochemical features and on its introduction into culture in the Institute of Botany, Bulgarian Academy of Sciences, Sofia. Several results have been published on the composition of the essential oil (Todorova et al. 2006), on the phenol and flavonoid contents of the roots (Staneva et al. 2009), on the dynamics and variability of salidroside content (Bozhilova 2011), and on the antioxidant activity of the root (Tasheva et al. 2011). Studies have focused on the callus culture of the slower growing species (Krashimira et al. 2011; Tasheva and Kosturkova 2011a,b) and attempts made in the evaluation of its cultivation possibilities as a new special crop (Platikova and Evstatieva 2008). Presently, semi-large-scale experimental cultivation exists in Bulgaria in size less than 1 ha, and the growers have difficulties with lower international prices (Liuba Evstatieva, 2011, pers. comm.).

4.5.4.4 Moldova

Natural populations of *R. rosea* exist in the eastern part of the Carpathian Mountains. Efforts have been focused on the research and utilization of roseroot in Moldavia as well, in collaboration with research institutes of Poland (Calugaru et al. 2007). A research project has been created for the selection and cultivation of *R. rosea* based on molecular, phytochemical, and physiological methods in cooperation with Polish researchers. The goal of the project was to provide selection and clone micropropagation of roseroot and study its cultivation possibilities under the conditions of Moldova (Dascaline et al. 2008).

4.5.4.5 Czech Republic

No significant research exists on *R. rosea* in Czech Republic. *In vitro* culture studies have been carried out on adaptogen species *Schizandra chinensis* and *R. rosea by* Martin et al. (2010). Several *Rhodiola* strains from European and Tatra Mountain origin are under observations in the Gene Bank of Olomuch (Karel Dusek, 2009, pers. comm.).

4.5.4.6 Slovak Republic

Among other adaptogenic medicinal plants (*Panax, Schisandra*, and *Rhaponticum*) preliminary agrobiological studies for the development of cultivation methods of *R. rosea* have been carried out in Slovakia, Klcov, Tatra Mountains (Jurcak and Suskova 1991).

4.5.5 OTHER COUNTRIES

4.5.5.1 Hungary

The interest for the use of roseroot as a new medicinal plant have resulted some university studies in Hungary (Czövek 1999; Konya 2008). Until now, attempts to grow roseroot under Hungarian climatic conditions have failed due to the high temperature conditions and the extreme dry summer, unsuitable for this arctic species. At the same time, active biotechnological research is carried out at Corvinus University of Budapest. The research has been started in Finland, with the dissertation made at the University of Oulu (György 2006) and being continued in the Department of Genetics and Plant Breeding of the Corvinus University. Several students (BSc, MSc, PhD) are participating in the research.

The research activity focused on the production of the main glycosides in *in vitro* callus cultures of *R. rosea* through biotransformation. Both rosin and rosavin (György et al. 2004, 2005; György and Hohtola 2009), salidroside (György 2006), and some new, earlier not detected, cinnamyl alcohol glycosides could be produced (Tolonen et al. 2004). Also, the biosynthesis routes of salidroside and cinnamyl alcohol glycosides are under study. So far, TyrDC have been isolated, and the expression of this gene was examined (György et al. 2009). Meanwhile, a transformation method was developed (Mirmazloum et al. 2012). The aim of this work is to establish an alternative system for the production of the pharmaceutically important glycosides with a high production rate.

The other direction of the research is the study of the genetic diversity of the European roseroot populations. This project aims to find promising genotypes for further biotechnological or breeding purposes. Populations from the Alps (Austria, Slovenia, Italy, and Switzerland), Carpathians (Romania, Slovakia), Pyrenees (Spain, Andorra), and from northern parts of Europe (Finland, Norway) have been explored. Genetic diversity is studied with ISSR and SSR markers. The chemical analysis of these samples is also studied with either HPLC or TLC and OPLC methods. Yet the results concerning the Finnish samples have been demonstrated in a conference paper (György et al. 2011).

4.5.5.2 Estonia

Due to the information transfer of the beneficial effects of roseroot from the former Soviet Union, some research and cultivation activities have been found in Estonia as well. During 1995–2000, with other adaptogenic medicinal plants, biological and agronomical questions of the cultivation methods of roseroot have been studied (Heintalu 2000). Presently, no significant plantations exist; the estimated area is less than 0.5 ha.

4.5.5.3 Great Britain

Only one study was published on roseroot in England. Biological observations have been carried out on the sex ratios of roseroot in natural population in the northern part of England. The ratio of the male to female individuals was 1.5:1 (Richards 1988).

4.5.5.4 Mongolia

The endemic roseroot population have been studied from 2007. Thirty-six constituents were identified in the oil. The main components in the oil were geraniol (32.3%), myrtenol (15.7%), octanol (13.7%), and *trans*-pinocarveol (11.6%). The essential oil composition has been compared to Finnish and Norwegian roseroot results (Shatar et al. 2007) and can be stated that the Mongolian roseroot oil had higher proportion of geraniol. In Norway, Rohloff (2002) found *n*-decanol (30.38%) and geraniol (12.49%). In Finland, the main compounds in the oil from a Norwegian origin cultivated roseroot strain were myrtenol (36.9%), geraniol (12.7%), *trans*-pinocarveol (16.1%), and dihydrocumin alcohol (12.1%).

The intraspecific variation of five roseroot populations of the east Mongolian Altai region has been studied (Magsar et al. 2011). The variation of the rosavin contents between the five populations was from 1.9% to 4.2% and within the most abundant population from 1.1% to 3.39%. The Mongolian researches cooperated with Polish universities in the roseroot research (Altansengent et al. 2007).

4.5.6 Present Status of *R. rosea* in Europe

The European Medicines Agency, Committee on Herbal Medicinal Products (HMPC), has published an assessment report on *R. rosea* rhizome and roots on July 12, 2011. (EMA/HMPC/232100/2011). The overall conclusion of the report is as follows: Herbal preparations of the underground organs of *R. rosea* are used in traditional medicine for centuries. In the former USSR (including countries which now belong to the European Union such as Estonia, Lithuania, and Latvia), a liquid extract (DER 1:1, extraction solvent ethanol 40%) was in medicinal use in 1975.

It was accepted as a so-called Temporary Pharmacopoeia Article, which allowed the large-scale production. Dry extracts (DER 1.5–5:1), extraction solvent ethanol 67%–70% v/v, are in medicinal use within the European Union since 1987. Therefore, the criteria for 30 years of medicinal use as defined for traditional herbal medicinal products in Dir. 2004/24 are fulfilled. The traditional use as an adaptogen for symptoms of asthenia such as fatigue and weakness is appropriate for traditional herbal medicinal products. The longstanding use as well as the outcome of the clinical trials supports the plausibility of the use of the mentioned herbal preparation in the proposed indication.

At the same time, "the published clinical trials exhibit considerable deficiencies in their quality." Therefore, "well-established use" cannot be accepted. The clinical trials as well as the traditional use do not give reasons for special safety concerns. No serious adverse events are reported. The *in vitro* observed inhibition of CYP3A4 and P-glycoprotein was not confirmed *in vivo*. Additionally, no case reports on interactions are published. Because of the limited duration of use of 2 weeks, the *in vitro* data seem to be of minor clinical relevance. Therefore, the overall benefit/risk balance is positive. Due to the missing published data on genotoxicity, the development of a Community List Entry could not be supported.

4.6 LIFE CYCLE AND CULTIVATION OF *R. ROSEA*

4.6.1 LIFE CYCLE

Detailed study of the development of long-lived perennial roseroot propagated from seed in the former Soviet Union (Nukhimovski et al. 1987; Nukhimovski 1974, 1976) identified four phases: (1) embryonic or seed phase, (2) virginal phase or first year of the growth, (3) juvenile phase, during the second year of the growth, and (4) maturity.

4.6.1.1 Seed Phase

The reddish brown seeds form capsules during the second half of the summer. Under natural conditions, the ripened seeds fall in August and September. During the winter, they undergo natural stratification.

Roseroot seeds are very small: on average, 1.8–2 mm long and 0.5–0.7 mm wide. The thousand seed weight is 0.20 g. In Tomsk area, Russia, Revina et al. (1976) have found the thousand seed weights of the third and fourth year old plants were 0.24 and 0.28 g. In Poland, Kozlowski and Szczyglewska (2001) measured the seed size. Thousand seed weight was 0.11–0.18 g. In Italy, the thousand seed weight of plants collected in the Alps at Trento, at 1840 m height, was 0.13 g (Aiello and Fusani 2004).

The seeds require stratification to germinate at a significant level. After harvest and under laboratory conditions, seeds show very low germination rates (4%–25%). A 21-day period of stratification between 0°C and 2°C has been shown to increase the rate to 76% (Nukhimovski 1974). Natural stratification by sowing the seeds outdoors during September–October can increase germination to 96%–100% (Nukhimovski 1976).

Seeds also require light to germinate. Using 100 mg GA3 treatment, the germination rates at 20°C and 24°C were 86.3% and 87.7%, respectively. Germination time was 7.2–8.9 days (Aiello and Fusani 2004). The seeds remained viable up to 10 years when stored at a temperature of about 5°C (Kozlowski and Szczyglewska 2001).

4.6.1.2   First Year of Growth

After natural stratification, seeds placed in a warm greenhouse conditions, in early spring, will germinate within 7–14 days. The first leaves are very small, and growth during the first year is very slow. During the second part of the summer, the young plants develop more leaves, and one to two shoots of 4–15 cm height. The underground parts consist of mainly hairy roots, and in this phase the rhizome is very small.

4.6.1.3   Second Year of Growth

In the beginning of the second year, new shoots begin to grow quickly. Two types of shoots develop: the majority is immature shoots, 10–22 cm height with no flowers. Some taller shoots (11–35 cm) develop flowers, and their proportion was determined in one study to be 3.8% (Nukhimovski 1976). The rhizome forms a small ball, larger than in the previous year.

4.6.1.4   Plant Growth from 3 to 40 Years

From the third year, plants develop numerous new shoots. Since roseroot is a dioecious species, the shoots have male and female flowers. The number of shoots depends on the growing conditions and age of the plants. In natural Altai Mountain conditions, Nukhimovski et al. (1987) reported a total of 249 shoots on a 200-year-old plant. Under cultivated conditions, the shoot number of 4-year-old plants has ranged from 226 to 260 per plant. The root and rhizome grow significantly, and fresh weight during the third to sixth growing years reaches on average 250–400, 300–600, and 500–1000 g/plant, respectively (Galambosi et al. 2003).

After 6 years, the central parts of the rhizomes begin to decay and root quality decreases significantly. No rhizome decay was observed in 4-year-old plants, but was observed in 6-year-old plants in Poland (Przybyl et al. 2008) as well as in a natural population in north Finland where 68% of older plants were reported dead (Galambosi et al. 2007). The older the plants, the higher was the proportion of dead parts in the rhizome. According to a study conducted by the author, the salidroside contents of the fresh and dead parts of a 12-year-old plant were, respectively, 0.530% and 0.045%, and the total rosavin contents were 1.233% and 0.189% (Galambosi et al. 2009).

According to several authors, the optimum life cycle of seed-propagated roseroot under cultivation seems to be 4–6 years (Galambosi et al. 2010b; Plescher et al. 2010; Przybyl et al. 2008). During this time, the plants develop sufficient root biomass for economical root yields.

4.6.1.5   Summer Growth Rhythm

In southern Finland, at Mikkeli (61°15′ N), we have observed the clear two-peak annual growth rhythm of cultivated roseroots. The snow-cover generally disappears in this region toward the end of April or the beginning of May, and roseroot starts to grow immediately. There is intense new shoot growth in spring, peaking at the end of June. The flowering period extends from the beginning of June until the end of July. The first growth phase ends with fruit ripening. After seed harvest, during August, new shoots begin to grow from the rhizome and continue to elongate until the end of September. The number of the new fall shoots is only about one-fifth of the spring shoots, and their height is only one-third to one-half, and they lack flowers. Both spring and autumn shoots die back during winter. Farther south, in Poland, a third, late autumn growth has even been observed (Zenon Weglarz, 2003, pers. comm.).

4.6.2   GENERAL ENVIRONMENTAL REQUIREMENTS

Roseroot is native to colder regions of the northern hemisphere. In mountainous areas, it grows close to the snow line. In natural habitats, it generally grows in moist soils (in sandy soils on tundra); in meadows; in pine forests and dunes; and on river-banks, cliffs, rock crevices, and rocky slopes (Komarov 1971).

4.6.3   FIELD CHOICE

Rhodiola will thrive under cultivation in deep, moderately rich, and well-drained sandy or sandy-loam soil that is neutral to slightly acidic (pH 5–7). Soil with a high peat content is not ideal, as it makes cleaning the roots difficult. Similarly, stony soil makes harvesting difficult. Under cultivation, the plant performs better in well-drained, moderately moist soil, especially in the critical establishment phase after transplantation, but established plants can tolerate drought. According to German experiences, it needs 500–600 mm precipitation for 2–3 t/ha dry root yield (Plescher et al. 2010). As high-altitude plants, they require special attention when grown at lower altitudes. They do not tolerate high summer temperatures and require regular water in spring. High humidity is very important for its growth. They are basically spring growers, and by midsummer are past their best.

Since roseroot will be growing for an average of 5 years at the same site, the soil must be well-prepared, cleaned of perennial weed roots, and free of competition from other weeds. The more thoroughly the soil has been worked prior to transplanting, the better the plants will do.

4.6.4   FERTILIZATION

In its natural habitat, roseroot grows on poor soil and on rocks covered with a very thin layer of soil. Its size under arctic and Nordic conditions is very small, in accordance with the poor nutrient supply the soil can provide. In field cultivation, roseroot develops much larger roots and produces more seeds. According to Stephenson (1994), the most important factor for strong and ongoing growth is very porous soil and compost. Additionally, the application of trace elements via watering promotes growth.

Little precise data are available concerning fertilization of roseroot. In Kola peninsula (Russia), a 6-year open-field plot experiment tests the effect of applications of manure, $CaCO_3$, and different doses of mineral elements (Elsakov and Gorelova 1999). The highest fresh root weights were measured using 50 t/ha manure and 5 t/ha $CaCO_3$, with additional NPK of 50–50–70 kg/ha (macroelements) and 0.3 kg/ha microelements. During the fourth to sixth growing years, the fresh root weights were 390, 677, and 810 g/plant. The salidroside content varied irregularly, 0.85%, 1.36%, and 1.15%, respectively.

An experiment with organic fertilizer was carried out from 1997 to 2000 in Finland (Galambosi et al. 2003). One-year-old seedlings were transplanted into plastic mulch at a density of 6 plants/m² in sandy moraine soil, with four replicates. Before transplanting, three doses of compost (5–10 and 20 t/ha) were incorporated into the soil (Biolan composted cow manure). No additional fertilization was performed during the fourth year of growth. The doses of N–P–K (in kg/ha) applied are presented in Figure 4.1.

The increased compost treatment increased shoot growth and fresh weight of the roots (the latter by 10%), but decreased fresh root dry matter content. The variation in fresh root weights was quite high, dry root yields were similar, and the results were not significant.

4.6.5 VARIETY

"Mattmark" is a synthetic variety of *R. rosea* launched in 2011. Four-year-old plants have an average individual dry weight of 99 g and contain 2.89% salidroside and 2.0% rosavins (Vouillamoz et al. 2011).

In Finland, in semi-large-scale cultivations, an accession originated from northern Norway was used. Its habit is compact and the shoot height is 30–40 cm. The average fresh weight of the 4–5-year-old plants ranged between 512 and 820 g/pot, with an extreme value of 1004 g/pot. The contents of salidroside (0.80%) and total rosavins (1.61%) are at medium level (Galambosi et al. 2007).

The search for a better quality strains resulted in an accession from North Lapland (Kilpisjärvi), with 2.02% salidroside and 2.25% total rosavin content

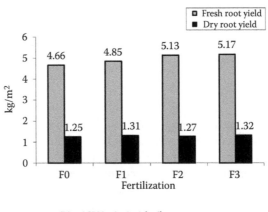

F0 = NPK= 0–0–0 kg/ha
F1 = NPK= 24–25–50 kg/ha
F2 = NPK= 50–50–100 kg/ha
F3 = NPK= 75–75–150 kg/ha

FIGURE 4.1 The effect of organic fertilization on *Rhodiola rosea* yield in Finland. From Galambosi, B., Galambosi, Zs., Valo, R., Kantanen, S., Kirjonen, H. 2003. Elaboration of Cultivation Methods for Roseroot (*Rhodiola rosea* L.) in Mikkeli, 1994–2002. In: Galambosi, B. (ed.) *Use and Introduction of Medicinal Plants with Adaptogen Effects in Finland.* Maa-ja elintarviketalous 37: 47–62. http:// www.mtt.fi/met/pdf/met37.pdf

(Galambosi et al. 2007). At the same time, in this study, dry rhizomes of 4-year-old accessions from Iceland, Germany, and Austria grown in Mikkeli, Finland, were analyzed as well. Their salidroside (0.208%–0.395%) and total rosavins content (1.201%–1.629%) proved to be not higher than those of the Finnish strain (Galambosi et al. 2009).

4.6.6 PROPAGATION METHODS

Roseroot can be propagated by seed, through crown, or by root division. Plants reproduced by vegetative propagation establish more rapidly and are initially more vigorous than those grown from seedlings. Methods of vegetative propagation are well suited for home-gardening purposes, small-scale cultivation, or propagation. For large-scale cultivation in the fields, transplanting seedlings is the most common and efficient method.

4.6.6.1 Root Division

The underground rhizome-like roots are cut into sections 1.5–15 cm long, with at least one bud and some hair roots on each cutting. Root division can be performed in early spring as well as in the summer, on crops in their third year. Cuttings should be transplanted into individual pots and placed in light shade in a cold frame or greenhouse until well-rooted. Larger cuttings with sufficient root hairs can be planted directly in the field.

4.6.6.2 Seed Propagation

R. rosea seeds require natural or artificial winter stratification to achieve a good germination rate. For natural stratification, seeds should be mixed with moist sand in autumn (during September–October), and should remain outdoors under a blanket of snow throughout the winter. In early spring, they should be transferred to a greenhouse for sowing.

The stratified seeds should be sown by hand in early spring on the substrate surface. For germination under greenhouse conditions, the optimum temperature is under 25°C in daytime and no less than 15°C at night. The seeds will germinate in about 7–14 days, and the small seedlings should be transferred into pots. In mass production of seedlings for mechanical sowing, a mixture of sand and stratified seed in a 20:1 ratio should be used.

Substrates for seedling generally contain peat and vermiculite as well as other additives, with a moderate dose of fertilizer. For larger root size, 2–4 small seedlings should be transplanted into one pot cell. These young seedlings remain in pots indoors for 3 months and possibly longer outdoor subsequently, depending on their growth rate. Transplantation of seedlings into the field can be performed in early summer or autumn.

4.6.6.3 Seedling Age

Seedling age at transplanting is an important factor, affecting both yields. From a practical point of view, the plant's 4–6 years cultivation period can be divided into two stages: potted seedling and field. Seedlings can be kept in pots for 6–12 months; after transplantation, the plant grows in the field over 3–4 years. The data in Figure 4.2 shows that the longer the seedling phase and the larger the transplanted seedlings, the better growth and yield will be in the field. Transplanted seedlings of 5–6 months

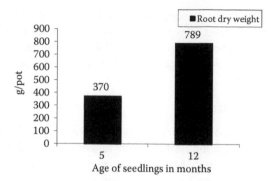

FIGURE 4.2 The effect of seedling age on the root weight of *Rhodiola rosea* after three growing seasons.

require 4–5 years in the field to reach maturity, whereas 12-month-old (or older) seedlings only require 3–4 years. Seedlings should therefore be kept in pots in the nursery for one or two growing seasons, to reduce production costs and improve yield.

4.6.6.4 Transplanting Seedlings

Seedlings 6–12 months old may be transplanted into the field in spring or autumn. Roseroot may grow in three-land systems: on flats in raised beds in plastic mulch. On flats, plants can be planted with 50–60 cm between rows and 15–20 cm within the rows. Row distance is determined by mechanical weed control machinery. In a raised or plastic mulch bed, the row distance is 30 cm within and between rows, three to four rows per bed or plastic.

The plastic mulch will normally be laid by a mulcher. The soil needs to be well prepared (free of large stones, roots, grass sods, etc.) and should not be wet for the mulcher to work properly. It is best if the plastic mulch is custom ordered and come with seedling holes punched at the correct spacing of 30 cm apart for hand transplanting. There are simple hand punchers for equal spacing of seedlings. Some type of the machines transplant the seedlings by own. There are systems with laying of drip irrigation tubes under the plastic mulch as well. During the first year, some hand weeding around the plants are necessary, especially for cleaning the weed dandelion. Between the plastic strips, the weeds are cut by a lawn mower, three to four times per growing season.

For large-scale production, an irrigation system is required. Transplanted seedlings should be irrigated immediately. The most critical phase in terms of water requirements is during the first and the second years. Well-established plants are quite drought tolerant.

4.6.7 Weed Control

During the first 2 years, roseroot grows quite slowly, and therefore competes poorly with weeds. No herbicide is currently registered for weed control. On flat cultivation, effective control requires several interrow cultivators and at least three manual weedings during a growing season. In smaller areas, mulching and moist soil are effective to control weeds, although reapplying mulch in each spring is costly.

TABLE 4.1

Effect of Seedling Age and Weeding System on Growth and Yield of *Rhodiola rosea* in Finland

Weeding System	Fresh Root Weight (g/pot)		Fresh Root Yield (g/m²)		Dry Root Yield (g/m²)	
Seedling age (months)	5	12	5	12	5	12
Hand weeding	168.7	475.1	675	1900	176	471
Black plastic mulch	246.1	657.7	984	2631	281	684
Hay mulch	206.4	686.6	827	2746	201	725
LSD 10%	n.s.	162.8				
		Field growing period: 3 years				

Source: Galambosi, B. et al. *Use and Introduction of Medicinal Plants with Adaptogen Effects in Finland.* Maa-ja elintarviketalous 37: 47–62. http://www.mtt.fi/met/pdf/met37.pdf, 2003.

Instead of time-consuming hand weeding, black plastic mulch seems to be the best weed control method, especially during the year of establishment, and also improves growth and retains moisture. In a 3-year experiment in Finland, the black plastic mulch increased the dry root yields by 53% and the black plastic mulch by 45% (Table 4.1).

4.6.8 PESTS AND DISEASES

Few pests and diseases have been reported for roseroot. According to Stephenson (1994), the greatest enemy of *Sedum* species in England is vine weevils (*Otiorhynchus sulcatus*), which feed on the edges of the newest, most succulent leaves and are difficult to eliminate. No pest or disease damage has been observed in Finland in the past 20 years, other than root damage occurred only once on two plants by *Armillaria mellea* fungi, which is common in forests. In Canada, serious damage has been caused by *Mysus persicae* on the green shoots (Ampong-Nyarko et al. 2010).

4.6.9 ROOT HARVEST

4.6.9.1 Harvesting Process

The harvest process varies according to conditions and available machinery. Steps include harvest from the field, predrying, and postdrying treatments.

1. Black plastic mulch removal (generally by hand).
2. Root harvest (vibrating shakers or diggers should be used).
3. Removal of the vegetative shoots (generally by hand).

4. Transportation of whole roots for predrying (over longer distances, by refrigerated truck).
5. Preliminary storage of the fresh root masses, if necessary.
6. Slicing and washing of whole roots (by hand or machinery).
7. Washing the sliced roots.
8. Drying of sliced roots with circulating warm air at 60°C.
9. Fractioning of dried roots by sieves with 3- and 10-mm holes.
10. Packaging dry roots in double paper–plastic sacks.
11. Storing the packaged material.

The main complication during the roseroot harvest process results from the presence of soil and small stones in the branching roots and rhizomes. These must be removed before slicing and washing.

Roots can be cleaned more effectively for a higher quality result by allowing the soil to dry for a week on the root before washing with water. According to the research results of Bacharov et al. (2009), this artificial water deficit shock has initiated some increases in the contents of the active metabolite in the rhizome.

4.6.9.2 Timing the Harvest

The roots can be harvested in their fourth or fifth year of growth. Generally, roots should be harvested in autumn (August and September). A springtime harvest (May) negatively affects some properties but improves others. According to the results of a 4-year-old plantation at Rovaniemi, Finland, the total dry root yield was about 60% lower when harvested in spring time (485 g/m²), as compared to autumn (761 g/m²) (Table 4.2). Salidroside and total rosavin contents were almost 60% and over 20% higher, respectively, when the root was harvested in spring. Figure 4.3

TABLE 4.2
Root Yield of 4-Year-Old *Rhodiola rosea* Effected by Different Harvest Times

Harvest (time)	Fresh Root Yield (g/m²)			Dry Root Yield (g/m²)		
	Root	Rhizome	Total	Root	Rhizome	Total
21st of May	796.2	1791.6	2587.8	177.6	392.4	570.0
29th of May	945.6	1638.6	2584.2	188.2	324.4	512.6
7th of June	770.4	1492.2	2262.6	138.7	234.3	373.0
Mean						
x_1	**837.4**	**1640.8**	**2478.2**	**168.2**	**317.0**	**485.2**
7th of June	867.0	1717.2	2584.2	212.4	461.9	674.3
21st of August	998.2	2226.6	3214.8	241.1	607.9	849.0
Mean						
x_2	**932.6**	**1971.9**	**2899.5**	**226.8**	**534.9**	**761.7**

Source: Galambosi, B. et al. Post harvest procedures of *Rhodiola rosea* in Finland. In: Abstracts of the 7th Annual Natural Health Product Research Conference: The Next Wave. *Pharmaceutical Biology* 48, S1, 6, 2010b.

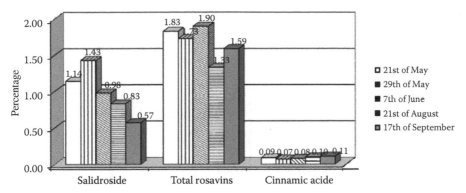

FIGURE 4.3 The effect of harvest time on the content of main compounds of *Rhodiola rosea*. (From Galambosi, B., Galambosi, Z., Uusitalo, M., Heinonen, A., Siivari, J. 2008. Ruusujuurta viljelyyn. *Maaseudun Tiede* 65, 3, 14).

shows that especially the salidroside contents were lower in autumn. This can, however, be compensated by higher root yields in autumn. Due to the higher root yields and better organization of the harvest process, from the grower's point of view, the autumn harvest is preferable.

4.6.9.3 Male–Female Plants

In a seed-propagated plantation, the male to female ratio is about 50:50. In a comparative study, quantitative and qualitative parameters of 4-year-old male and female individuals of five different roseroot varieties were compared (Galambosi et al. 2009). No significant differences were found in rhizome diameter or root length, but Figure 4.4 shows that the total weight of fresh roots of male plants was significantly higher, 18%–20%, than that of female plants. The average salidroside content of male plants was 40% higher (female: 0.29%, male: 0.40%), and total rosavins content was 10% higher (female: 1.20%, male: 1.36%), but these differences were not significant. In the light of these experimental results, male plants achieved higher productivity. Male plants seem to be heavier and of superior quality internally, but propagating more male plants could be far more expensive than using seed-propagated seedlings.

4.6.9.4 Predrying Treatments

In commercial production, large quantities of fresh root mass must be processed rapidly, while avoiding damage to the raw material through inappropriate storage temperature or injury by mechanical equipment.

In a series of experiments in Finland, several possible events that may occur during harvesting have been simulated. First, the whole fresh roots were stored: one in a refrigerator (5°C) and the other at room temperature (20°C), for 2, 4, and 8 days. Second, the effects of frozen storage were examined by slicing the rhizomes, freezing them at –20°C for 7 days, and comparing them to sliced rhizomes kept at room temperature for 2 hours. Finally, the effects of cell damage were analyzed by damaging whole roots with tractor wheels and comparing them with undamaged ones.

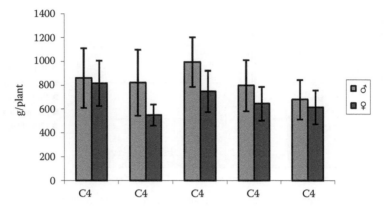

FIGURE 4.4 Total fresh root weight of male and female plants of five *Rhodiola rosea* accessions in Finland.

The results (Table 4.3) show that tyrosol ($x = 0.03\%$), cinnamic alcohol ($x = 0.07\%$), and salidroside (1.04%–1.49%) were unaffected by the treatments. Instead, the concentrations in cold storage at 5°C for two, four, and eight days appeared to increase the salidroside content in the roots (1.25%–2.69%). Roots stored in cool for 4 days produced the highest value. Salidroside levels in the samples pretreated by slicing, freezing, and thawing were the second highest (2.15%).

The variation of the content of total rosavins in this experiment was quite low (0.95%–1.64%). Damaging the roots resulted in the lowest rosavin content (0.85%). Storage at room temperature resulted in the lower average content ($x = 0.99\%$) compared to the cold storage ($x = 1.39\%$). The process of slicing, freezing, and thawing have resulted in the highest rosavin contents (1.52%–1.64%). This result may indicate future utilization in the industrial process of fresh root yields.

Before drying, the roots have to be cleaned and washed. The way of washing can vary according to the conditions of the farm. A simple solution is to soak the root masses into a water tank or basin, or use a common concrete mixer. For washing of large root masses, drum washers are suitable, with conveyor belt equipped with water pipe. Before washing, the root masses should be sliced. Slicing of the roots is necessary for different reasons. The sliced roots can be washed more effectively. The stones can be separated from the roots, and the roots can be dried more effectively.

The drying of the roots of *Rodiola* is generally very slow, and the slicing is absolutely necessary. Without slicing, the roots dry for more than 5 days; the sliced roots are dried after 44 hours in an electrical drier.

For slicing, different machinery can be used. The pieces of fresh roots should not be too small or too thick. Especially, crushing the fresh root should be avoided. In a processing experiment in Finland, the hand-cut particles of fresh roots with 4–8-mm thickness were compared to a bush–chopper-sliced root masses with thin particles (1–2 mm). The results presented in Table 4.4 show that the concentrations of salidroside and tosal rosavins in the machine-slicing of roots decrease significantly, nearly by 50% (Table 4.4).

TABLE 4.3
Contents of Salidroside and Total Rosavins of *Rhodiola rosea* Roots Effected by Different Preliminary Storage Treatments

Treatment	Root Type Control	Storage Length	Storage Temperature (°C)	Salidroside %	Total Rosavins (%)
A	(Whole)	0 hour	0	1.12	0.95
1B	Whole root	2 days	5	1.95	1.47
1C	Whole root	4 days	5	2.69	1.30
1D	Whole root	8 days	5	1.25	1.40
1E	Whole root	2 days	20	1.39	0.78
1F	Whole root	4 days	20	1.33	1.28
1G	Whole root	8 days	20	1.14	0.91
2B	Sliced rhizomes	2 hours	20	1.49	1.10
2C	Sliced rhizomes	7 days	−20	1.15	1.64
2D	Sliced rhizomes	7 days	−20[a]	2.15	1.52
3B	Damaged roots	0 hours	0	1.04	1.32
3C	Damaged roots	24 hours	20	1.26	0.85

[a]Thawing.

Source: Galambosi, B. et al. Post harvest procedures of *Rhodiola rosea* in Finland. In: Abstracts of the 7th Annual Natural Health Product Research Conference: The Next Wave. *Pharmaceutical Biology* 48, S1, 6, 2010b.

TABLE 4.4
Effect of Slicing on the Active Metabolites of *Rhodiola rosea* in Finland

Treatment	Salidrosid (%)	Total Rosavin (%)	Tyrosol (%)	Cinamic Alcohol (%)
Hand sliced	0.96	2.11	0.02	0.10
Machine sliced	0.47	1.00	0.05	0.12

Source: Galambosi, B. et. al. Post harvest procedures of *Rhodiola rosea* in Finland. In: Abstracts of the 7th Annual Natural Health Product Research Conference: The Next Wave. *Pharmaceutical Biology* 48, S1, 6, 2010b.

According to Dubichev et al. (1991), the decrease of the metabolites in the fresh roots starts soon after 30 minutes of beginning of processing. The white color of sliced roots turns to brownish or reddish color after 1–2 hours. The excessively fine-grained fresh root material seems to be sensitive to enzyme activity and autofermentation.

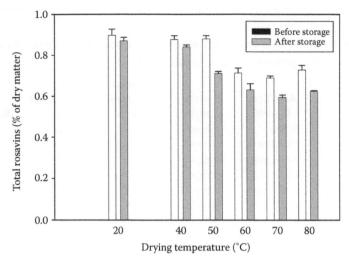

FIGURE 4.5   Total rosavin content of *Rhodiola rosea* in relation to drying temperature and one-year storage time.

Therefore, slicing into very small particles should be avoided. The optimum thickness of the fresh root particles should be 5–10 mm. To avoid this negative autofermentation, before drying, the storing time of sliced fresh root in room temperature should be as short as possible. The dryer has to be preheated for 60°C before any fresh root masses are placed into the dryer.

4.6.9.5   Drying Temperature

Russian experimental results have demonstrated that there is a moderate variation in the contents of rosavin and salidroside in different drying regimes. The salidroside content of the rhizome at drying temperatures of 20°C–50°C–80°C were 0.8%, 0.5%, and 0.8%, respectively. The contents of rosavidin were 3.22%, 2.67%, and 3.25%, respectively. The authors proposed the optimum drying temperatures for roseroot between 70°C and 80°C (Kurkin et al. 1989). Nearly similar results have been achieved in a recent study, where rhizomes of a 4-year-old Norwegian population were dried at different temperature regimes (Thomsen et al. 2011). Figure 4.5 shows that the highest total rosavin contents (0.90%) were measured at drying temperatures of 20°C, 40°C, and 50°C. The higher drying temperatures at 60°C–70°C–80°C resulted in about 20% decrease in the total rosavin contents. For drying roseroot, a temperature between 50°C and 60°C is recommended.

4.6.9.6   Postdrying Process

The harvested roseroot yield consisted of rhizomes and hairy roots, with a proportion of rhizomes generally about 70%–80% (Przybyl et al. 2008). The whole dried root material must be processed to achieve a consistent final quality. Sand and dust (dried hairy root particles) must be separated from the thick roots and rhizomes.

TABLE 4.5

Content of Active Compounds of Different Fractions of *Rhodiola rosea* Dry Roots after Separation

Fraction (dust percentage)	Salidroside	Rosavin	Rosarin	Rosaridin	Flavonoid
A/Whole root	0.454	1.051	0.067	0.315	0.186
Dust A (17.8%)	0.222	0.553	0.058	0.087	0.167
B/Rhizome	0.382	2.292	0.132	0.298	0.091
Dust B (8.9%)	0.268	0.746	0.07	0.106	0.119
C/Hairy-root	0.172	0.737	0.044	0.134	0.187
Dust C (16.4%)	0.103	0.601	0.043	0.033	0.212

Source: Unpublished data.

In Table 4.5, we present the separation results of a commercially produced root yield in Mikkeli, Finland. The proportion of the dust was the highest (17%–18%) of the whole and the hairy root fractions, while the rhizomes contained only 8% dust. The dust-containing particles of the hairy roots seem to be a valuable raw material as well. The total flavonoid contents of the hairy root fraction were nearly two times higher than that of the thick rhizomes. From the economical point of view of growers, this fraction should be utilized, for example, for extraction.

The dried root can be packed into double sacks (paper and plastic) for a longer storage period. According to the Russian Pharmacopeia, the dry roots can be stored for 3 years without quality deterioration. In a 1-year storage experiment, presented in Figure 4.3, dry rhizomes were stored for 12 months at room temperature in paper bags, resulting in a slight reduction in total rosavin content (10%–15%), but these changes are not drastic (Thomsen et al. 2011).

4.6.10 YIELD POTENTIAL

The root yield of a roseroot plantation depends on various factors, including plant density in initial pots and in the field, plant age at harvest, and environmental conditions. Most available data concerns small-scale production.

Nukhimovski et al. (1987) have reported on an experimental plantation in Moscow district, Russia. The fresh underground masses of the 4- and 5-year-old transplanted plants were 233 and 388 g/plant, respectively. At a distance of 60 cm × 20 cm between plants (a density of 8 plants/m²), total fresh root yield ranged from 1.86 to 3.1 kg/m².

In an experimental plantation in Finland, the fresh and dry root yields of 4-year-old plants ranged from 1.8 to 2.8 kg/m² and from 0.4 to 0.6 kg/m², respectively. The dry root yield of 5-year-old plants was 1.2–1.4 kg/m². In a 5-year-old semi-large-scale plantation in black plastic, with a plant density of 6/m² and cultivation surface of 550 m², total dry root yield was 334 kg (1.65 kg/m²). In Germany, the expected yield level for field cultivation is about 2000 kg/ha dry root yield (Plescher et al. 2010).

4.6.11   SEED HARVEST

Plants generally flower in June. Ripened seeds can be collected once the flowers have turned brown. In natural growing, first flowers appear only in the seventh and eighth years, due to the plant's slow growth. In the High Altai areas, between 1750 and 2200 m up on sea level, Polozhii and Reviakina (1976) found 560–960 seeds in one flowering stem. Under cultivation conditions, plants develop more quickly, produce larger flowers, and more seeds. In the Tomsk Botanical Garden (Russia), the seed quantity of the 2- and 3-year old plants were 780–3620 seeds per flowering stem (Revina et al. 1976). In south Finland, Mikkeli, the seed quantity in a 4-year-old cultivated plantation was 0.085–0.100 g/stem. The measured seed yield was 2.07–4.53 g/plant. As the plant becomes older, the seed yield may increase up to 6–10 g/plant.

Seeds yield can be collected from cultivated crops every summer from the second year of growth. In July–August, when the color of the seed capsule is deep brown, it will naturally open and release seeds when shaken. Manual collection should be carried out in the morning, when the capsules are still moist. The cut flower heads with short stem should be collected in boxes or bags. After drying for a week, the dry capsule mass must be shaken vigorously with care.

Seeds can be kept in dry storage, but little data is available on the viability of stored seeds. In Finland, the seeds are stored at room temperature for 3 years, and a fresh supply is collected regularly. According to Kozlowski and Szczyglewska (2001), seeds remain viable as long as 10 years, when stored at a temperature of about 5°C.

4.7   SUMMARY AND FUTURE TRENDS

Mankind has turned to medicinal and aromatic plants throughout history. Presently, there are over 150 species, with well-elaborated growing technologies in Europe (Schippmann et al. 2006). *R. rosea* is a relatively new medicinal crop, whose domestication process dates back to only about 40 years and is ongoing. Scientific study of the adaptogenic properties and medicinal potential of this species began in the former Soviet Union after the Second World War, and the first cultivation experiments were set up during the 1970s.

On the basis of available research on roseroot cultivation, we may make some general observations:

- Soon after the pharmacological studies in nearly all countries, where roseroot is an endemic species in the flora, the domestication experiments are started. These countries are Sweden, Finland, Norway, Poland, Italy, and Switzerland.
- The main subjects of these researches were to obtain the basic agrobiological observations of the species for elaborating the first growing technologies (Poland, Finland, Germany, and Switzerland), the quality analyses of the endemic populations (Finland, Norway, Italy, Romania, Bulgaria, and Mongolia), and the development of analytical skill in some countries (Germany, Switzerland, and Finland). After these activities, collection of local accessions and studies on its genetic variability and selection have been started (Norway, Finland, Sweden, Switzerland). The selection work results the first bred variety in Switzerland.
- Contrary to these intensive research and development activities, the available data are demonstrated, that the present cultivation areas in all

European countries remain small, estimated totally about 8–12 ha (Sweden < 1 ha, Norway < 0.5 ha, Finland 2–4 ha, Italy < 0.5 ha, Switzerland 1 ha, Germany 2–3 ha, Poland < 0.5 ha, Bulgaria < 0.5 ha, and Estonia < 0.5 ha).
- This situation emphasizes the most critical features of the roseroot cultivation and production: the 4–6 year-long life cycle, the lack of special machinery, and the high labor production costs in the small-scale farming.
- However, the cheap raw material collected from the nature is still very attractive. Presently, there is quite active, legal, or illegal collection and partly this determines the world price level of the roseroot raw material. The low price level has no inspiration for long-term investments in field cultivation. The continuing wild collection results more threaten the status of the natural *R. rosea* populations.
- Observing these difficulties in the cultivation of roseroot, in many countries, intensive research have been started on the production of the main chemical compounds of roseroot by *in vitro* callus cultures or through biotransformation (Bulgaria, Poland, Finland, and Hungary).

From the above-mentioned situation, the author is convinced that the continuous or increasing raw material demand of *R. rosea* can be supplied only from cultivation. The domestication research has to be continued, and it has to be focused for the following directions:

- To improve the present field technologies, increasing the root biomass production and the yields by different agronomical means.
- To develop and improve the mechanization of the field technologies for increasing the efficacy and decreasing the production costs.
- To develop the extraction methods of the fresh root yield for avoiding the high drying and processing costs of the large root biomasses (Siivari and Galambosi 2007).
- Additionally, it has to be continuing the selection and breeding activities of roseroot for improving the inner quality of the cultivated populations.
- Additionally, basic research should be needed for studying using possibilities of modern and new technological methods for reducing the long life cycle of roseroot, for example, all-year-round cultivation in "photo culture" or in hydroponics.

REFERENCES

Aiello, N., Bontempo, R., Vender, C. 2010. Seed germination tests on Arnica Montana and *Rhodiola rosea* L. wild populations. *Book of Abstracts*, First International Symposium on Medicinal, Aromatic and Nutraceutical Plants from Mountainous Areas, July 6–9, 2011, Saas-Fee, Switzerland, p. 85.

Aiello, N., Fusani, P. 2004. Effetti della prerefrigerazione e dell' acido gibberellico sulla germinaxione del seme di *Rhodiola rosea*. *Sementi Elette*, Bologna, Italia, 4:33–5.

Aiello, N., Scartezzini, F., Vender, C., Cangi, F., Mercati, S., Fulceri, S. 2011. Experimental activity carried out on golden root (*Rhodiola rosea* L.) in the province of Trento. *Book of Abstracts*, First International Symposium on Medicinal, Aromatic and Nutraceutical Plants from Mountainous Areas, July 6–9, 2011, Saas-Fee, Switzerland, p. 32.

Alm, T. 1996. Bruk av rosenrot (*Rhodiola rosea*) mot skorbut. *Polarflakken* 1:29–32.

Alm, T. 2004. Ethnobotany of *Rhodiola rosea* (Crassualceae) in Norway. *SIDA* 21, 1:321–44.

Altantsetseg, K., Przybyl, J. L., Weglarz, Z., Geszprych, A. 2007. Content of biologically active compounds in roseroot (*Rhodiola* sp.) raw material of different derivation. *Herba Polonica* 4:20–5.

Ampong-Nyarko, K., Cole, D., Sloley, D., Zhang, Z. 2010. Potential and attainable yield of cultivated *Rhodio rosea* in Alberta. In: Abstracts from the 7th Annual Natural Health Product Research Conference: The Next Wave. *Pharmaceutical Biology* 48, S1:pp. 5,6.

Anon, 1992. Nordens ginseng. *Hälsa* Nr. 7–8. p. 17.

Arnajörg L. J., 1992. Burnirot (*Rhodiola rosea*) In: Islenskar Laekninghurtit. Islensk *Natura IV*. p. 25.

Asdal, A., Galambosi, B., Olsson, K., Wedelsbäck, B. K., Þorvaldsdottir, E. 2006. *Rhodiola rosea* L. (Roseroot). In: *Spice and Medicinal Plants in the Nordic and Baltic Countries.* Conservation of genetic resources: Report from a project group at the Nordic Gene Bank. Alnarp; Nordic Gene Bank. pp. 94–104, 150–2.

Bacharov, D., Volodina, S., Poletaeva, I., Volodin, V. 2009. Effect of water deficiency on the accumulation of salidroside and rosavine in *Rhodiola rosea* (Crassulaceae). *Rastitelnyje Rresursi* 45(3):94–102.

Battistelli, M., De Sanctis, R., De Bellis, R., Cucchiarini, L., Dacha, M., Gobbi, P. 2005. *Rhodiola rosea* as antioxidant in red blood cells: Ultrastructural and hemolytic behavior. *Acta Physiologica Latino Americana* 32:277–85.

Borodin, A. M. (ed.). 1985. *The Red Book of the USSR*. Lesnaya Promyshlennost 1, Moscow, p. 392.

Bozhilova, M. 2011. Salidroside contennt in *Rhodiola rosea* L., dynamics and variability. *Botanica Serbica* 35 (1):67–70.

Brown, R. P., Gerbarg, P. L., Ramazanov, Z. 2002. *Rhodiola rosea*. A phytomedicinal overview. *HerbalGram* 56:40–52.

Buchwald, W., Mscisz, A., Krajewska-Patan, A., Furmanowa, M., Mielcarek, S., Mrozikiewicz, P. M. 2006. Contents of biologically active compounds in *Rhodiola rosea* roots during the vegetation period. *Herba Polonica* 52(4):39–43.

Calugaru, T., Dascaliuc. A., Ivanova, R. 2007. Total polyphenolic content and radical scavenging activity of extracts from *Rhodiola rosea l.* callus. *Romanian Biological Sciences* V(1–2):23–4.

Calugaru, T., Ivanova, R., Ralea, T., Ciocarlan, A., Mrozikiewicz, P. M., Krajewska-Patan, A., Tkachenko, A., Dascaliuc, A. 2007. Prespectives of cultivation *Rhodiola rosea* L. for medicinal use. *Herba Polonica* 53(2):138.

Costica, M., Costica, N., Toma, O. 2007. Phytocoenological, histo-anatomical and biochemical aspects in *Rhodiola rosea* L. species from Romania. *Analele Ştiinţifice ale Universităţii," Alexandru Ioan Cuza," Secţiunea Genetică şi Biologie Moleculară* TOM VIII. pp. 119–21.

Czövek, E. 1999. *Drosera es Rhodiola* fajok kemiaja es farmakologiaja. Szent-Györgyi Albert Orvostudomanyi Egyetem, Gyogynöveny-es Drogismereti Intezet, Szakdolgozat. p. 35.

D'Ambrosio, M., Guerriero, A., Mari, A., Vender, C. 2008. Characterization of wild and cultivated accessions of *Rhodiola rosea* L. from the Alpine region by analyses of their marker compounds. In: *Book of Abstracts*, 12th International Congress Phytopharm 2008, July 2–4, 2008, St. Petersburg, Russia, p. 10.

Darbinyan, V., Kteyan, A., Panossian, A., Gabrielian, E., Wikman, G., Wagner, H. 2000. *Rhodiola rosea* in stress induced fatigue: A double blind cross-over study of a standardized extract SHR-5 with a repeated low-dose regimen on the mental performance of healthy physicians during night duty. *Phytomedicine* 7(5):365–71.

Dascaline, A., Calugaru-Spatatu, T., Ciocarlan, A., Costica, M, Costica, N., Krajewska-Patan, A., Dreger, M., Mscisz, A., Furmaniova, M., Mrozikiewicz, P. M. 2008. Chemical composition of golden root (*Rhodiola rosea* L.) rhizome of Carpathian origin. *Herba Polonica* 54(4):17–27.

Dragland, S. 2001. Rosenrot, Botanikk, Innhalstoff, dyrking og bruk. *Planteforsk, Gronn forskning 09/2001*. Oppdatert utgave 3.3.2004.

Dragland, S., Galambosi, B. 1996. Roserot (*Rhodiola rosea* L.). In: *Produksjon og forsteforedling av medisinplanter*. Forskningsparken i Ås, 143–5.

Dubichev, A. G., Kurkin, V. V., Zapesochnaya, G. G., Vorontsov, E. D. 1991. Chemical composition of the rhizomes of *Rhodiola rosea* by the HPLC method. *Chemistry of Natural Compounds* 27 (2):161–4.

Egger, P., Ambrosio, M. D., Aiello, N., Contriini, C., Fusani, P., Scartezzini, F., Vender. C. 2007. Active constituents profiling of *Rhodiola rosea* L. *Planta Medica* 73:269.

Elameen, A., Klemsdal, S. S., Dragland, S., Fjellheim, S., Rognl, I. O. A. 2008. Genetic diversity in a germplasm collection of roseroot (*Rhodiola rosea*) in Norway studied by AFLP. *Biochemical Systematics and Ecology* 36:706–15.

Elsakov, G. V., Gorelova, A. P. 1999. Fertilizer effects on the yield and biochemical composition of rose-root stonecrop in North Kola region. *Agrokhimiya* 10:58–61.

EMA 2011. Assessment report on *Rhodiola rosea* L., rhizoma et radix. The European Medicines Agency, Committee on Herbal Medicinal Products (EMA/HMPC/232100/2011).

Fjelldal, E., Svenske, M., Martinussen, I., Volodin, V. L., Galambosi, B. 2010. Geographic variation in chemical composition in roseroot (*Rhodiola rosea*) in Finnmark County. In: *Book of Abstracts, Circumpolar Agricultural and Land Use Resources—Prospects and Perspectives for Circumpolar Productions and Industries*, The 7th Circumpolar Agricultural Conference Alta, Finnmark, Norway, September 6–8, p. 35.

Frisk, P. 2008. Frälst på rosenrot? *Köp den på rot! Hälsä* 12:20.

Frolov, Y., Poletaeva, I. 1998. *Rhodiola rosea* in the European north-east. *UrD RAS*, Ekaterinburg. p. 192.

Furmanowa, M., Kedzia, B., Hartwich, M., Kozlowski, J., Krajevska-Patan, A., Mscisz, A., Jankowiak, J. 1999. Phytochemical and pharmacological properties of *Rhodiola rosea* L. *Herba Polonica* 45(2):108–13.

Furmanowa, M., Hartwich, M., Alfermann, A. W., Koźmiński, W., Olejnik, M. 1999. Rosavin as a product of glycosylation by *Rhodiola rosea* (roseroot) cell cultures. *Plant Cell, Tissue and Organ Culture* 56(2):105–10.

Furmanowa, M., Oledzka, H., Michalska, M., Sokolnicka, I., and Radomska, D. 1995. *Rhodiola rosea* L. (Rosenroot): *In vitro* regeneration and the biologial activity of roots. In: Bajaj, Y. P. S. (ed.). *Biotechnology in Agriculture and Forestry*, Volume 33, Medicinal and Aromatic plants VIII, Springer-Verlag, Berlin, Heidelberg, pp. 412–26.

Galambosi, B. 2004. Coltivazione della pianta. In: Ramazanov, Z., Ramazanov, A. (eds.). *Rhodiola Rosea. Le origini e la storia. Fitochimica e farmacologia*. Aboca, Italy, pp. 94–100.

Galambosi, B. 2005. *Rhodiola rosea* L., from wild collection to field production. *Medicinal Plant Conservation* 11(1):31–5.

Galambosi, B. 2006. Demand and availability of *Rhodiola rosea* L. raw material. In: Bogers, R. J., Craker, L. E., and Lange, D. (eds.). *Medicinal and Aromatic Plants: Agricultural, Commercial, Ecological, Legal, Pharmacological and Social Aspects*. Wageningen UR Frontis Series, Volume 17, pp. 223–36. Springer.

Galambosi, B., Galambosi, Z., Valo, R., Kantanen, S., Kirjonen, H. 2003. Elaboration of cultivation methods for roseroot (*Rhodiola rosea* L.) in Mikkeli, 1994–2002. In: Galambosi, B. (ed.) *Use and Introduction of Medicinal Plants with Adaptogen Effects in Finland*. Maa-ja elintarviketalous 37: 47–62. http://www.mtt.fi/met/pdf/met37.pdf.

Galambosi, B., Galambosi, Z. S., Heinonen, A., Uusitalo, M. 2010b. Post harvest procedures of *Rhodiola rosea* in Finland. In: Abstracts of the 7th Annual Natural Health Product Research Conference: The Next Wave. *Pharmaceutical Biology* 48, S1:6.

Galambosi, B., Galambosi, Z. S., Hethelyi, E., Szöke, E., Volodin, V., Poletaeva, I., Iljina, I. 2010a. Importance and quality of rosenroot (*Rhodiola rosea* L.) growing in the European North. *Zeitschrift für Arznei & Gewürzpflanzen* 15(4):160–9.

Galambosi, B., Galambosi, Z. S., Slacanin, I. 2007. Comparison of natural and cultivated roseroot (*Rhodiola rosea* L.) roots in Finland. *Zeitschrift für Arznei- & Gewürzpflanzen* 12(3):141–7.

Galambosi, B., Galambosi, Z. S., Uusitalo, M., Heinonen, A. 2009. Effects of plant sex on the biomass production and secondary metabolites in roseroot (*Rhodiola rosea* L.) from the aspect of cultivation. *Zeitschrift für Arznei- & Gewürzpflanzen* 14(3):114–21.

Galambosi, B., Galambosi, Z. S., Uusitalo, M., Heinonen, A., Siivari, J. 2008. Ruusujuurta viljelyyn. *Maaseudun Tiede* 65, 3(13.10.2008):14. http://www.mtt.fi/maaseuduntiede /pdf/mtt-mt-v65n03s14.pdf.

Galambosi, B., Valo, R. 2006. Ruusujuuri (*Rhodiola rosea* L.). In: Ahokas, H., Galambosi, B., Airikko, H., Kallela, M., Sahramaa, M., Suojala-Ahlfors, T., Valo, R., Veteläinen, M. (eds.). *Suomen kansallisten kasvigeenivarojen pitkäaikaissäilytysohjeet: Vihannes-, yrtti- ja rohdoskasvit.* Maa-ja elintarviketalous 85, pp. 73–81. http://www.mtt.fi/met /pdf/met85.pdf.

Ghiorghita, G., Hartan, M., Maftei, D.-E., Nicuta, D. 2011. Some considerations regarding *in vitro* culture of *Rhodiola rosea* L. *Romanian Biotechnological Letters*, Bucarest, 16(1 Supplement):79–85, 5902–8.

György, Z. 2006. Glycoside production by *in vitro Rhodiola rosea* cultures. *Acta Universitatis Ouluensis C Technica* 244, Oulu University Press, Oulu.

György, Z., Derzsó, E., Galambosi, B., Pedrycz, A. 2011. Genetic diversity of Finnish *Rhodiola rosea* populations based on SSR and ISSR analysis. *Acta Horticulturae* (in press).

György, Z., Hohtola, A. 2009. Production of cinnamyl glycosides in compact callus aggregate cultures of *Rhodiola rosea* through biotransformation of cinnamyl alcohol. In: Mohan, J. and Saxena P. K. (eds.). *Protocols for In Vitro Cultures and Secondary Metabolite Analysis of Aromatic and Medicinal Plants.* The Humana Press, Inc., Springer, New York, pp. 305–12.

György, Z., Jaakola, L., Neubauer, P., Hohtola, A. 2009. Isolation and expression analysis of a *Rhodiola rosea* tyrosine decarboxylase cDNA. *Journal of Plant Physiology* 166 (14):1581–6 IF 2,44.

György, Z., Tolonen, A., Neubauer P., Hohtola, A. 2005. Enhanced biotransformation capacity of *Rhodiola rosea* callus cultures for glycosid production. *Plant Cell Tissue and Organ Culture* 83:129–35. IF 1,113.

György, Z., Tolonen, A., Pakonen, M., Neubauer, P., Hohtola, A. 2004. Enhancement of the production of cinnamyl glycosides in CCA cultures of *Rhodiola rosea* through biotrans-formation of cinnamyl alcohol, *Plant Science* 166(1):229–36. IF 1,389.

Hartwich, M. 2010. The importance of immunological studies on *Rhodiola* rosea in the new effective and safe herbal drug discovery. *Central European Journal of Immunology* 35(4):263–6.

Hedman, S. 2000. *Rosenrot: nordens mirakelört.* Mikas Förlag. Ölandstryckarna. 80 pp.

Hegi, G (ed.). 1963. *Rhodiola*, Rosenwurz. In: *Illustrierte Flora von Mitteleuropa.* Zweite völlig neubearbeitete Auflage. Band IV/2, Lieferung 2/3. Paul Parey, Hamburg/Berlin, pp. 99–102.

Heintalu, A. 2000. Biology and cultivation technics of medicinal plants, like *Rhodiola rosea L., Rhaponticum carthamoides* (Willd) Iljin and *Allium suvorovii Regl. x Allium giganteum Regl.* in Estonian soils. Tallin. ISBN 998 60786-4. 259 s.

Hellum, B. H., Tøsse, A., Høybakk, K., Thomsen, M., Rohlof, J., Nilsen, O. G. 2010. Potent *in vitro* inhibition of CYP3A4 and P-Glycoprotein by *Rhodiola rosea*. *Planta Medica* 76:331–8.

Héthelyi, B. É., Varga E., Hajdú Z., Szarka S., Galambosi, B. 2004. Vadon termő és Finnországban termesztett *Rhodiola rosea* L. hatóanyagai, farmakológiai hatása és fitokémiai vizsgálata (GC, GC/MS). *Olaj, Szappan Kozmetika*, 53(6):236–46.

Hethelyi, É. B., Korany, K., Galambosi, B., Domokos, J., Palinkas, J. 2005. Chemical composition of the essential oil from rhizomes of *Rhodiola rosea* L. grown in Finland. *Journal of Essential Oil Research* 17(6):628,9.

Hjaltalin, O. J. 1830. *Isländs botanik*. Hins islezka bokmenntafelags. Köpenhamn. 128.

Hoeg, O. A. 1976. *Sedum rosea* (L.)Scop. (*Rhodiola rosea* (L) Rosenrot, pp. 595–7. In: *Planter og tradisjon, Universitetsforlaget*, Oslo-Bergen-Tromso, p. 751.

Hördur, K. 1987. Roseroot (*Rhodiola rosea*) In: *A Guide to the Flowering Plants and ferns of Iceland*. Örn og Örlygurs Puiblishing House, Reykjavik, p. 158.

Iljina, I. 1997. *Komi folk medicine*. Syktyvkar, Komi khisnoje isdatelstvo. p. 118.

Jurcak, S., Suskova, J. 1991. Investigation of cultivation technology of tonic plants. In: *Book of Abstracts*, Pestovanie zber a spracovanie liecivych rastlin. CSER, VysokeTatry, Nova Lubovna, June 4–7, p. 10.

Kedzia, B., Furmanova, M., Krajewska-Patan, A., Holderna-Kedzia, E., Mscisz, A., Wojcik, J., Buchwald, W., Mrozikiewicz, P. M. 2006. Studies on the toxicity and adaptogenic activity of extract obtained from underground parts of selected *Rhodiola* sp. *Herba Polonioca* 52(4):117–32.

Kim, E. F. 1976. Experience of cultivation of the drug plant *Rhodiola rosea* in low-mountain area of the Altai. *Rastitelnije Resurssi*. 12(4):583–90.

Komarov, V. L. (ed.). 1939. Genus 698. *Rhodiola* L. In: *Flora of the U.S.S.R.* Volume IX, *Rosales* and *Sarraceniales*. Izdatel'stvo Akad. Nauk SSSR, Moskva-Leningrad. Israel Program for Scientific Translation, Jerusalem 1971. pp. 20–36.

Konya, L. 2008. *Rhodiola rosea* termesztestechnologiaja es egeszsegmegörzö szerepe. *Pannon Egyetem Georgikon Mezögazdasagtudomanyi Kar, Keszthely*, Szakdolgozat, p. 40.

Kotiranta, H., Uotila, P., Sulkava S, Peltonen. S. L. (eds.). 1998. *Red Data Book of East Fennoscandia*. Ministry of the Environment, Finnish Environment Institute and Botanical Museum, Helsinki, p. 351.

Kovaleva, N. P., Dolgushev, V. A., Tikhomirov, A. A. 1997. Viability and germination rate of roseroot (*Rhodiola rosea* L.) seeds obtained under artificial illumination. *Russian Agricultural Sciences* 12:11–3.

Kovaleva, N. P., Tikhomirov, A. A., Dolgushev, V. A. 2003. Specific characteristics of *Rhodiola rosea* growth and development under the photoculture conditions. *Russian Journal of Plant Physiology* 50(4):527–31.

Kozlowski, J., Szczyglewska, D. 2001. Seed germination biology of medicinal plants. Part XXII. Species of the family Crassulaceae: *Rhodiola rosea* L. *Herba Polonica* 47(2):137–41.

Kozyrenko, M., Gontcharova, S. B., Gontcharov, A. A. 2011. Analysis of the genetic structure of *Rhodiola rosea* (Crassulaceae) using inter-simple sequence repeat (ISSR) polymorphisms. *Flora* 206:691–6.

Krajewska-Patana, A., Furmanova, M., Dreger, M., Gorska-Pukszta, M., Lowicka. A., Mscisz, A., Mielcarek, S., Baranika, M., Buchwald, W., Mrozikiewicz, P. M. 2007. Enhancing the biosynthesis of salidroside by biotransformation of p-tyrosol in callus culture of *Rhodiola rosea* L. *Herba Polonica* 53(1):55–64.

Krasimira, M., Tasheva, K. Kosturkova, G. 2011. *Rhodiola rosea in vitro* plants morphophysiological and cytological characteristics. *AC Romanian Biotechnological Letters* 16(6 Supplement):79–85.

Kucinskaite, A., Sawicki, W., Briedis, V., Sznitovska, M. 2007. Fast disintegrating tablets containing *Rhodiola rosea* L. extracts. *Acta Poloniae Pharmaceutica–Drug Research* 64(1):63–7.

Kump, A. 1991. Some remarks to *Rhodiola rosea* L. and salidroside. In: *Book of Abstracts*, Pestovanie zber a spracovanie liecivych rastlin. CSER, VysokeTatry, Nova Lubovna, June 4–7, p. 8.

Kurkin, V. A., Zapesochnaya, G. G., Gorbunov, Y. N., Nukhimovskii, E. L., Shreter, A. I., Shchavlinskii, A. N. 1986. Chemical investigation of some species of the genera *Rhodiola* L. and *Sedum* L. and problems of their chemotaxonomy. *Rastiteln'nye resursy* 22(3):310–9.

Kurkin, V. A., Zapesochnaya, G. G., Kir'yanov, A. A., Bondarenko, L. T., Vandyshev, V. V., Mainskov, A. V., Nukhimovskii, E. L., Klimakhin, G. I. 1989. Quality of *Rhodiola rosea* L. raw material. *Khimiko-Farmatsevticheskii Zhurna* 23(11):1364–7.

Kurkin, V. A., Zapesochnaja, G. G., Tsavlinckij, E. L., Hukhimovskii, E. L., Vandisev, V. V. 1985. Method for determination of rootstock of *Rhodiola rosea*. *Chimiko Farmatsevticheskii Zhurnal* 3:185–90.

Kylin, M. 2010. Genetic diversity of Roseroot (*Rhodiola rosea* L.) from Sweden, Greenland and Faroe Islands. Master's thesis in Biology, SLU. Alnarp. p. 58.

Lei, Y., Gao, H., Tsering, T., Shi, S., Zhong, Y. 2006. Determination of genetic variation in *Rhodiola* crenulata from the Hengduan Mountains region, China using inter-simple sequence repeats. *Genetics and Molecular Biology* 29(2):339–44.

Magsar, J., Sharkuu, A., Baczek, K., Przybyl, J. L., Weglarz, Z. 2011. Infraspecific variation pf rosenroot (*Rhodiola rosea* L.) naturally occurring in Mongolian Altai. *Book of Abstracts*, First Int. Symposium on Medicinal, Aromatic and Nutraceutical Plants from Mountainous Areas, July 6–9, 2011, Saas-Fee, Switzerland. p. 17.

Makarov, V. G., Zenkevich, I. G., Shikov, A. N., Pimenov, A. I., Pozharitskaya, O. N., Ivanova, S. A., Galambosi, B. 2003. Comparative analysis of *Rhodiola rosea* of Scandinavian and Russian origin. Proceedings of Congress Phytopharm 2003. Actual problems of creation of new medicinal preparations of natural origin. St. Petersburg-Pushkin, Russia, July 3–5, pp. 570–4.

Mäkitalo, I. and Jankkila, H. 1994. *Lapin luonnontuotealan kehittämishankkeet 1996–1998*. Rovaniemi, Lapin Maaseutukeskusry. p. 62.

Malnoe, P., Carron, C. A., Vouillamoz, J. F., Rohloff, J. 2009. L'orpin rose (*Rhodiola rosea* L.), une plante alpine anti-stress. *Revue suisse de Viticulture Arboriculture Horticulture* 41(5):281–6.

Martin, J., Pomahačová, B., Dušek, J., Dušková, J. 2010. *In vitro* culture establishment of *Schizandra chinensis* (Turz.) Baill. and *Rhodiola rosea* L., two adaptogenic compounds producing plants. *Journal of Phytology* 2010, 2(11):80–8.

Martinussen, I., Volodin, V., Rothe, G., Jakobsen, K., Nilsen, H. 2011. Effect of climate on plant growth and level of adaptogenic compounds, in maral root (*Leuzea carthamoides* (Willd.) DC., crowned snow-wort (*Serratula coronata* L.) and rosenroot (*Rhodiola rosea* L.) *The European Journal of Plant Science and Biotechology 5*. (Special Issue). 5(1):72–7.

Mattioli, L., Funari, C., Perfumi, M. 2008. Effects of *Rhodiola rosea* L. extract on behavioural and physiological alterations induced by chronic mild stress in female rats. *Journal of Psychopharmacology* 23:130–42.

Mattioli, L., Perfumi, M. 2007. *Rhodiola rosea* L. extract reduces stress- and CRF induced anorexia in rats. *Journal of Psychopharmacology* 21:742–50.

Mattioli, L., Perfumi, M. 2011. Evaluation of *Rhodiola rosea* L. extract on affective and physical signs of nicotine withdrawal in mice. *Journal of Psychopharmacology* 25(3):401–10.

Mirmazloum, I., Forgács, I., Zok, A., György, Z. 2012. Establishing transgenic *Rhodiola rosea* callus cultures. XVIII. Növénynemesítési Tudományos Napok, 2012. Összefoglalók, Szerk, Veisz O., MTA, Budapest p. 90.

Muehlhoff, J. 2010. Development of a *Rhodiola rosea* based herbal medicine for stress: Vitango® – A case study. *Pharmaceutical Biology*, Volume 48, Supplement 1. Abstracts from the 7th Annual Health Products Research Conference. The Next Wave. 6–7.

Nukhimovski, E. L. 1974. Ecological morphology some of the medicinal plants in natural conditions of growing. 2. *Rhodiola rosea* L. *Rastitel'nye Resursy.* 10(4):348–55.

Nukhimovski, E. L. 1976. Initial stages of biomorphogenesis of *Rhodiola rosea* L. cultivated in the Moscow region. *Rastitel'nye Resursy*, 12(3):348–55.

Nukhimovski, E. L., Yurtseva, N. S., Yurtsev, V. N. 1987. Biomorphological characteristics of *Rhodiola rosea* L. cultivation (Moscow district). *Rastitel'nye-Resursy* 23(4):489–501.

Panossian, A. 2003. Adaptogens: Tonic herbs for fatigue and stress. *Alternative and Complementary Therapies* 9:327–32.

Panossian, A., Wikman, G., Wagner, H., 1999. Plant adaptogens: New concepts on their mode of action. *Phytomedicine* 6:1–14.

Panossian, A., Wagner, H. 2005. Stimulating effect of adaptogens: An overview with particular reference to their efficacy following single dose administration. *Phytotherapy Research* 19:819–38.

Panossian, A., Wikman, G., and Sarris J. 2010. Rosenroot (*Rhodiola rosea*): Traditional use, chemical composition, pharmacology and clinical efficacy. *Phytomedicine* 17:481–93.

Pautova, I. A., Tkachenko, K. G. 2004. *Rhodiola* genus (Crassulaceae DC.) in a culture into north-west Russia. In: Proceedings of 8th International Congress Phytopharm 2004. Mikkeli, Finland, June 21–23, pp. 496–8.

Perfumi, M., Mattioli, L. 2007. Adaptogenic and central nervous system effects of single doses of 3% rosavin and 1% salidroside *Rhodiola rosea* L. extract in mice. *Phytotherapy Research* 21:37–43.

Petkov, V., Yonkov, D., Mosharoff, A., Kambourova, T., Alova, L., Petkov, V. V., Todorov I. 1986. Effects of alcohol aqueous extract from *Rhodiola rosea* L. roots on learning and memory. *Acta Physiologicaet Pharmacologica* Bulgaria 12(1):3–16.

Platikanov, S., Evstatieva, L. 2008. Introduction of wild golden root (*Rhodiola rosea* L.) as a potential economic crop in Bulgaria. *Economic Botany* 62(4):621–7.

Plescher, A., Holzapfel, C., Hannig, H. 2010. Inkulturnahme und Pilotanbau von Rosewurz (*Rhodiola rosea* L.) In: *20th Bernburger Winterseminar für Arznei- und Gewürzpflanzen*, February 23–24, 2010. (Powerpoint presentation).

Polozhii, A. V., Reviakina, N. V. 1976. Developmental biology of *Rhodiola rosea* L. in the Katujn range (Altai Mts.). *Rastitel'nye Resursy* 12(1):53–9.

Popov, K. 1874. Zyryane i zyryanskiy kray. Izvestiya Imperatorskogo Obshchestva Lyubiteley yestestvoznaniya, antropologii i etnografii. XIII(2) Moscow, p. 18.

Przybyl, J., Weglarz, Z., Geszprych, A., Pelc, M. 2005. Effect of mother plant age and environmental factors on the yields and quality of roseroot (*Rhodiola rosea* L.) seeds. *Herba Polonica* 51:5–12.

Przybyl, J. L., Weglarz, Z., Geszprych, A. 2008. Quality of *Rhodiola rosea* cultivated in Poland. *Acta Horticulturae* 765:143–50.

Revina, T. A., Krasnov, E. A., Sviridova, T. P., Stepaniuk, G. I., Surov, I. P. 1976. Biological characteristics and chemical composition of *Rhodiola rosea* L. cultivated in Tomsk. *Rastitel'nye Resursy* 12(3):355–60.

Richards, A. J. 1988. Male predominant sex ratios in holly (*Ilex aquifolium* L., Aquifoliaceae) and Roseroot (*Rhodiola rosea* L., Crassulaceae). *Watsonia* 17:53–7.

Rohloff, J. 2002. Volatiles from rhizomes of *Rhodiola rosea* L. *Phytochemistry* 59:655–61.

Sandberg, F. 1998. *Herbal Remedies and Herb Magic*. Det Bästa, Stockholm, Sweden, p. 223.

Sandberg, F., Bohlin L. 1993. *Fytoterapi: wäxbaserade läkemede*. Hälsokosträdets förlag AB, Stockholm, Sweden, p. 131.

Saratikov, A. S., Krasnov, E. A. 1987. Chapter VIII: Clinical studies of *Rhodiola*. In: Saratikov, A.,S., Krasnov, E., A. (eds.). Rhodiola rosea *Is a Valuable Medicinal Plant (Golden Root)*. Tomsk State University Press, Tomsk, Russia, pp. 216–27.

Scartezzini, F., Aiello, N., Vender, C., Cangi, F., Mercati, S., Fulgeri, S. 2010. Quantitative and qualitative performance of two golden root (*Rhodiola rosea* L.) accessions grown at different altitude in northern Italy. *Book of Abstracts*, First International Symposium on Medicinal, Aromatic and Nutraceutical Plants from Mountainous Areas, July 6–9, 2011, Saas-Fee, Switzerland, p. 80.

Schittko, U. 2004. Pflanzenporträt: Rosenwurz (*Rhodiola rosea* L.). *Z. Arznei- Gewurzpflanzen*, 9(1):42,3.

Schippmann, U., Leaman, D., Cunningham, A. B. 2006. A comparison of cultivation and wild collection of medicinal and aromatic plants udder sustainable aspects. In: Bogers, R. J., Craker, L. E. and Lange, D. (eds.). *Medicinal and Aromatic Plants: Agricultural, Commercial, Ecological, Legal, Pharmacological and Social Aspects*. Wageningen UR Frontis Series, Volume 17. Springer, Dordrecht, The Netherlands, pp. 75–92.

Shatar, S., Adamns, R. P., Koenig, W. 2007. Comparative studies of the essential oil of *Rhodiola rosea* L. from Mongolia. *Journal of Essential Oil Research* 19(3):215–7.

Shevtsov, V. A., Zholus, B. I., Shervarly, V. I., Vol'skij, V. B., Korovin, Y. P., Khristich M. P., Roslyakova, N. A., Wikman G. 2003. A randomized trial of two different doses of a SHR-5 *Rhodiola rosea* extract versus placebo and control of capacity for mental work. *Phytomedicine* 10:95–105.

Siivari, J., Galambosi, B. 2007. Processing of *Rhodiola rosea* and *Bergenia crassifolia* raw materials for dry extracts. In: Beneficial health substances from berries and minor crops. *NJF Report 3*, 1:20. http://www.njf.nu/filebank/files/20070409$210108$fil$3BGHMD xHaPL0C49DUdoA.pdf.

Skarpanska-Stejnborn, A., Pilaczynska-Szczesniak, L., Basta,P., Deskur-Smielecka, E. 2011. The influence of supplementation with *Rhodiola rosea* L. extract on selected redox parameters in professional rowers. *International Journal of Sport Nutrition Exercise Metabolism* 21(2):124–34.

Slacanin, I. 2011. Development of extraction and rapid analytical methods for the selection of medicinal plants. *Book of Abstracts*, First International Symposium on Medicinal, Aromatic and Nutraceutical Plants from Mountainous Areas, July 6–9, 2011, Saas-Fee, Switzerland, p. 29.

Small, E., Catling, M. P. 1999. *Rhodiola rosea* (L.) Scop. Roseroot. In: *Canadian Medicinal Crops*. NRC Research Press, Ottawa, Ontario, Canada. pp. 134–9.

Sokolov, S. Y., Ivashin, V. M., Zapesochnaya, G. G., Kurkin, V. A., Shavlinskiy, A. N. 1985. Study of neurotropic activity of new substances isolated from *Rhodiola rosea*. *Khimiko-Farmatsevticheskii Zhurnal* 19:1367–71.

Spasov, A. A., Mandrikov, V. B., Mironova, I. A. 2000a. The effect of the preparation rhodiosin on the psychophysiological and physical adaptation of students to an academic load. *Eksperimental'naia i Klinicheskaia Farmakologiia* 63(1):76–8.

Staneva, J., Todorova, M., Neykov, N., Evstatieva, L. 2009. Ultrasonically assisted extraction of total phenols and flavonoids from *Rhodiola rosea*. *Natural Product Communications* 4(7):935–8.

Stephenson, R. 1994. *Sedum: Cultivated Stonecrops*. Timber Press, Inc., Portland, O. R. p. 335.

Tasheva, K., Kosturkova, G. 2011a. Establishment of callus cultures of *Rhodiola rosea* Bulgarian ecotype. *Book of Abstracts*, First International Symposium on Medicinal, Aromatic and Nutraceutical Plants from Mountainous Areas, July 6–9, 2011, Saas-Fee, Switzerland p. 37.

Tasheva, K., Kosturkova, G. 2011b. The role of biotechnology for conservation and biologically active substances production of *Rhodiola rosea*: Endangered medicinal species. *The ScientificWorld Journal*, 2012, Article ID 274942, 13.

Tasheva, K., Trendafilova, A., Kosturkova, G. 2011. Antioxidant activities of Bulgarian Golden root- endangered medicinal species. Book of Abstracts, First International Symposium on Medicinal, Aromatic and Nutraceutical Plants from Mountainous Areas, July 6–9, 2011, Saas-Fee, Switzerland, p. 76.

Thomsen, M. G. 2012. Breeding and selection of high quality plants of *Rhodiola rosea*. *Reviews of Clinical Pharmacology and Plant Drug Therapy*. TOM 10/2012/2. Abstracts Phytopharm 2012. M 206.

Thomsen, M. G., Galambosi, B., Galambosi, Z., Uusitalo, M., Heinonen, A. 2011. Post harvest treatment of *Rhodiola rosea*. Book of Abstracts, First International Symposium on Medicinal, Aromatic and Nutraceutical Plants from Mountainous Areas, July 6–9, 2011, Saas-Fee, Switzerland, p. 43. (*Acta Horticulturae*, in press).

Tikhonova, V. L., Kruzhalina, N., Shugayeva, E. V. 1997. The effect of freezing on the viability of seeds of some cultivated medicinal plants. *Rastitel'nye Resursy* 33(1):68–74.

Todorova, M., Evstatieva, L., Platicanov, S., Kuleva, L. 2006. Chemical composition of essential oil from Bulgarian *Rhodiola rosea* L. rhizomes. *Journal of Essential Oil-Bearing Plants* 9(3):267–71.

Tolonen, A. 2003. Analysis of secondary menabolites in plant and cell culture tissue of *Hypericum perforatum* L. and *Rhodiola rosea* L. Department of Chemistry, University of Oulu. (Dissertation) http://herkules.oulu.fi/isbn9514281610/.

Tolonen, A., György, Z., Jalonen, J., Neubauer, P., Hohtola, A. 2004. LC/MS/MS identification of glycosides produced by biotransformation of cinnamyl alcohol in *Rhodiola rosea* compact callus aggregates. *Biomedical Chromatography*, 18:550–8. IF 1,069.

Tolonen, A., Pakonen, M., Hohtola, A., Jalonen, J. 2003. Phenylpropanoid glycosides from *Rhodiola rosea*. *Chemical and Pharmaceutical Bulletin* 51(4):467–70.

Toma, O. 2009. Contributions to *Rhodiola rosea* L. biochemical matrix of soluble proteins identification. Analele Ştiinţificeale Universităţii' AlI Cuza'dinIaşi (SerieNouă), Secţiunea IIa. *Geneticăşl Biologie Moleculară* 10(1):53–6.

Tsikov, P. S. (ed.). 1980. Atlas of areas and resources of medicinal plants of Soviet Union. Kartographia, Moskva. p. 340.

Tuominen, L., Tuominen, M., Galambosi, B. 1999. *Luonnon yrttien viljelyopas*. Arktiset Aromit ry. p.64.

Väre, H., Ohenoja, M., Halonen, P. 1998. Ruusujuuri ja muuta mukavaa Pudasjärven Ruskeakalliolla. *Lutukka* 14:115–8.

Virey, J. J. 1819. *Traité de pharmacie*. Chez Rémont – Paris, 1819. (cit: Hedman, 2000).

Vouillamoz, J. E., Carron, C.-A., Baroffo, C. A., Varlen, C. 2011. *Rhodiola rosea* L. "Mattmark," the first synthetic variety is launched in Switzerland. *Book of Abstracts*, First International Symposium on Medicinal, Aromatic and Nutraceutical Plants from Mountainous Areas, July 6–9, 2011, Saas-Fee, Switzerland. p. 26.

Weglarz, Z., Przybyl, J. L., Geszprych, A. 2008. Roseroot (*Rhodiola rosea* L.): Effect of internal and external factors on accumulation of biologically active compounds. In: Ramawat, K. G., Merillon, J. M. (eds.). *Bioactive Molecules and Medicinal plants*. Springler, Berlin, Heidelberg. pp. 297–315.

Xia, T., Chen, S., Chen, S., Zhang, D., Zhang, D., Gao, Q., Ge, X. 2007. ISSR analysis of genetic diversity of the Qinghai-Tibet Plateau endemic *Rhodiola chrysanthemifolia* (Crassulacea). *Biochemical Systematics and Ecology* 35:209–14.

Yan, T.-F., Zu, Y.-G., Yan, X.-F., and Zhou, F.-J. 2003. Genetic structure of endangered *Rhodiola sachalinensis*. *Conservation Genetics* 4(2):213–8.

Zapesochnaya, G. G., Kurkin, V. A., Boyko, V. P., Kolkhir, V. K. 1995. Phenylpropanoids– promising biologically active compounds of medicinal plants. *Khimiko-Farmatsevticheskii Zhurnal* 29:47–50.

Zini, E., Piovan, S., Filippini, R., Komjanc, M., Caniato, R. 2007. DNA analysis in *Rhodiola rosea* populations from North Italy's Alps and in callus culture. *Book of Abstracts*, 11th International Congress Phytopharm 2007, June 27–30, Leiden, The Netherlands, p. 157.

Zych, M., Furmanova M., Krajenska-Patan, A., Lowicka A., Dreger, M, Mendlewska, S. 2005. Micropropagation of *Rhodiola kirillovii* plant using encapsulated axillary buds and callus. *Acta Biologica Cracoviensia. Series Botanica* 47/2:83–7.

FIGURE 5.2 *Rhodiola rosea* plant growing in Edmonton, Alberta, one week after bud break.

FIGURE 5.14 Adult *Rhodiola* scolytine *Dryocoetes krivolutzkajae* Mandelshtam.

(a) (b)

(c)

FIGURE 6.5 Crown rot caused by *Fusarium* spp. on *Rhodiola*, leading to wilting of the leaves and discoloration (a) and discoloration of the root tissue (b). Crown rot caused by *Sclerotinia* is shown in (c).

FIGURE 6.7 *Rhodiola* seedlings with whitish mycelium and spores of the powdery mildew fungus covering the young leaf surfaces.

(a)   (b)   (c)

FIGURE 6.8 *Botrytis cinerea* causes the development of gray-brown lesions on the leaves of *Rhodiola* (a). The fungus first colonizes damaged or senescing plant tissue, such as old flowers, from which it spreads to healthy tissue (b and c).

FIGURE 6.12 Purplish reddening along the leaf margins and tips of a *Rhodiola* plant infected with aster yellow phytoplasma.

(a)

(b)

FIGURE 6.15 Phytoplasma infection of *Rhodiola* causes distortion and virescence (greening) of ray and disk florets (a). A normal flower is shown in (b).

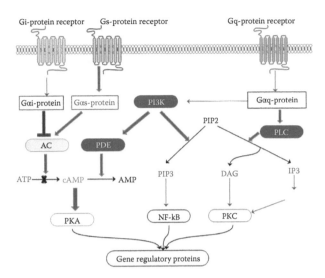

FIGURE 9.2 Hypothetic molecular mechanisms by which *Rhodiola* activate adaptive stress response pathways. Neurons normally receive signals from multiple extracellular stressors that activate adaptive cellular signaling pathways, for example, many neurotransmitters activate GTP-binding protein coupled receptors (GPCR). The receptors in turn activate kinase cascades including those that activate protein kinase C (PKC), protein kinase A (PKA), and phosphatidylinositol-3-kinase (PI3K). Effect of *Rhodiola* on G-protein-coupled receptors pathways: upregulated genes are represented in red, downregulated in blue. The Gs alpha subunit (or Gs protein) stimulates the cAMP-dependent pathway by activating adenylate cyclase. Gi alpha subunit (or Gi/G0 or Giprotein) inhibits the production of cAMP from ATP. DAG, diacylglycerol; IP3, inositoltriphosphate; PLC, phospholipase C.

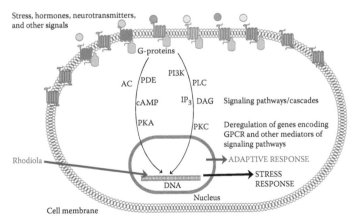

FIGURE 9.3 Evidence suggests that *Rhodiola* SHR-5 extract may initiate the activation or suppression of some genes that encode the expression of some GPCRs and key mediators of GPCR-signaling pathways. By reducing the expression of specific GPCRs, SHR-5 decreases cellular sensitivity to stress and increases stress resilience when the individual is exposed to different kinds of stressors, including emotional, physical, heat, chemical, toxic, infectious, malignant, etc.

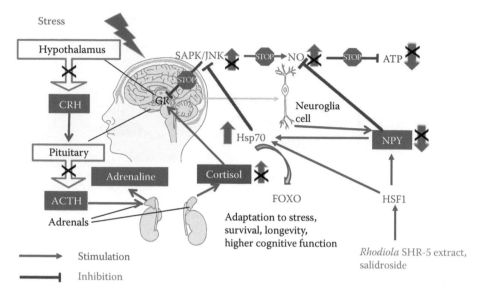

Stress

Hypothalamus

SAPK/JNK — STOP → NO — STOP → ATP

CRH

GR

Neuroglia cell

Hsp70

NPY

Pituitary

Adrenaline

Cortisol

ACTH

FOXO

HSF1

Adrenals

Adaptation to stress, survival, longevity, higher cognitive function

→ Stimulation

⊣ Inhibition

Rhodiola SHR-5 extract, salidroside

FIGURE 9.6 Hypothetical neuroendocrine mechanism of stress protection by *Rhodiola* SHR-5 extract and salidroside. Stress induces CRH release from the hypothalamus followed by ACTH release from the pituitary, which simulates release of adrenal hormones and neuropeptide Y (NPY) to mobilize energy resources and cope with the stress. Feedback regulation of overreaction is initiated by cortisol release from the adrenal cortex, followed by binding to glucocorticoid receptors (GR) in the brain. This signal stops the further release of brain hormones and brings the stress-induced increase of cortisol down to normal levels. While brief and mild stress (eustress/challenge) is essential to life, severe stress (distress/overload) is associated with extensive generation of oxygen-free radicals, including nitric oxide (NO), which is known to inhibit ATP formation (energy providing molecules). Stress-activated protein kinases (SAPK/JNK/MAPK) inhibit GR, consequently feedback downregulation is blocked and cortisol content in the blood remains high during fatigue, depression, impaired memory, impaired concentration and other stressful conditions. Adaptogens normalize stress-induced elevated levels of cortisol and other extra- and intracellular mediators of stress response, such as elevated NO, SAPK via upregulation of expression of NPY, heat shock factor (HSF-1) and heat shock proteins Hsp70, which are known to inhibit SAPK. Consequently, NO generation is reduced and ATP production is no longer suppressed. Hsp70 functions intracellularly to enhance anti-apoptotic mechanisms protect proteins against mitochondria-generated oxygen-containing radicals, including nitric oxide, and superoxide anion. The released Hsp70 acts as an endogenous danger signal and plays an important role in immune stimulation. While released NPY plays a crucial role in the HPA axis and maintains energy balance, both NPY and Hsp70 are directly involved in cellular adaptation to stress, increased survival, enhanced longevity and improved cognitive function. Hsp70 inhibits the FOXO transcription factor, playing an important role in adaptation to stress and longevity. These pathways contribute to adaptogenic effects: antifatigue, increased attention and improved cognitive function.

5 *Rhodiola rosea* Cultivation in Canada and Alaska

Kwesi Ampong-Nyarko

CONTENTS

5.1 Introduction 126
5.2 Morphology, Growth, and Development of *Rhodiola* Plant 126
    5.2.1 Anatomical Structure 126
    5.2.2 Biomass Partitioning in *Rhodiola* Plant 127
5.3 Climatic Effects on *Rhodiola* Growth 128
    5.3.1 Day Length 128
    5.3.2 Temperature 130
    5.3.3 Hail Storms 131
5.4 Ecophysiology 132
    5.4.1 Photosynthesis 132
    5.4.2 Transpiration 133
    5.4.3 Respiration 133
    5.4.4 Water Use 134
5.5 Crop Nutrition and Fertility Management 135
    5.5.1 Soils 135
    5.5.2 Effect of Soil pH 136
    5.5.3 Nutrient Uptake and Removal 137
    5.5.4 Fertilizer Use and Placement 137
5.6 Production Systems 138
    5.6.1 Seeding and Propagation Methods 138
    5.6.2 *In Vitro* Propagated Plants 138
    5.6.3 Land Preparation 138
    5.6.4 Plasticulture 139
    5.6.5 Bare Soil Production 139
    5.6.6 Container Production 139
    5.6.7 Time of Transplanting 140
    5.6.8 Demand for Labor 140
5.7 Weeds and Weed Control 140
    5.7.1 *Rhodiola*–Weed Competition 140
    5.7.2 Critical Period of Weed Competition 142
    5.7.3 Weed Control 143

5.8   Time to Harvest 143
5.9   Pest of *Rhodiola* 145
        5.9.1   Aphids.. 145
        5.9.2   *Rhodiola Scolytine*.. 146
        5.9.3   Black Vine Weevil 147
        5.9.4   Meadow Spittlebug 147
5.10   Post Harvest... 147
        5.10.1 Drying.. 147
        5.10.2 Economics of Production.. 149
5.11   Research Priorities... 149
        5.11.1 Constraints to High Yields under Cultivation 149
Acknowledgments.. 150
References.. 150

5.1   INTRODUCTION

Rhodiola rosea in commerce is traditionally sourced from wild crafting. In recent years, attempts have been made to bring the crop into cultivation (Galambosi 2006). In North America, Alberta, and Alaska, it has been at the forefront of this emerging crop industry. In 2012, *Rhodiola* under cultivation occupied about 20 ha in the province of Alberta, Canada, and about 3 ha in Alaska, United States. Besides its importance as natural health product, *Rhodiola* provides full or part-time employment to over 150 growers in Alberta and over 20 growers in Alaska. In Alberta, *Rhodiola* is grown in the Boreal Plains and the northern part of the Prairies ecozone. In Alaska, growers are located in the MatSu Valley, Bethel, Delta Junction, Trapper Creek, Homer, Chickaloon, Fairbanks, Nenana, and Anchor Point (Illig 2011). The *Rhodiola* growing areas in Alberta have a continental, semiarid climate. Precipitation from May to August varies from slightly below 200 mm to more than 325 mm in the mountains. Temperatures are generally higher in southern than northern Alberta. The July average daily temperature ranges from warmer than 18°C in the south to cooler than 13°C in the Rocky Mountains and the north. Arctic air masses in the winter produce extreme minimum temperatures varying from cooler than −54°C in northern Alberta to −46°C in southern Alberta. Annual bright sunshine totals range between 1900 and 2500 h/year (Chetner et al. 2003). The normal climate for Edmonton and High Level in Alberta and Palmer Alaska is given in Table 5.1.

5.2   MORPHOLOGY, GROWTH, AND DEVELOPMENT OF *RHODIOLA* PLANT

5.2.1   ANATOMICAL STRUCTURE

Anatomical structure of the rhizome and stem has been described by Costică et al. (2007). Structure of rhizome in cross section shows that the protective tissue is represented by six to eight layers of cork and just one to two layers of phelloderm. The cortical parenchyma is formed of 12–20 layers of oval or less tangential elongated cells.

TABLE 5.1
Normal Climate for Three Locations in Alberta and Alaska (1971–2000)

	Edmonton Alberta			High Level (Alberta)			Palmer (Alaska)		
	Mean Low (°C)	Mean High (°C)	Precipi-tation (mm)	Mean Low (°C)	Mean High (°C)	Precipi-tation (mm)	Mean Low (°C)	Mean High (°C)	Precipi-tation (mm)
Jan	−19.1	−8	22.7	−27.2	−16	20.3	−14.4	−5.5	22.3
Feb	−16.3	−4.7	13	−24	−10.8	17.7	−12.2	−2.7	20
Mar	−9.9	1	16	−16.8	−2.4	17.3	−8.3	1.6	16.7
Apr	−2.2	10.7	26.3	−4.9	8.8	15.6	−2.2	7.7	12.4
May	3.4	17.4	49.9	2.3	16.8	36.4	3.3	14.4	17.8
Jun	7.7	20.5	87.4	7.4	21.1	58.3	7.7	18.3	33.7
Jul	9.5	22.2	95.2	9.5	22.8	65.5	9.4	19.4	53.3
Aug	8.3	21.7	70.3	7.4	21	50.6	8.3	18.3	58.9
Sep	3.3	16.9	47.1	1.8	14.9	34.4	4.4	13.3	63.2
Oct	−2.4	10.9	19.8	−4.3	5.7	29.7	−3.3	5.5	39.3
Nov	−11	−0.4	17.7	−16.9	−7.7	27.4	−10.5	−2.2	26.9
Dec	−16.7	−5.9	17.3	−25.2	−14.2	20.9	−12.7	−4.4	28.2

Edmonton Alberta Latitude: 53°19′00.000″ N; Longitude: 113°35′00″ W; Elevation: 723.30 m.
High Level (Alberta) Latitude: 58°37′17.000″ N; Longitude: 117°09′53″; W Elevation: 338.30 m.
Palmer (Alaska) Latitude: 61°35′59″ N; Longitude: 149°6′46″ W; Elevation: 72.8 m.

The cells from the first three to five layers have the walls thicker than the cells from the inner cortex. The central cylinder has secondary structure. The conductive tissues are disposed on a ring, being separated by narrow parenchyma rays. Conductive elements of phloem and xylem are relatively less represented, being radial spread in a mass of conductive parenchyma. In the center of the organ, there is extensive parenchyma pith. For the aerial stalks, the cross section has an oval contour. The epidermis is unilayered, having cells with outer walls covered by a thin cuticle. The cortex is a thick, homogenous parenchyma from 12 to 14 layers of cells. The first one to three layers of cells have their inner and outer walls thicker than lateral ones. Conductive tissues are on a ring. The phloem is formed of conductive elements of small dimensions. The secondary xylem is formed from conductive vessels with thick, lignified walls. The primary xylem has less conductive elements with less lignified walls. The pith, in the center of the vegetative organ, is in a very large proportion disorganized. The upper and lower epidermises are composed of irregular shaped cells with numerous simple pits in the walls. The leaf lamina is amphystomatic, having stomata of anizocitous type. The stem is straight, 10–35 cm in height with only a few leaves, cylindrical in shape.

5.2.2 BIOMASS PARTITIONING IN *RHODIOLA* PLANT

Plants that naturally occur on nutrient-poor sites have inherently lower growth rates when compared to plants from more favorable habitats (Lambers et al. 1995). Genetic variation in plant growth rate is predominantly due to variation in the pattern of

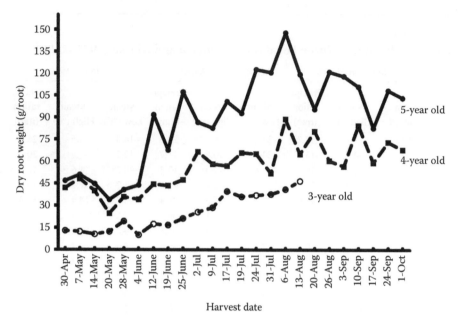

FIGURE 5.1 Biomass of three-, four-, and five-year-old *Rhodiola rosea* plants throughout the growing season.

biomass allocation in leaves, stems, and roots: fast-growing genotypes have more leaf area per unit plant weight, but the leaves have a low biomass density. *Rhodiola* root growth in the first 2 years is slow. In field trials in Alberta, in third-year plants, 74%–80% of the root yield occurred by the end of June. Consequently, this period contributes most to final yield. In fourth- and fifth-year plants, an additional growth phase for root dry matter accumulation was identified. In this growth phase, root dry matter declined by 30% (Figures 5.1 and 5.2). Root dry matter then started to increase during a 3-week constant growth phase until it reached 75% of its final dry biomass at maturity. Organic food materials stored in rhizomes from previous year's growth are mobilized to support the rapid growth in spring. This accounts for the root mass reduction in weight every spring as it is used as a food source for new spring growth before it begins to increase again. If environmental stress adversely affects the foliage, new growth might not be able to compensate for the reduction in root mass.

5.3   CLIMATIC EFFECTS ON *RHODIOLA* GROWTH

5.3.1   DAY LENGTH

The natural day length or photoperiod affects the growth of *R. rosea* plants. The day length pattern during the *Rhodiola* growing season varies most in high latitudes in Palmer and High Level (Figure 5.3). Optimum temperature together with long photoperiod favors the growth of *R. rosea*. Short photoperiods induced growth cessation and the formation of resting buds in 3- or 11-month-old plants (Figure 5.4). Long photoperiods resulted in immediate growth activation of dormant buds. A photoperiod of 18 hours was

FIGURE 5.2 *Rhodiola rosea* growth stages during the season at Edmonton, Alberta (a) April 22; (b) May 08; (c) May 21; (d) June 16; (e) July 09; (f) July 21.

the optimum for root, leaf biomass, and leaf area expansion. Plants exposed to 24-hour photoperiod on the other hand had high number of dormant buds, reduced leaf area, root, stem, and leaf biomass. Phytochrome, a chromoprotein, is influenced by light, and it is mainly produced in darkness (Sengbusch 2002). *R. rosea* needs a dark period during which the pool of phytochrome is replenished. High temperature of 25°C also induced dormancy in plants (Figure 5.5). In *Sedum telephium*, no chilling was required for dormancy release, even in plants induced to dormancy and maintained at high temperature

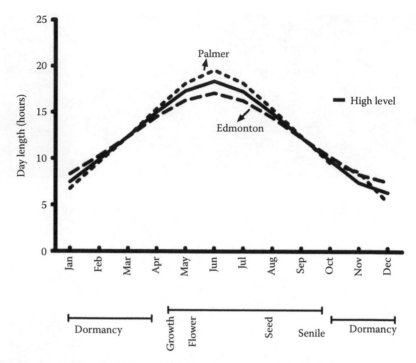

FIGURE 5.3 Day length at Edmonton, High Prairie, and Palmer and *Rhodiola rosea* crop growth cycle at Edmonton.

(21°C) for more than 3 months (Heide 2001). The critical photoperiod for dormancy release was about 15 hours, a minimum of four long-day cycles (24 hours) being required. The ecophysiological significance of photoperiodic control of dormancy is that the mechanism ensures stability of winter dormancy, even under conditions of climatic warming.

5.3.2 TEMPERATURE

Temperature regime greatly influences the growth duration and growth pattern of the *R. rosea* plants. Plant metabolic rates are dependent on temperature and explain how plant growth rates vary with temperature. Table 5.2 shows the optimal temperatures for germination, bud break, and photosynthesis in *R rosea*. After bud break, *Rhodiola* is able to tolerate freezing temperatures of –2.2°C as the average low temperatures in April in central Alberta is –4.9°C. The cold resistance of perennials of temperate and polar regions undergoes seasonal changes, with higher resistance in the winter months than in summer. The increase in hardiness that occurs in the autumn is stimulated by falling temperatures and shortening day length. High temperatures above the optimum occur periodically in the summer months. The thermal death point of a plant tissue of temperate plants generally follows exposure for a few hours at temperatures between 45°C and 55°C (Wickens 2001). In many tuber crops, low night temperature favors root formation and high temperature during the day favors vegetative development. However, this has not been studied in *R. rosea*.

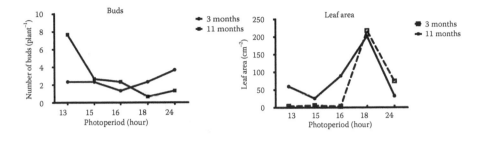

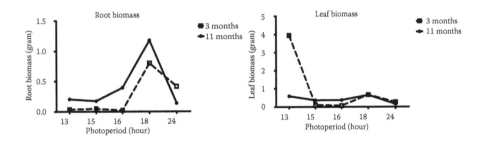

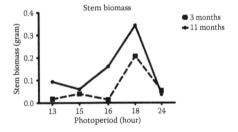

FIGURE 5.4   Effect of photoperiod on *Rhodiola rosea* leaf area, number of buds, and biomass.

5.3.3   HAIL STORMS

Some areas of production in Alberta lie on the edge of Alberta's hailstorm alley and may be prone to damaging hailstorms every few years. Alberta gets more hail than anywhere else in Canada, with most hailstorms forming over the foothills of the Rocky Mountain. The three key ingredients for hailstorms are soil moisture, surface heating, and a triggering mechanism, such as an approaching weather system or a dry breeze that flows down from the mountains, clashing with moisture over the foothills to trigger a storm (Alberta Agriculture and Rural Development 2011). Hail can destroy crops, particularly seedlings, slicing plants to ribbons in a matter of minutes. Crops so damaged recover and put up new shoots if it occurs early in the season and will grow again the following year.

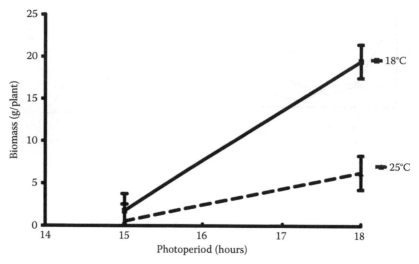

FIGURE 5.5   Effect of temperature and photoperiod on *Rhodiola rosea* growth.

TABLE 5.2
Response of the *Rhodiola rosea* Plant to Temperature at Various Growth Stages

Growth Stage	Low	High	Optimum Temperature
Seed germination	4°C	30°C	
Seedling emergence and establishment		25°C	18°C
Bud break	–2.2°C	4°C	10.7°C
Photosynthesis	5°C–7°C	25°C–30°C	8°C–18°C

5.4   ECOPHYSIOLOGY

5.4.1   PHOTOSYNTHESIS

Crassulacean plants are adapted during evolution to the shortage of water, high temperature, and high insolation. Several adaptive mechanisms protect the photosynthetic apparatus from injury by adverse environmental factors. These mechanisms include structural and functional traits, such as reduction of leaf surface area, the occurrence of thick cuticle, the presence of specialized water-retaining tissue, small-sized chloroplasts, low content of photosynthetic pigments, and specific photosynthetic metabolism known as Crassulacean acid metabolism (CAM) (Luttge 2004; Mamushina and Zubkova 2005). The leaf mesophyll in *R. rosea* is clearly differentiated into palisade and spongy tissues. The content of chlorophyll was determined as 5–7 mg/g dry weight and carotenoids at 1.5–2.0 mg/g dry weight in *R. rosea* leaves. The rate of CO_2 net uptake in Crassulacean species depends on the mesostructure and correlates with the content of pigments and soluble carbohydrates. The photosynthetic rate in *R. rosea* under optimal irradiance and temperature attained the value of 40 mg/g dry weight. The temperature optimum for photosynthesis of *R. rosea* was observed at 8°C–18°C. At chilling temperatures (5°C–7°C), the leaves of *R. rosea* retained 50% of their maximal photosynthetic rate. In *R. rosea*, the rate of photosynthetic electron flow was depressed at high irradiances and temperatures that were supraoptimal for net photosynthesis. It was concluded that the photosynthetic apparatus of *R. rosea* species is well adapted to moderate and chilling temperatures. The content of soluble carbohydrates in *Rhodiola* leaves accounted for about 8% of leaf dry matter (Golovko et al. 2008). The mesophyll differentiation and comparatively small fractional content of chlorophylls integrated into light-harvesting complex (less than 50%) is the basis to classify *Rhodiola* as a light-demanding plant. At the same time, a small number of large chloroplasts, as well as low values of light compensation point, indicate the shade tolerance of this plant species (Golovko et al. 2008). Crassulacean plants, especially *Rhodiola*, photosynthesize actively at chilling and moderate temperatures, which ensures the metabolism level required for pursuing their life strategy under cold climate conditions (Golovko et al. 2008).

5.4.2 TRANSPIRATION

Young rapidly expanding leaves manifested a high respiration rate, which usually decreased with plant age. The age-dependent changes in the respiration rate are generally related to the decrease in the activity of the cytochrome pathway in mature leaves. The alternate pathway contributed 30%–40% to the respiration of young leaves and the twice-lower value in the mature leaves (Golovko and Pystina 2001). Growth form is regarded as a proxy for transpirational water loss (Korner et al. 1979). Morphological, anatomical, and physiological characters, specific to each growth form, have a direct influence on the rate at which water moves along the soil–plant–atmosphere continuum and the consequent rate of transpirational water loss. A study summarizing maximum leaf conductance values of 246 plant species found that succulent plants (including *Sedum* spp.) had the lowest maximum leaf conductance values of 13 morphologically and/or ecologically distinct plant groups (Korner et al. 1979). CAM is also associated with various anatomical and morphological features that minimize water loss, including thick cuticles, low surface-to-volume ratios, large cells and vacuoles with enhanced water storage capacity (i.e., succulence), and reduced stomata size and frequency (Zotz and Hietz 2001). Many species may exhibit facultative CAM metabolism, thus transpiration rates may depend greatly on soil moisture conditions (Villarreal and Bengtsson 2005).

5.4.3 RESPIRATION

Respiration, as a source of metabolites and energy, is closely related to growth and plant tolerance to adverse environmental factors. Therefore, the study of respiration in arctic plants is important for resolving a number of ecological and physiological problems, such as productivity of wild and cultivated plants grown in northern latitudes and carbon turnover under low temperature conditions (Semikhatova et al. 2007). Respiration rate in *R. rosea* was measured as 0.76 nmol CO_2/g fresh weight/s and followed the Stocker's rule, that is, respiration rates of identical species were equal at average temperatures of their natural habitats (Semikhatova et al. 1992). The high respiration rate in plants native to cold habitats is caused largely by structural changes in leaves: enlarged cells, comparatively small specific volume of vacuoles, enlarged mitochondria, and elevated frequency of mitochondria (Semikhatova et al. 1992; Larcher 2004). Plant respiration can consume an appreciable amount of the carbon fixed each day during photosynthesis over and above the losses due to photorespiration (Siedow et al. 2010). Attempts to establish a quantitative relationship between respiratory energy metabolism and the various processes going on in the cell have led to a breakup of respiration into growth respiration and maintenance respiration (Lambers 1985). Utilization of energy by maintenance respiration is not well understood, but estimates indicate that it can represent more than 50% of the total respiratory flux. Although numerous questions remain regarding these issues, there are several empirical examples of relations between plant respiration rates and crop yield (Wilson and Jones 1982). The effect of growth conditions and plant age on the relationships between respiratory pathways was investigated in *R. rosea* by Golovko and Pystina (2001). The alternative pathway contributed 0%–50% to the leaf respiration and 15%–20% to the respiration of mature leaves, and in the young rapidly expanding leaves, the contribution was twice higher. It was concluded that a high alternative pathway contribution to the respiration of leaves correlates with their rapid growth and that a high supply of respiratory substrates is one of prerequisite for the alternative pathway activation (Golovko and Pystina 2001).

Rhodiola is a perennial rhizomatous biomorph with annual growth cycles in which 7-month periods of formation and functioning of the aboveground organs regularly alternate with long periods characterized by the absence of any external manifestations of growth. Some researchers are inclined to believe that geoephemeroids have no period of dormancy in the literal sense. For instance, in the genus *Fritillaria*, hidden growth and developmental processes in the vegetative disseminules (bulbs, tubers) usually do not cease during this period. The preliminary formation of the vegetative or monocarpic shoot of the next year (so-called preformation) takes place, which is most typical of tundra and alpine plants (Billings and Mooney 1968; Skripchinskii and Skripchinskii 1976). All plants of cold climates exhibit substantial respiration at 0°C and need temperatures of –5°C to –10°C before rates approach zero (Körner 2003).

5.4.4 WATER USE

The formative water in plant is obtained mainly from the soil through absorption by the plant roots. Water stress affects almost all plant functions including photosynthesis and roots ability to take up nutrients. In general, Sedum has been found to have great drought tolerance. Many species tolerated up to 1 month without precipitation, and a few *Sedum* spp. endured as long as 4 months without water in greenhouse experiments (Durhman et al. 2004, 2006; VanWoert et al. 2005). In green

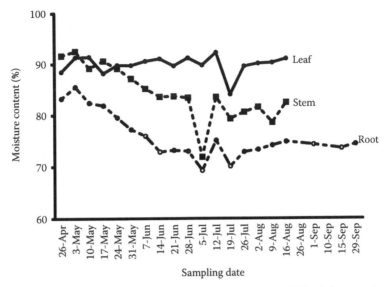

FIGURE 5.6 Plant water content in different plant structures of *Rhodiola rosea* in annual crop cycle.

roof microcosm studies, *R. rosea* survived the 2-month experiment when plants were watered at 24 days interval (Wolf and Lundholm 2008). *Rhodiola* plant water content varies within various plant structures and varies during the growth period (Figure 5.6). One field experiment was established in Alberta to determine supplementary irrigation requirements for *R. rosea* production. Plastic mulch appeared to conserve water in the plots and *R. rosea* benefited from supplementary irrigation only when no mulch was used. *Rhodiola* seedling establishment was reduced by 53% and there was significant reduction in root biomass when there was no supplementary irrigation in the non-mulched plots. At planting, it is important to have moist soil in order to obtain a good seedling establishment after transplanting. Where the normal rainfall during the growing season is about 175 mm or less, and distribution is erratic, supplementary irrigation may be necessary.

5.5   CROP NUTRITION AND FERTILITY MANAGEMENT

5.5.1   Soils

R. rosea is grown on a variety of soils ranging from poorly to well drain. The typical prairie soil, chernozemic soil order, and luvisolic soils are the most commonly used for growing *Rhodiola* in Alberta (Alberta Agriculture and Rural Development 2005). Gelisols describes the majority of soil in the Alaskan tundra where *Rhodiola* cultivation occurs. As the soil provides water, air, nutrients, and stability, its physical and chemical properties determine its potential for *R. rosea* production. Specific chemical fertility and physical conditions are required by each crop to maximize yield potential (Mahloch 1974). Soil texture, sand, silt, and potassium were functionally the most important soil parameters affecting *R. rosea* root yield in Alberta (Ampong-Nyarko et al. 2011). Texture, sand, and silt had a positive relationship with *R. rosea* root yield. Texture refers to the size distribution of the primary mineral particles in the soil and is quantified by the proportion of sand, silt, and clay. The effect of soil texture on crop growth and quality is complex. Structural pores between soil particles influence soil compressibility, root penetration, and therefore root growth (Dexter and Tanner 1973). Soil texture provides a measure for permeability and to some extent, for water retention capacity, water infiltration, aeration, rooting ability, and workability. It further affects crops growth and nutrient absorption (Wang and Zhan 2008). Potassium levels in *R. rosea* roots were between 0.93% and 1.49% with a mean of 1.11% of the dry weight. Potassium plays a vital role in photosynthesis and the regulation of plant stomata and water use among several other functions (Zhao et al. 2001). In Alberta, *Rhodiola* plants start growth early in spring, normally when soil temperatures are still low (3.2°C). *Rhodiola* can be grown on a variety of soils. Generally, *Rhodiola* prefers a fertile, well-drained, open loam soils (Evans 1983; Ampong-Nyarko et al. 2011) though it can tolerate some damp conditions. Fertilizer requirements vary with soil type, location, and production system.

5.5.2 Effect of Soil pH

A pH range of approximately 6 to 7 promotes the most availability of plant nutrients. In a greenhouse study, *R rosea* growth was not affected in slightly acidic to moderately alkaline soils (pH 6.6–7.5) (Figure 5.7). There were no significant differences in bioactives between different pH levels.

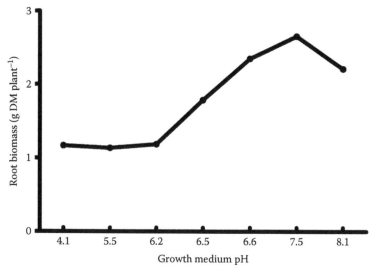

FIGURE 5.7 Effect growth medium pH on *Rhodiola rosea* root biomass.

5.5.3 Nutrient Uptake and Removal

The six macronutrients N, K, Ca, Mg, P, and S either are involved in the structure of proteins and nucleic acids or are key cations involved in charge stabilization (Evans and Edwards 2001). The average elemental composition of dried *R. rosea* roots is given in Table 5.3. The process of nutrient uptake at different growth stages is a function of climate, soil properties, amount, and method of fertilizer application. Nutrient uptake rate among others depends on root surface area, uptake properties of root surface, nutrient concentration in soil, and plant demand determined by growth rate. Nutrient uptake increases with increasing nutrient supply to some maximum levels. Estimates of crop nutrient removal are useful as a guide for *Rhodiola* nutrient demand. *Rhodiola* considering its native habitat is adapted to infertile soils. Plants from infertile soils have a low capacity to absorb phosphate (Bloom 1985). Plant strategies in infertile soils include increased absorption efficiency, increased root growth, and greater redistribution of ions within plants to produce new organs. Similarly, storage is well developed in perennial plants that occupy low resource environments both as insurance against catastrophic tissue loss and as a support for rapid growth during brief periods when conditions are favorable (Chapin et al. 1990).

5.5.4 Fertilizer Use and Placement

R. rosea did not significantly respond to nitrogen, phosphorus, potassium, or sulfur fertilizers applied to soils deficient in nitrogen and marginal in phosphorus in field trials (Figure 5.8). However, there was significant increase in root yield with foliar application of fertilizers. Nutrient use efficiency of foliar fertilizer computed as rhizome biomass produced per unit of nitrogen applied was 213. While there is a need to fine-tune a fertilizer program for *R. rosea*, it is profitable to apply foliar fertilizer annually. A foliar fertilizer concentration of 2000 ppm nitrogen, phosphorus, and potassium applied once within the first 6 weeks of spring emergence was most effective. This approach synchronizes nutrient supply

---

TABLE 5.3

Average Elemental Composition of Dried *Rhodiola rosea* Root (Observed Ranges in Parentheses) after 5 Years Growth

Element	Amount	Root Yield (kg/ha)	Estimated Nutrient Removal (kg/ha)
Nitrogen (%)	2.0 (1.5–2.8)	5,800	116
Phosphorus (µg/g)	3,286 (2,660–4,900)	5,800	19
Potassium (µg/g)	11,133 (9,280–14,900)	5,800	65
Calcium (µg/g)	7,461 (230–9,410)	5,800	43
Magnesium (µg/g)	1,171 (768–2,080)	5,800	7
Sulfur (%)	0.12 (0.10–0.15)	5,800	7
Sodium (µg/g)	123.9 (4.7–322)	5,800	0.7

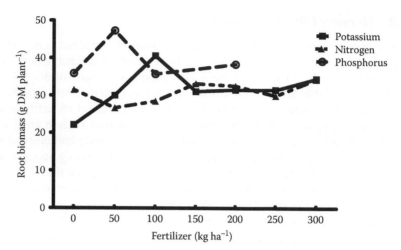

FIGURE 5.8 *Rhodiola rosea* responses to nitrogen, phosphorus, and potassium fertilizer applied to the soil.

with the plant nutrient requirements. However, the effect of fertilizer on root quality and bioactive concentration that add value to the harvested root has not been established.

5.6 PRODUCTION SYSTEMS

5.6.1 Seeding and Propagation Methods

Rhodiola can be propagated from seed or from crown division of the basal shoot. Seed germination rates can be improved by winter stratification or by pretreating seeds with gibberellic acid. Vegetative propagation has the advantages of producing a larger, fuller plant faster. It gives plants that are genetically similar to the parent, unchanged perpetuation of naturally cross-pollinated or heterozygous plants. Newly developing buds that are removed from a growing plant will root readily. The mother plant regenerates and produces new buds.

5.6.2 *In Vitro* Propagated Plants

Established *in vitro* schemes are suitable for mass propagation of *R. rosea* (Tasheva and Kosturkova 2012). Chapter 7 discusses biotechnology in detail. The whole process could take about 3 months. For this period, more than 100 regenerated plantlets could be propagated from one seedling of *Rhodiola*. Each regenerant could give three to five explants for further clonal propagation. The main drawback is the cost. It can be used for the production of stock plants that are then used for seed multiplication.

5.6.3 Land Preparation

The purpose of land preparation is to provide conditions to enhance seedling transplant establishment and to control perennial and annual weeds. Once the planting site has been selected, proper preparation is needed. This step is critical in order to reduce the labor required for in-crop weed management and to address nutrient requirements for the duration of the crop growth (Spencer et al. 2007). *Rhodiola* needs to be grown on land that is as free as possible from weeds and their seeds, especially persistent and competitive perennial weeds. Hard-to-kill perennial weeds can easily jeopardize a *Rhodiola* stand or make it very difficult to manage over the life of the crop. Land preparation includes cultivation and leveling, and bed formation and laying of plastic mulch. The soil is ripped, disked, and cultivated; finally, the land is leveled with tractor-pulled implement. Cultivation turns over the topsoil and loosens the compacted soil below to achieve a good tilth for laying plastic mulch and provides medium where storage root growth is not impeded. To reduce land preparation cost without sacrificing, farmers should limit tillage to a minimum (Spencer et al. 2007).

5.6.4 PLASTICULTURE

Field planting of *Rhodiola* is done on bare soil or by using plasticulture. Plastic mulch consists of a thin film of plastic (typically around 1.2 m in width) that is laid across a soil surface using specialized equipment. Mulch is used primarily as a tool to prevent weed growth. Mulch may be installed some time before planting once soil preparation is completed or just prior to planting. Some equipment will lay mulch and plant at the same time. Between row spacing should be adjusted to accommodate equipment that will be used for interrow cultivation. Generally, increasing planting density increases total yields per unit area. Planting densities and spacing may be limited by the equipment used for planting. Many growers plant using within-row spacing of 30 cm and 30 cm between the rows. Planting may be carried out by hand or mechanically by tractor-mounted transplanters. When transplanted by tractors, insufficient root contact with soil or wrong placement will reduce seedling establishment. Growers use transplants grown outdoors or in the greenhouse using plug trays. Plug plant is a general term for seedlings that have been started in trays of individual cells. Transplants establish better in the field because roots are not damaged in pulling.

5.6.5 BARE SOIL PRODUCTION

Production can be on flat or raised beds. *Rhodiola* plant plugs are usually planted 30 cm in rows and 75 cm between rows. Ten centimeters of clean straw mulch, wood chips, or saw dust should be applied after transplanting to control weeds (Spencer et al. 2007). The competition that results in a bare soil planting can lead to low seedling survival, and it is difficult to differentiate between *Rhodiola* seedlings and weeds, resulting in an unacceptable level of *Rhodiola* plant damage and subsequent low yields. Several types of transplanters are available to mechanize the field transplanting process.

5.6.6 CONTAINER PRODUCTION

Using container systems to grow *R. rosea* are practiced by some growers. The plants are grown in 13-L containers using high plant densities until they reach a marketable size and bioactive levels in 3 or 4 years. It offers several advantages including greater control of the environment and other agronomic production factors. This is suitable for small operations.

5.6.7 TIME OF TRANSPLANTING

Studies on dates of field transplanting of *R. rosea* were conducted in central Alberta. Seedlings were transplanted at monthly intervals from May to October. The number of *Rhodiola* plants was counted the following June after the winter. There were no significant differences in the number of plants that survived the winter. Plants can therefore be transplanted anytime during the year. It was also observed that plants transplanted in late June and early July during the hottest and driest parts of the season require frequent irrigation to achieve good plant establishment. Though *Rhodiola* can be transplanted any time during the year, there are two transplanting windows recommended. The first is in April and early May (spring), and the second in mid-August to September (autumn). Seedlings transplanted in August to September require no weeding in the year of transplanting. Spring transplanting requires extra effort in weed control and run the risk of poor establishment if transplanting is delayed. The early spring transplanting takes the advantage of moisture and melting snow. Growers observed heaving of transplanted seedlings when the crops were planted late autumn. Plant density will vary for the different systems, mechanization, and weed control methods adopted.

5.6.8 DEMAND FOR LABOR

R. rosea production is labor intensive from the point of view of weed control. Labor cost (both paid and unpaid) for production represents over half of variable costs. Weeding makes up about 80% of labor with the remainder going to planting, spraying, and irrigating (Bauer and Chaudhary 2011). Maintenance activities include irrigation, and controlling pests and diseases. Figure 5.9 shows the annual crop production cycle and associated labor demands. The drudgery of hand weeding and labor shortages has made *Rhodiola* farming unattractive for some.

5.7 WEEDS AND WEED CONTROL

5.7.1 RHODIOLA–WEED COMPETITION

R. rosea is adapted for survival in severe climate of the alpine biome. Its success depends on the ability to tolerate conditions that would kill other plants. Most alpine plants can grow in sandy and rocky soil. When they are brought under cultivation without the harsh conditions to which they are adapted, they are easily swamped by bigger plants. Competition between *Rhodiola* and weeds occurs when one of the limiting environmental resources falls short of the combined requirements of both. The degree of *Rhodiola*–weed competition depends on rainfall, *Rhodiola* plant, soil factors, weed density, length of time of *Rhodiola* and weed growth, crop age when weeds started to compete, and nutrient resources, among other variables (Ampong-Nyarko

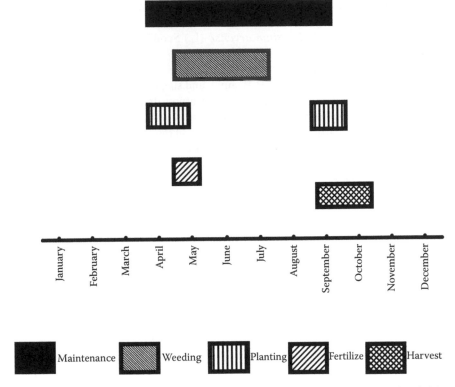

FIGURE 5.9 The annual *Rhodiola rosea* crop production cycle and associated labor demand.

TABLE 5.4
Average Dry Weight of 10 Weeds and 10 Established *R. rosea* in Black Plastic Mulch—2007 Trial Had 2 Years of Weed Pressure and 2008 Was under 3 Years of Weed Pressure

	Weed Top Dry Biomass/Plant (g)	Weed Root Dry Biomass/Plant (g)	*Rhodiola* Root Dry Biomass/Plant (g)
Weed free	0	0	184.9
Dandelion	67.2	78.4	15.0
Canada thistle	79.2	11.3	10.6
Annual weeds	23.9	13.1	20.6

and De Datta 1991). Weed management can make or break *Rhodiola* production; transplants are very small seedlings and they are slow to establish. *Rhodiola* will not compete effectively with weeds over its entire 3- to 5-year life cycle.

Yield reductions caused by uncontrolled weed growth throughout a *Rhodiola* crop have been estimated to be from 44% to 94%, depending on the culture (Table 5.4).

Weeding also increased percentage of total root bioactives by 46% in two seasons of growth. Perennial weeds such as common dandelion (*Taraxacum officinale* F.H. Wigg.) and Canada thistle (*Cirsium arvense* (L.) Scop.) can be devastating to production of *R. rosea*. Other weeds of importance in *Rhodiola* production include field sowthistle (*Sonchus arvensis* L.), lambs quarters (*Chenopodium album* L.), barnyard grass (*Echinochloa crus-galli* (L.) P. Beauv.), and shepherd's-purse (*Capsella bursapastoris* (L.) Medik). Common dandelion has features that make it a successful weed wherever *Rhodiola* is grown. These characteristics include active growth period in spring and fall, rapid growth rate, rapid seed spread rate, high seedling vigor, the tap root system, and its ability to regrow after harvest. Dandelion seeds are blown in the wind and are collected in the seedling holes of the *Rhodiola* transplants where they establish in close proximity.

5.7.2 CRITICAL PERIOD OF WEED COMPETITION

Competition between weed seedlings and transplanted seedlings of *Rhodiola* usually begins as competition for light. The effect of competition on *Rhodiola* growth is most severe during the first 8 weeks of the season where 50% of yield losses occur (Figure 5.10). The so-called critical period of weed competition lies between 21 and 35 days when weed competition is most damaging to *Rhodiola*. In the third, fourth, and fifth year, *Rhodiola* is usually among the first crop to start growing in the spring and has a head start on many annual weeds. However, root yield loss was similar in the second and third years when weeds were allowed to compete with *Rhodiola* throughout the season.

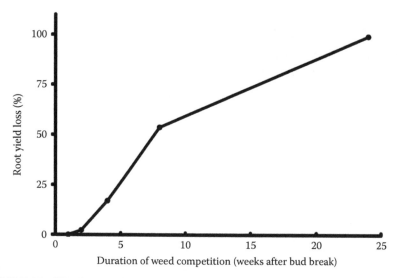

FIGURE 5.10   Weed competition and *Rhodiola rosea* root yield.

5.7.3 Weed Control

Planning is important in making appropriate decisions on weed control. Often, the decision to control is not made until the problem has become serious, when control may be uneconomical, ineffective, or even impossible. Future weed problems can be obtained by surveying and recording the weed species in a *Rhodiola* field after emergence, and at mid season. This record is useful in planning for weed control. Cultural control methods include transplanting bigger and older seedlings, transplanting in the fall, land preparation, fertilization, plasticulture, and plant population. Cultural control involves preventing the introduction, establishment, and spread of weed seeds that will establish in subsequent years. A basic principle of cultural control is to increase the competitive ability of *Rhodiola* to enable it to suppress weed growth. A vigorous *Rhodiola* crop competes more effectively with weeds than does a less vigorous crop. Age of seedlings at transplanting affects the crop competitive ability. The older the seedlings, the more competitive they are. With inadequate weed control, older seedlings are more desirable than younger seedlings. Manual weed control includes hand pulling and mechanical hand weeding. Nevertheless, manual methods are slow, unattractive, and tedious. Hand pulling controls weed seedlings growing in the same hole as *R. rosea* or near the plants. Plasticulture contributes to weed suppression. However, weeds are still a problem within the plant holes where it matters most. It has been observed that frequent removal of these weeds within the plant holes causes mechanical damage to *Rhodiola* roots and reduces yields. Herbicide use for weed control is not practiced because most growers grow *Rhodiola* as naturally grown or organic. Organic herbicides such as corn meal gluten and blend of acetic and citric acids were evaluated but were found to be ineffective (Kimmel et al. 2008). It is feasible to use directed application of glyphosate to control weeds in nonorganic production systems. The economic benefit of weed control must exceed its cost.

5.8 TIME TO HARVEST

Rhodiola plants under cultivation need 4–6 years for optimal development and accumulation of biologically active substances in the rhizomes (Brown et al. 2002; Platikanov and Evstatieva 2008). Letchamo et al. (1998) studied plants under cultivation and from natural populations in the wild. Higher root yield was obtained from cultivated plants. In a time to maturity studies in Alberta, growth of a *R. rosea* rhizome over 7 years followed a sigmoidal curve consisting of a slow starting phase in the first and second years (Figure 5.11). This was followed by a phase of rapid root expansion between the third and fifth years. After the fifth year, it reached an asymptote as a result root yield was not significantly different between the fifth, sixth, and seventh years (Figure 5.11). The mean weight of oven-dried rhizomes with roots increased to 120 g per plant in the fifth year of plant vegetation in both Alberta and central Poland. In the sixth year of vegetation, the symptoms of plant aging were noticed. The oldest, central parts of rhizomes decayed and the rhizomes divided into smaller, autonomic parts (Przybył et al. 2008). The accumulation of the main biologically active compounds, rosavin, rosarin, rosin, and salidroside, in the underground organs of *Rhodiola* plants generally followed a pattern similar to root biomass reaching optimum levels around the fifth year (Figure 5.12). However, salidroside was very high in seedling stage in the first year but fell to low levels in the second and then started to increase again with plant age (Figure 5.12). The variations in salidroside, rosarin, rosavin, and rosin concentration in *R. rosea* roots in a growing season is shown in Figure 5.13. The rhizomes were characterized by higher contents of salidroside, rosavin, rosarin, and *trans*-cinnamic alcohol in comparison with roots (Przybył et al. 2008). The results of several years of study carried in

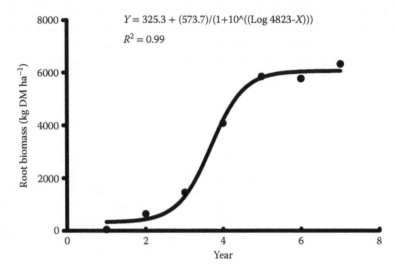

FIGURE 5.11   Effect of plant age on *Rhodiola rosea* rhizome biomass as an indicator of plant maturity.

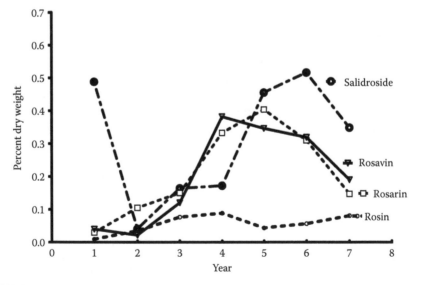

FIGURE 5.12   Effect of *Rhodiola rosea* plant age on the concentrations of rosavins, salidroside, rosarin, and rosin in rhizomes and roots.

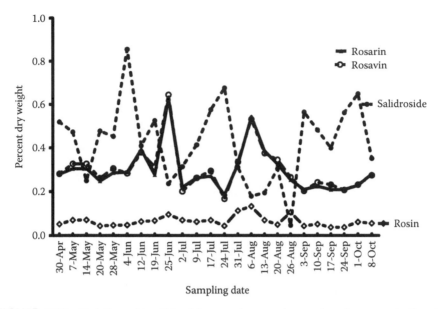

FIGURE 5.13 Variations in salidroside, rosarin, rosavins, and rosin concentration in *Rhodiola rosea* rhizomes and roots in a growing season.

Poland indicate a high intraspecific variability concerning the accumulation of these compounds. It was also stated that both the weight of the underground organs of roseroot and the content of active compounds change during plant development. The yield and quality of *R. rosea* raw material was also significantly affected by climatic and soil conditions (Węglarz et al. 2000).

In Alberta, *R. rosea* harvesting is usually done by machines, normally by a modified potato harvester. It is essential to harvest the crop when the bioactives are at their peak in the harvest year. Normally, the later in the fall, the better and harvest is normally completed early enough to avoid the ground frost. Farmers judge their harvest timing decisions by sampling *Rhodiola* roots for bioactive levels and biomass.

5.9 PEST OF *RHODIOLA*

5.9.1 APHIDS

Aphids (Insecta: Hemiptera: Aphididae) have been observed on *Rhodiola* shoots and constitute the major pest for production in Alberta. Aphid colonies are found on the undersides of the leaves or new growth and flower heads. Leaves attacked by aphids coil up; have spotty yellow discolorations, usually on the undersides; and the leaves may later dry out and wilt. Heavy infestations result in the senesce of *Rhodiola* plants, though such plants are able to regrow the following year. Potential yield loss estimate of 25% in rhizome have been estimated in trials at the Crops Diversification Centre North in Edmonton. Aphids also are constraint in seedling production. In Alberta, they are normally observed in June. In severe cases of infestation, some plants have been stunted in growth and led to early senescence. Peach aphids (*Myzus persicae*), foxglove aphid (*Aulacorthum solani*), and melon or cotton aphid (*Aphis gossypii*) have all been reported by growers. Root aphids have also been observed by growers especially in *Rhodiola* seedling production. Some of the many species of aphids attack the roots of plants and these can stunt the growth or even kill them, if the infestation is large enough.

5.9.2 *Rhodiola* Scolytine

In 2009, an interesting taxonomic problem with a new insect pest was found in native *R. rosea* in Atlantic Canada, in the specimens from Simons Cove, New Brunswick, and Labrador. The specimen was identified as *Dryocoetes krivolutzkajae* Mandelshtam by Prof. Anthony Cognato, the director of the A.J. Cook Arthropod Research Collection, Michigan State University. This new species breeds in the roots of *R. rosea*. It was first described in 2001 as a new species of bark beetles from the northeastern part of Russian Far East (Mandelshtam 2001). This is the first description in North America. There are over 135 members of the genus *Dryocoetes*, which include *D. autographus* (bark beetle), *D. betulae* (birch bark beetle), and *D. confusus* (western balsam bark beetle). Under bad infestation, *Rhodiola* rhizomes that looked healthy from outside were completely decayed when sliced open (Figure 5.14). The current approach is to prevent the pest from moving into the areas where they do not occur.

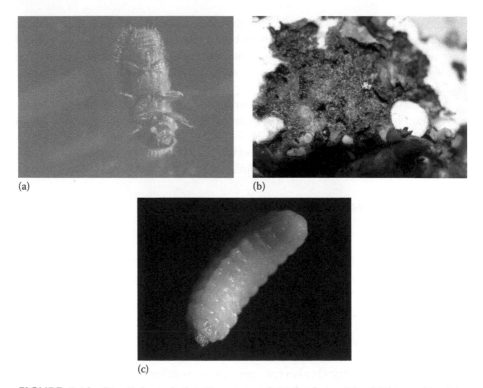

(a)   (b)

(c)

FIGURE 5.14 Rhodiola scolytine *Dryocoetes krivolutzkajae* Mandelshtam. (a) adult, (b) Rhodiola rhizomes damaged showing larva (c) Larva.

5.9.3 BLACK VINE WEEVIL

The black vine weevil (*Otiorhynchus sulcatus* F. (Coleoptera: Curculionidae) feeds on a variety of plant species and is a serious pest of ornamental nursery crops. The larval stage has a more restricted diet than the adult stage, but the larvae are more damaging because they feed on roots and often stunt or kill their hosts (Reding 2008). Vine weevil attacks wide range of plants. Far more serious is the damage caused by the soil-dwelling larvae, which are plump, white, legless grubs up to 10 mm long with pale brown heads. These larvae eat the roots and bore into tubers and stem bases of succulents, devastating many pot plants. The black vine weevil has not been found in *Rhodiola* crops in Alberta. A major problem in combating weevil attack is monitoring and timing of control measures. Because of the night activity of the adult weevils, growers do not observe the emerging weevils in a timely manner and oviposition often starts before effective control measures are taken. There is a report of field-active attractants for *O. sulcatus*, which holds promise for the development of new monitoring strategies for growers in the near future (Tol et al. 2012). Placing trap boards under *Rhodiola* plants can help in detecting their presence. A biological control of the larvae is available as a pathogenic nematode (*Steinernema kraussei*). It is applied when the soil temperature is warm enough for the nematode to be effective (5°C–20°C) and before the vine weevil grubs have grown large enough to cause serious damage.

5.9.4 MEADOW SPITTLEBUG

The meadow spittlebug, *Philaenus spumarius* (Hemiptera: Cercopidae), is occasionally found in *Rhodiola*. The nymphs are yellow to orange and cause the damage. They are found behind leaf sheaths, in folded leaves, or on the leaves and stems in masses of froth or spittle during late April, May, and early June. Both the nymphs and the adults may be found on a wide variety of weeds and plants. When abundant, spittlebugs stunt plant growth by sucking the sap. Spittlebug populations infrequently need control.

5.10 POST HARVEST

5.10.1 DRYING

Like many other medicinal crops, *Rhodiola* have to be dried before storage. Drying is the most common and fundamental method for postharvest preservation of medicinal plants because it allows for the quick conservation of the medicinal qualities of the plant material in a simple manner (Müller and Heindl 2006). Drying represents 30%–50% of the total costs in medicinal plant production (Qaas and Schiele 2001). Washing and drying of *Rhodiola* rhizomes and roots results in modifications that significantly affect their physiochemical properties. Drying stabilizes the roots for long-term storage, prior to further processing. Research has been conducted to determine the optimum drying temperatures for *Rhodiola* roots and crowns, as well as to determine the effect of drying operations on the bioactive levels of the roots and crown tissues. Normally, the higher the temperature, the more volatilization and change there is to the bioactive components. Cutting and bruising of the roots subject them to accelerated enzymatic activity and other deteriorating processes. Finally, longer drying periods increase the changes resulting from these higher temperatures and the cutting and bruising. Generally then, the roots should be dried at the lowest practical temperature, with minimal disturbance to the flesh of the root, in the shortest possible time. Drying behavior of medicinal plants during convective drying is mainly influenced by the conditions of drying air such as temperature, relative humidity, and velocity (Müller and Heindl 2006). Figure 5.15 shows the results of drying experiments with thin layers (20 mm) of *R. rosea* in a laboratory dryer with continuous measuring of sample mass. The optimum drying temperatures for the crown to obtain optimum total rosavins was between 41.1°C and 49.8°C, and for the root component it was between 54.3°C and 57.1°C. Drying temperatures also affected the visual quality of the product, with 60°C producing the best root color, that is, pinkish coloration. Drying enabled the roots to reach equilibrium moisture content by 50 hours for roots and about 60 hours for the crown.

The optimum drying temperature varied among the chemical compounds. For rosavin and rosarin, the optimum temperature was 40°C, while it was 60°C for salidroside and rosin. Drying at 25°C gave the lowest concentration among all phenylpropenoid components. For rosavin, the difference between drying at 40°C (0.2795% dry weight) and 60°C (0.2485% dry weight) was significant (Figure 5.16). In rosarin, there were no significant differences between 40°C and 60°C; both 25°C and 80°C were significantly less. Drying at 60°C gave significantly higher rosin than all the other drying temperatures. To achieve increased dryer capacity, drying temperature should be chosen as high as possible without reducing the quality of the product. Exceeding the thresholds for microbial count is the most frequent reason for rejection of medicinal plant material from growers by the pharmaceutical industry (Baier and Bomme 1996). Official thresholds of colony-forming units in terms of aerobic bacteria, molds, enterobacteria, and *Escherichia coli* are postulated in the European Pharmacopoeia (2005). When the bulk of harvested material is not ventilated, autoheating due to respiration activity provides favorable conditions for microorganism growth, in terms of temperature and humidity (Böttcher and Günther 1995).

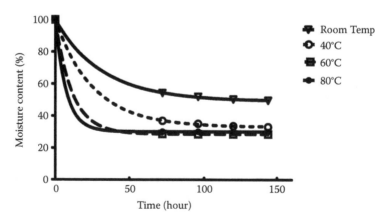

FIGURE 5.15   Influence of drying air temperature on drying time of *Rhodiola rosea* rhizomes.

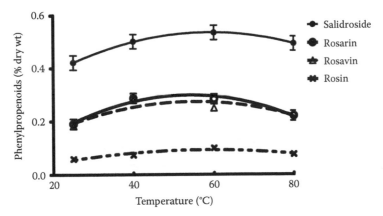

FIGURE 5.16 Effect of drying air temperature on salidroside, rosarin, rosavin, and rosin content of *Rhodiola rosea* rhizomes.

5.10.2 ECONOMICS OF PRODUCTION

Production targets are focused on high yield along with high levels of bioactive compounds in the rhizomes. Both of these factors will be driving influences on grower's economic return. The total costs of production for *Rhodiola* includes the costs of such items as seedlings, land preparation, planting, fertilizer, harvesting, root washing, and other activities. It is becoming increasingly apparent that weeding costs, in particular, may be substantial for this crop if organic status of the crop is to be achieved.

5.11 RESEARCH PRIORITIES

5.11.1 CONSTRAINTS TO HIGH YIELDS UNDER CULTIVATION

A gap exists between the potential and the actual farm yield of *R. rosea*. This gap exists because farmers use factors of production that result in lower yields than those possible on their farms (De Datta 1981). The factors used and the reasons why they are used need to be identified. The biological explanation of yield gap indicates that potential yields could approach 250 g dry rhizomes and roots mass per plant if they would use higher yielding variety, apply maximum-yield levels of fertilizer and insecticide to control aphids, control weeds, correct existing soil problems, and use the best cultural practices (Figure 5.17). Plant to plant variability in *Rhodiola* yields exists not only among farmers' but also among researchers' plots. Current agronomic research will continue to look at labor-saving alternatives to control weeds economically. Phenylpropanoids contribute to all aspects of plant responses toward biotic and abiotic stimuli. They are not only indicators of plant stress responses but are also key mediators of the plant's resistance toward pests (Camera et al. 2004). The environmental factors that affect the content of rosavin, rosarin, rosin, *trans*-cinnamic alcohol, salidroside, and tyrosol in the raw material need to be determined. A long-term goal is to establish *R. rosea* breeding program to release new varieties with improved yield, quality, and supply consistency of biologically active compounds.

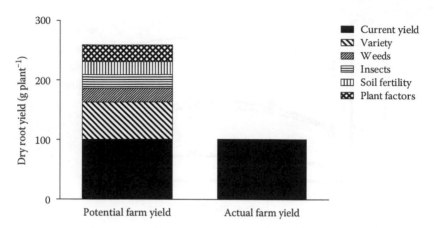

FIGURE 5.17   Yield gap between current on-farm yield and potential yield and biological constraints.

ACKNOWLEDGMENTS

The work reported here was part of the *Rhodiola rosea* commercialization project funded by the Alberta Agriculture and Rural Development and AVAC Ltd. I thank members of the *Rhodiola rosea* commercialization team, especially Dr. Susan Lutz, Dr. Jim Calpas, Dan Cole, Lawrence Papworth, Dr. Mohyuddin Mirza, Nabi Chaudhary, Dr. Brian Sloley, Paul Laflamme, John Brown, Dr. Christine Murray, Zhixiong Zhang, Cameron Stevenson, Robert Spencer, Shirzad Chunara, Steven Williams, Dr. Hugh Semple, Dean Dyck, and Jim Jones. I also thank the Alberta *Rhodiola Rosea* Growers Organization (ARRGO) members who participated in the study. I thank Dr. Petra Illig for information on production in Alaska.

REFERENCES

Alberta Agriculture and Rural Development. 2005. Agricultural Land Resource Atlas of Alberta Soil Groups of Alberta. http://www1.agric.gov.ab.ca/$department/deptdocs.nsf/all /agdex10307/$file/pg_14_soil_groups.pdf?OpenElement; accessed February 12, 2012.
Alberta Agriculture and Rural Development. 2011. Hail season approaches. http://www1 .agric.gov.ab.ca/$department/newslett.nsf/all/agnw18130; accessed February 12, 2012.
Ampong-Nyarko, K. and De Datta, S. K. 1991. *A Handbook for Weed Control in Rice*. Manila: International Rice Research institute.
Ampong-Nyarko, K., Lutz, S., Sloley, B. D., Piquette, K., and Zhnag, Z. 2011. Assessments of soil productivity for roseroot (*Rhodiola rosea* L.) cultivation in Alberta. *Z. Arznei-Gewurzpfla* 16: 156–62.
Baier, C. and Bomme, U. 1996. Impurities of medicinal plants—Current situation and future prospects. *Z Arznei-Gewurzpfla* 1: 40–8.
Bauer, J. and Chaudhary, N. 2011. The economic feasibility of *Rhodiola rosea* production. Alberta Rhodiola Growers Organization Spring Forum, Thorsby, Alberta.
Billings, W. D. and Mooney, H. A. 1968. The ecology of arctic and alpine plants. *Biological Review* 43: 481–529.
Bloom, A. J. 1985. Wild and cultivated barleys show similar affinities for mineral nitrogen. *Oecologia* 65: 555–7.

Böttcher, H. and Günther, I. 1995. Nachernteverhalten und Nacherntephysiologie von Arznei- und Gewürzpflanzen. *Herba Germanica* 3: 47–66.

Brown, R. P., Gerbarg, P. L., and Ramazanov, Z. 2002. *Rhodiola rosea*: A phytomedicinal overview. *Herbal Gram* 56: 40–52.

Camera S. L., Gouzerh, G., Dhondt, S., Hoffmann, L., Frittig, B., Legrand, M., and Heitz, T. 2004. Metabolic reprogramming in plant innate immunity: The contributions of phenyl-propanoid and oxylipin pathways. *Immunological Review* 198: 267–84.

Chapin, F. S. III, Schulze, E.-D., and Mooney, H. A. 1990. The ecology and economics of stor-age in plants. *Annual Review of Ecology and Systematics* 21: 423–47.

Chetner, S. and the Agroclimatic Atlas Working group. 2003. *Agroclimatic Atlas of Alberta, 1971–2000.* Edmonton: Alberta Agriculture and Rural Development.

Costică, M., Costică, N., and Toma, O. 2007. Phytocoenological, histo-anatomical and biochemical aspects in *Rhodiola rosea* L. species from Romania. *Analele Ştiinţifice ale Universităţii "Alexandru Ioan Cuza," Secţiunea Genetică şi Biologie Moleculară* 8: 119–21.

De Datta, S. K. 1981. *Principles and Practices of Rice Production.* New York: Wiley.

Dexter, A. R. and Tanner, D. W. 1973. The response of unsaturated soils to isotropic stress. *Journal of Soil Science* 24: 491–502.

Durhman, A. K., Rowe, D. B., and Rugh, C. L. 2006. Effect of watering regimen on chloro-phyll fluorescence and growth of selected green roof plant taxa. *HortScience* 41: 1623–8.

Durhman, A. K., VanWoert, N. D., Rowe, D. B., Rugh, C. L., and. Ebert-May D. 2004. Evaluation of crassulacean species on extensive green roofs, pp. 504–17. Proceedings of 2nd North American Green Roof Conference: Greening Rooftops for Sustainable Communities, Portland, OR.

Europäisches Arzneibuch (Ph.Eur. 5.00) 2005. *European Pharmacopoeia.* Stuttgart: Deutscher Apotheker Verlag.

Evans, R. L. 1983. *Handbook of Cultivated Sedums.* Northwood: Science Reviews Limited.

Evans, J. R. and Edwards, E. 2001. Nutrient uptake and use in plant growth. http://five.dsm.usb.ve/bibliografia_articulos_tareas/tomanutrientes.pdf; accessed February 19, 2012.

Galambosi, B. 2006. Demand and availability of *Rhodiola rosea* L. raw material. In: R. J. Rogers, L. E. Craker, and D. Lange (Eds). *Medicinal and Aromatic Plants.* The Netherlands: Springer. pp. 223–36.

Golovko, T. K., Dalke, I. V., and Bacharov D. S. 2008. Mesostructure and activity of photosyn-thetic apparatus for three crassulacean species grown in cold climate. *Russian Journal of Plant Physiology* 55: 603–12.

Golovko, T. K. and Pystina, N. V. 2001. The alternative respiration pathway in leaves of *Rhodiola rosea Ajuga reptans*: Presumable physiological role. *Russian Journal of Plant Physiology* 48: 733–40.

Heide, O. M. 2001. Photoperiodic control of dormancy in *Sedum telephium* and some other herbaceous perennial plants. *Physiology Plant* 113: 332–7.

Illig, P. 2011. Alaska *Rhodiola* growers report. Annual Report, Anchorage.

Kimmel, N., Cole, D., Wood, T., and Stevenson, C. 2008. *Rhodiola rosea* weed management project progress report. Edmonton: Alberta Agriculture and Rural Development.

Korner, C. H., Scheel, J. A., and Bauer, H. 1979. Maximum leaf diffusive conductance in vas-cular plants. *Photosynthetica* 13: 45–82.

Körner, C. 2003. *Alpine Plant Life: Functional Plant Ecology of High Mountain Ecosystems.* Berlin: Springer.

Lambers, H. 1985. Respiration in intact plants and tissues. Its regulation and dependence on environmental factors, metabolism and invaded organisms. In: R. Douce and D. A. Day (Eds). *Higher Plant Cell Respiration (Encyclopedia of Plant Physiology, New Series, vol. 18).* Berlin: Springer. pp. 418–73.

Lambers, H., Nagel, O. W., and van Arendonk, J. J. C. M. 1995. The control of biomass parti-tioning in plants from "favourable" and "stressful" environments: A role for gibberellins and cytokinins. *Bulgarian Journal of Plant Physiology* 21: 24–32.

Larcher, W. 2004. *Physiological Plant Ecology.* Berlin: Springer.

Letchamo, W., Kireeva, T., and A. Shmakov, A. 1998. A comparative study of *Rhodiola rosea* under cultivation and natural growing conditions. *HortScience* 33(3): 482.

Luttge, U. 2004. Ecophysiology of crassulacean acid metabolism (CAM). *Annals of Botany* 93: 629–52.

Mandelshtam M.Ju. 2001. A new species of bark-beetles (Coleoptera: Scolytidae) from Russian Far East. *Far Eastern Entomologist* N 105: 11–12.

Mamushina, N. S. and Zubkova, E. K. 2005. CAM photosynthesis: Occurrence, ecological, and physiological aspects. *Botanicheskij Zhurnal* 90: 1641–50.

Mahloch, J. L. 1974. Multivariate techniques for water quality analysis. *Journal of Environmental Engineering Division* 100: 1119–32.

Müller, J. and Heindl, A. 2006. Drying of medicinal plants. In: R. J. Bogers, L. E. Craker, and D. Lange (Eds). *Medicinal and Aromatic Plants.* Wageningen: Springer. pp. 237–52.

Platikanov, S. and Evstatieva, I. 2008. Introduction of wild golden root (*Rhodiola rosea* L.) as a potential economic crop in Bulgaria. *Economic Botany* 62: 621–7.

Przybył, J. L., Węglarz, Z., and Geszprych, A. 2008. Quality of *Rhodiola rosea* cultivated in Poland. *Acta Horticulturae (ISHS)* 765: 143–50.

Qaas, F. and Schiele, E. 2001. Influence of energy costs on the profitability of the dehydration of medicinal and aromatic plants. *Zeitschrift für Arznei- und Gewürzpflanzen* 6: 144–5.

Reding, M. E. 2008. Black vine weevil (Coleoptera: Curculionidae) performance in container and field-grown hosts. *Journal of Entomological Science* 43: 300–10.

Semikhatova, O. A., Gerasimenko, T. V., and Ivanova, T. I. 1992. Photosynthesis, respiration, and growth in the Soviet Arctic. In: F. S. Chapin III, R. L. Jefferies, J. F. Reynolds, G. R. Shaver, and J. Svoboda (Eds). *Arctic Ecosystems in a Changing Climate.* San Diego, CA: Academic Press. pp. 169–92.

Semikhatova, O. A., Ivanova, T. I., and Kirpichnikova, O. V. 2007. Comparative study of dark respiration in plants inhabiting arctic (Wrangel Island) and temperate climate zones. *Russian Journal of Plant Physiology* 54: 582–8.

Sengbusch, P. V. 2002. Botany online—The Internet hypertextbook in the www network of information and knowledge comments, opinions and recommendations. http://www.biologie.uni-hamburg.de/b-online/www/botany_online.htm; accessed March 12, 2012.

Siedow J. N., Møller I. M., Allan G., and Rasmusson A. G. 2010. Does respiration reduce crop yields? *Plant Physiology*, Fifth Edition Online. http://www.plantphys.net; accessed April 2, 2012.

Skripchinskii, V. V. and Skripchinskii, V. L .V. 1976. Morphobiological bases of ontogeny in ephemeroid geophytes and the problem of its establishment in the course of evolution (problems in ecological morphology of plants). *Trudy Moskovskogo obshchestva ispytatelei prirody* 62: 167–85.

Spencer, R., Ampong-Nyarko, K., and Cole, D. 2007. *Preliminary Guidelines for the Cultivation of Rhodiola rosea in Alberta.* Edmonton: Alberta Agriculture and Rural Development.

Tasheva, K. and Kosturkova, G. 2012. The role of biotechnology for conservation and biologically active substances production of *Rhodiola rosea*: Endangered medicinal species. *The Scientific World Journal* 2012, Article ID 274942, 13 pp.

Tol, V., Obert, W. H. M., Bruck, D. J., Griepink, F. C., and De Kogel, W. J. 2012. Field attraction of the vine weevil *Otiorhynchus sulcatus* to kairomones. *Journal of Economic Entomology* 105: 169–75.

VanWoert, N. D., Rowe, D. B., Andresen, J. A., Rugh, C. L., and Xiao, L. 2005. Watering regime and green roof substrate design affect Sedum plant growth. *HortScience* 40: 659–64.

Villarreal, E. L. and Bengtsson, L. 2005. Response of a sedum green-roof to individual rain events. *Ecological Engineering* 25: 1–7.

Wang, W. J. and Zhan, H. H. 2008. Effect of soil texture on starch accumulation and activities of key enzymes of starch synthesis in the kernel of zm 9023. *Agricultural Sciences in China* 7: 686–91.

Węglarz, Z., Przybył, J. L., and Geszprych, A. 2008. Roseroot (*Rhodiola rosea* L.): Effect of internal and external factors on accumulation of biologically active compounds. In: K. G. Ramawat and J. M. Merillon (Eds). *Bioactive Molecules and Medicinal Plants*. Koenigstein, Germany: Koeltz Scientific Books. pp. 297–315.

Wickens, G. E. 2001. *Economic Botany: Principles and Practices*. Dordrecht, The Netherlands: Kluwer Academic Publishers.

Wilson, D. and Jones, J. G. 1982. Effect of selection for dark respiration rate of mature leaves on crop yields of *Lolium perenne* cv. S23. *Annals of Botany* 49: 313–20.

Wolf, D. and Lundholm, J. T. 2008. Water uptake in green roof microcosms: Effects of plant species and water availability. *Ecological Engineering* 33: 179–86.

Zhao, D., Oosterhuis, D. M., and Bednarz, C. W. 2001. Influence of potassium deficiency on photosynthesis, chlorophyll content, and chloroplast ultra structure of cotton plants. *Photosynthetica* 39: 103–9.

Zotz, G. and Hietz, P. 2001. The physiological ecology of vascular epiphytes: Current knowledge, open questions. *Journal of Experimental Botany* 52: 2067–78.

6 Diseases of Wild and Cultivated *Rhodiola rosea*

Sheau-Fang Hwang, Stephen E. Strelkov,
Kwesi Ampong-Nyarko, and Ron J. Howard

CONTENTS

6.1 Introduction 155
6.2 Infectious Diseases of *Rhodiola* 156
    6.2.1 Seed Decay and Seedling Blight 156
    6.2.2 Root Rot and Winter Kill 159
    6.2.3 Powdery Mildew 161
    6.2.4 Gray Mold, Flower, and Leaf Blight, *Botrytis cinerea* 163
    6.2.5 Alternaria Leaf Spot 163
    6.2.6 Sclerotinia Stem Rot 164
    6.2.7 Aster Yellows 166
    6.2.8 Rusts, *Puccinia* sp. 169
6.3 Disease Management Recommendations 169
    6.3.1 Seed Decay, Seedling Blight, and Root Rot Winter Kill 169
    6.3.2 Foliar Diseases 170
    6.3.3 Aster Yellows 171
References 171

6.1 INTRODUCTION

Rhodiola rosea L. (syn. *Sedum rosea* (L.) Scop.), a member of the Crassulaceae family, produces several compounds that have been termed as "adaptogens," which are capable of enhancing the physical and mental performance of people in stressful situations, contain antioxidants, and may also act as anti-inflammatory agents (Kelly 2001; Brown et al. 2002). Because of this, *Rhodiola* has long been used in the traditional medicines of eastern Europe and Asia, and has recently gained popularity in the health food market in North America (Ampong-Nyarko 2004; Galambosi 2005).

    Rhodiola has a circumpolar distribution and grows on well-drained, alpine soils in its native habitat, such as China and Russia. However, the plant can be cultivated in more hospitable environments, for example, it has been grown for many years in rock gardens as an ornamental, resulting in one of its common names, "rose stonecrop." Alberta, Canada, represents an ideal subarctic environment for the cultivation of *Rhodiola* because of pristine environment, relatively high altitude, and rigorous climate, all of which act to promote accumulation of increased levels of the active ingredient(s) of many medicinal herbs (http://www.ricola.com/en-us/World-of-Herbs/Ricola-Herbology/Herb-drops-in-production). In Alberta, *Rhodiola* seedlings are typically planted in a nursery and are later transplanted to the field.

As with other plant species, *R. rosea* is susceptible to attack by various pathogenic microorganisms that cause infectious diseases, and can also be negatively impacted by a wide variety of environmental stresses (noninfectious diseases). There have been relatively few plant pathological studies on *Rhodiola* because, until recently, this species has not been extensively cultivated as a crop. Since *Rhodiola* requires a 4- to 5-year commitment of time, land, and labor before reaching harvestable maturity, disease-associated losses in production and quality may represent a major roadblock to its adoption as a crop by agricultural producers. Root rots are of particular concern, since *Rhodiola* derives its economic value from its roots. The successful development of *Rhodiola* as a crop in Alberta depends not only on producer acceptance but also on the ability to maximize the quantity and quality of the bioactive compounds. To achieve this quality, the economically important plant diseases on this crop must be identified and effective control measures developed.

Early detection of diseases can help prevent their spread and thus reduce the resulting damage to the standing crops. This chapter provides descriptions of the major infectious diseases of *Rhodiola* as observed in central Alberta. In addition, management recommendations are provided for those diseases that pose a serious threat to commercial *Rhodiola* production.

6.2   INFECTIOUS DISEASES OF *RHODIOLA*

6.2.1   Seed Decay and Seedling Blight

One of the major constraints to the cultivation of *Rhodiola* in Alberta has been poor seed germination and seedling establishment. Seedling damping-off is caused by various soil-borne, plant pathogenic fungi, including *Pythium* spp., *Rhizoctonia solani* Kühn, *Fusarium* spp., and *Alternaria* spp. (Hwang et al. 2009a). These pathogens can occur in infested growing media and cause significant losses under cool, wet conditions, which often prevail in greenhouses during the spring (Figures 6.1 and 6.2). Moreover, healthy transplants may become infected in the field by indigenous populations of these pathogens (Figure 6.3). Diseased plants first exhibit a blackening of the root tips, followed by browning of the root and stem, shrinking of the cotyledons, and twisting of the young leaves. Finally, the seedlings wilt and collapse onto adjacent plants. Under conditions favorable for disease development, such as cool-to-warm, moist soils, this collapse may occur quite suddenly (3–5 days) and results from the disintegration of root and stem tissues due to the necrotizing effects of the fungal infections. Seedling blight inoculum, including fungal spores, sporangia, mycelia, and/or hyphal fragments, can spread with infested potting media, soil, roots, or plant material. Interactions between these root pathogens likely occur in the field, and they are frequently isolated from diseased root tissues as a complex of disease-causing agents. For example, it is not uncommon to recover *R. solani* and *Fusarium avenaceum* (corda ex Fr.) Sacc., from a single symptomatic root.

(a)

(b)

FIGURE 6.1   (a) Healthy *Rhodiola* seedlings compared with (b) seedlings showing symptoms of seedling blight and damping-off.

Pythium spp. cause stunting of the seedlings and browning of the root tips. Following severe infection, the plants may wilt and die. *Pythium* spp. may also cause pre- or postemergence damping-off, if the growing substrate is saturated for prolonged periods.

At the seedling stage, *F. avenaceum, Pythium* sp., and *R. solani* may all reduce stand establishment and cause poor plant growth (Liu and Cheng 2011; Bai et al.

FIGURE 6.2 A *Rhodiola* seedling with a healthy root system (left) compared with two seedlings infected by *Fusarium* (center) and *Pythium* (right). Note the shrunken, necrotic roots of the infected seedlings and their smaller size.

(a)

(b)

FIGURE 6.3 (a) Healthy young *Rhodiola* plants at 6 months of age compared with (b) seedlings of the same age affected by root rot.

2012). *Rhodiola* seedlings infected by these pathogens may have reduced plant height and shoot vigor even without the appearance of any obvious disease symptoms such as discolored root tissues.

Surveys of *Rhodiola* plantations in central Alberta have confirmed that there is a great potential for severe economic loss as a consequence of seedling blight. Isolates of *Pythium* spp., *R. solani,* and *Fusarium* spp. were commonly recovered from the roots of diseased *Rhodiola* plants. All are destructive soil-borne pathogens, which can cause severe preemergence damping-off and postemergence seedling root rot.

6.2.2 ROOT ROT AND WINTER KILL

Root and crown rot of *Rhodiola* are caused by *R. solani, Fusarium verticillioides* (Sacc.) Nirenberg (*Gibberella fujikuroi* [Sawada] Wollenw.), *Fusarium oxysporum* (Schlecht) Snyder et Hansen, *F. avenaceum,* and *Pythium* spp. Root rot may become serious over the 4- to 5-year period from planting to harvest and reduce the yield and quality of the harvested product. The complex of fungi associated with root rot often kills the plant after 2 years, typically during winter or early spring, and its impact on the vigor of surviving plants increases with plant age. Roots injured by root rot may decline in both volume and quality, thereby reducing the yield and quality of the crop.

Infection by *R. solani* first affects the leaves and stems near the ground, causing the formation of dark brown lesions on the roots and shoots and wilting of the lower leaves (Figure 6.4). The wilting symptoms spread rapidly over the entire plant, with the leaves becoming grayish brown in color and taking on a water-soaked appearance. Finally, the infected leaves turn black. In addition to the foliar symptoms, sunken lesions may occur at the soil line. In severe infections, defoliation and damping-off also may result and plant morbidity is common. The anastomosis grouping of *R. solani* strains associated with root rot of *Rhodiola* is AG-4 in both North America and China (Jiang et al. 1996; Bai et al. 2012).

Fusarium spp. are the most commonly encountered fungi in infected crowns and roots, usually as a result of invasion through injured tissues in the crown or root areas. *F. oxysporum* has been isolated from discolored vascular bundles of roots and crowns, while *F. solani* has been found in the cortex. Other species of *Fusarium,* such as *F. avenaceum* and *F. equiseti* (Corda) Sacc., have also been isolated from diseased roots. Infection starts at the base of the shoot and root and extends upward into the shoot and downward into the roots. Symptoms may include vascular discoloration and wilting of the plant (Figure 6.5). As the disease progresses, brown lesions expand on the lateral and main roots, the aboveground parts of the plant senesce and the plants finally die. The disease is most severe at high temperatures (24°C–30°C) and high humidity.

It is very important to examine the roots of plants showing foliar symptoms. *Rhodiola* roots may appear to be healthy, but a cross-sectional cut through the crown area can reveal discolored vascular tissues, indicating infection by soil-borne pathogens (Meng et al. 1994; Li et al. 2003, 2005; Li and Lu 2009).

Good winter survival is critical for the successful cultivation of *Rhodiola* (Figure 6.6). The plants are not completely winter hardy and require a mulch of straw for winter survival in Alberta. Unfortunately, this increases the likelihood that soil-borne pathogens will survive the winter and become active in the following spring,

FIGURE 6.4 Adult *Rhodiola* plant showing symptoms of Rhizoctonia crown rot. Note the necrosis at the base of the stem.

(a)   (b)

(c)

FIGURE 6.5 **(See color insert.)** Crown rot caused by *Fusarium* spp. on *Rhodiola*, leading to wilting of the leaves and discoloration (a) and discoloration of the root tissue (b). Crown rot caused by *Sclerotinia* is shown in (c).

(a)

(b)

FIGURE 6.6 Winter injury on *Rhodiola* can range from partial dieback (a) to total plant mortality (b).

resulting in the development of root rot. Winter kill of *Rhodiola* can be directly caused by low-temperature injury or by a progressive decay of the root tissues caused by soil-borne root rot pathogens. Low-temperature injury and root rot operate synergistically. The decay of the root tissues interferes with the accumulation of nutrients and can predispose *Rhodiola* plants to low-temperature injury in the following winter. Low-temperature injury may also predispose plants to fungal invasion and the development of root rot during the growing season. Root rot and winter kill may become major limiting factors for *Rhodiola* production.

6.2.3 POWDERY MILDEW

Powdery mildew is caused by an obligate plant-parasitic fungus belonging to the family Erysiphaceae. This pathogen grows on the surface of the leaves and enters host epidermal cells with specialized penetration structures called appresoria. Once inside the host cells, the fungus differentiates into feeding structures called haustoria. The haustoria absorb nutrients from the plant cells, which are then used for growth and reproduction of the fungus. The velvety white mildew symptoms that are visible on the upper surfaces of infected leaves result from the presence of a combination of whitish vegetative mycelium and asexual spores called conidia, which are borne in chains on upright conidiophores. Wind-dispersed conidia can germinate in the absence of free water under high relative humidity, and the disease is often severe when environmental conditions are humid at night, the temperature nears the dew point, and warm and dry conditions persist during the day. Small black structures, called cleistothecia, which are the sexual fruiting bodies of the powdery mildew fungus, are often visible as black specks among the white, powdery, mycelial growth in mature infections. The spores within the cleistothecia, known as ascospores, may be genetically more diverse than conidia, facilitating the development of new strains of the fungus. Powdery mildew is seen commonly on *Rhodiola* seedlings grown in greenhouses. The mildew generally begins on cotyledonary leaves and spreads to true leaves very rapidly. The white powdery coating that covers the leaves eventually causes affected plant parts to turn purple or bluish-green and then brown (Figure 6.7) due to the stress caused by the fungus as it parasitizes the host plant.

The disease can spread very quickly throughout a greenhouse because of the extremely high asexual reproductive potential of the fungus. These infections cause severe losses in productivity, mainly as a result of a loss of photosynthetic capacity and photosynthates by the host. The airborne conidia of the powdery mildew fungus spread particularly rapidly in enclosed areas such as greenhouses. Infected plants lose vigor, their growth is impaired, and subsequent field performance may be poor. Heavily infected leaves and even entire plants can die prematurely. Since greenhouse-grown plugs are very expensive and a critical component of a successful *Rhodiola* industry, this disease represents a serious threat to *Rhodiola* production at multiple steps in the production chain.

FIGURE 6.7 (**See color insert.**) *Rhodiola* seedlings with whitish mycelium and spores of the powdery mildew fungus covering the young leaf surfaces.

Under field conditions, powdery mildew tends to occur late in the growing season. It does not appear to cause significant damage, but the leaves can become covered by white mycelium and spores.

In other crops, powdery mildew–resistant varieties have been developed through traditional breeding efforts and represent the best defense against powdery mildew, so long as the prevalent races of the fungus are controlled. Unfortunately, no powdery mildew–resistant genotypes of *Rhodiola* have yet been identified, so breeding for resistance has not yet been possible.

6.2.4   GRAY MOLD, FLOWER, AND LEAF BLIGHT, *BOTRYTIS CINEREA*

B. cinerea (de Bary) Whetzel is a fungus that commonly infects damaged or senescing plant tissue, such as old flowers (Figure 6.8). From these tissues, the fungus moves into healthy stems and leaves in rapidly spreading lesions, causing a damaging blight and producing profuse gray mycelium laden with spores. These spores are dispersed by air currents, thus helping the pathogen spread long distances. Gray mold is favored by cool, wet conditions and the presence of senescing tissue. Under warm, dry conditions, the disease develops slowly and forms small to large, gray, brown, or black lesions. Latent infections may also occur.

6.2.5   ALTERNARIA LEAF SPOT

Alternaria spp. infect the leaves, causing small dark brown- to black-colored spots (Figure 6.9), which can coalesce to form large lesions under moist conditions. The inner portion of lesions/spots may fall out, leaving holes in the leaves. The fungus may also infect young shoots and small flowering heads, resulting in dieback symptoms (Figure 6.9). In addition, *Alternaria* spp. can infect the seed and become a seedborne pathogen. Occasionally, the pathogen infects the roots, crowns, and leaves of young seedlings, causing root rot and leaf spots.

(a)   (b)   (c)

FIGURE 6.8   (See color insert.) *Botrytis cinerea* causes the development of gray-brown lesions on the leaves of *Rhodiola* (a). The fungus first colonizes damaged or senescing plant tissue, such as old flowers, from which it spreads to healthy tissue (b and c).

(a)   (b)

FIGURE 6.9   Symptoms of Alternaria leaf spot on *Rhodiola* leaves include the formation of concentric lesions that are darker on the outside and lighter near the center (a). Leaf tissue in the center of the lesions may fall out, resulting in the formation of holes in the leaves (b).

6.2.6   SCLEROTINIA STEM ROT

Stem rot, caused by *Sclerotinia sclerotiorum* (Lib.) de Bary, is characterized by the formation of dark sclerotia (fungal survival structures of varying size) that can remain dormant in the soil for up to 5 years. Under favorable weather conditions, the sclerotia germinate to produce hyphae that can directly invade plants, and/or apothecia, which produce sexual, airborne ascospores that germinate on plant leaves and stems (Figure 6.10). Moist conditions are required for establishment and increased severity of stem rot. The disease can be very destructive and develops quickly after a few wet days. No studies have examined the pathogenicity of *S. sclerotiorum* on *Rhodiola* in North America. Common symptoms include yellowing of the lower leaves, followed by wilting and death of the rest of the plant. A white, cottony mass of mycelium grows around the crown or on the soil near the crown, serving to distinguish stem rot from other diseases (Figure 6.5). One of the most important diagnostic features of this disease is the formation of the sclerotia, which are visible on the outside and inside of infected stems as reddish-brown, seed-like structures. Dark brown to black lesions may also form at and above the soil line (Figure 6.11). The crowns and roots can become rotted and black (Figure 6.5), and diseased plants are easily pulled from the soil.

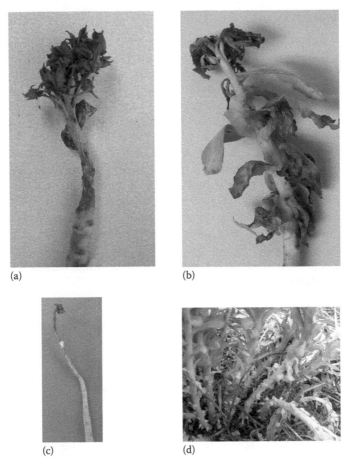

(a)   (b)

(c)   (d)

FIGURE 6.10 Airborne spores of *Sclerotinia sclerotiorum* can land on the upper stem, leaves, and floral structures of *Rhodiola* plants, resulting in necrosis of these tissues. Symptoms and signs of infection include dieback of the stem above the lesion (a), dieback of the leaves (b), the presence of immature sclerotia on necrotic stems (c), and wilting of stems and death of plants (d).

FIGURE 6.11 *Sclerotinia sclerotiorum* can invade the lower stems of *Rhodiola* plants, causing the development of dark brown to black lesions at or above the soil line.

6.2.7 ASTER YELLOWS

Aster yellows of *Rhodiola*, a disease caused by a phytoplasma, was first identified in
Alberta in 2007 (Hwang et al. 2009b). Phytoplasmas are plant-parasitic prokaryotes
that lack a cell wall and are found in the sugar-conducting tissue (phloem) of host
plants. Affected *Rhodiola* plants typically show symptoms such as a purplish reddening along the leaf margins and tips (Figure 6.12), and discoloration and stunting of the
plant (Figure 6.13). The bright yellowing of the leaves, resulting from phytoplasma
infection, is particularly apparent in the early spring, when young shoots emerge
from the ground. Plants infected with yellows sometimes have extremely numerous,
small, branched axillary shoots that proliferate from the stem nodes. These develop
leaflets with shortened internodes so that the plants eventually develop a bunchy or
"witches' broom" appearance (Figure 6.14). At later stages of disease development,
infected flowers may develop into vegetative, leaf-like structures, symptoms that are
commonly referred to as virescence (greening of the floral tissues) and phyllody
(replacement of the floral structures with leaf-like organs) (Figure 6.15). The affected
flowers do not produce seeds. Yellows can also reduce winter hardiness, which often
kills infected plants.

FIGURE 6.12   (See color insert.) Purplish reddening along the leaf margins and tips of a
Rhodiola plant infected with aster yellow phytoplasma.

FIGURE 6.13 Discoloration and stunting of a *Rhodiola* plant infected with aster yellow phytoplasma (right).

FIGURE 6.14 Phytoplasma infection of *Rhodiola* can cause a proliferation of the axillary shoots. This gives infected plants a bunchy or "witches' broom" appearance.

(a)

(b)

FIGURE 6.15   (**See color insert.**) Phytoplasma infection of *Rhodiola* causes distortion and virescence (greening) of ray and disk florets (a). A normal flower is shown in (b).

Phytoplasmas, which can cause more than 200 diseases in addition to yellows, can only survive and reproduce in living plant tissue, or within some of their insect vectors. They cannot be isolated or cultured using conventional laboratory methods. This characteristic makes them difficult to detect and identify. Transmission electron microscopy has been used to observe the ultrastructural characteristics of phytoplasmas and to demonstrate their presence in the phloem of host plants (not shown). Examination of ultrathin sections of leaf veins from affected leaves and flowers may reveal large numbers of phytoplasma bodies in the phloem. Recent advances in molecular biology have facilitated the identification of phytoplasmas. The phytoplasma associated with aster yellows of *Rhodiola* are a member of the subgroup 16SrI-B (aster yellows subgroup B).

Phytoplasmas are transmitted between plants primarily by phloem-feeding leafhoppers. The leafhopper vector feeds on the phloem of aster yellow-infected plants by inserting a needle-like mouthpart, called a stylet, into the cell. Once the phytoplasma is acquired by the leafhopper, an incubation period follows in which it multiplies within the insect and then moves to the salivary glands. At this point, the phytoplasma can be transmitted to a new host through the saliva as the leafhopper feeds. Within 8–24 hours after the introduction of a phytoplasma to healthy leaf tissue, the pathogen moves out of the leaf and into the phloem of the host plant. Leafhoppers feed on a variety of plants and can spread phytoplasma pathogens from wild to cultivated plants and vice versa. Infected biennial and perennial plants may allow phytoplasmas to persist from one year to the next in the fields and natural habitats. In some circumstances, phytoplasmas may overwinter within an insect vector.

6.2.8 RUSTS, *PUCCINIA* SP.

The term "rust" refers to both the disease and the pathogen causing the disease. Rust fungi are specialized obligate parasites that can complete their life cycle on one (monoecious) or more (heteroecious) host species. Symptoms of rust infection include production of rust-colored spores or gelatinous structures in powdery pustules on leaves or stems. The surrounding tissue is discolored and yellowed, and plants are often stunted. Although limited information is available regarding rusts of *Rhodiola*, the rust fungus *Puccinia umbilici Guepin* has been reported on *Rhodiola* in Russia and Norway (Gjærum 1989). In an early report, Arthur and Cummins (1934) observed the production of only telia (fruiting bodies or sori that produce a spore type called teliospores) by *P. umbilici* on *Rhodiola integrifolia* Raf. (*Sedum Rhodiola*) in Colorado and Europe. Although rust has not yet been reported on *Rhodiola* in Alberta, the ability of rust spores to travel great distances makes this disease a potential threat to commercial production.

6.3 DISEASE MANAGEMENT RECOMMENDATIONS

6.3.1 SEED DECAY, SEEDLING BLIGHT, AND ROOT ROT WINTER KILL

- Sterilize seeds with 1% bleach solution for 3–5 minutes and rinse them with sterile water before seeding.

- Use a pathogen-free growing medium and do not overwater seedlings, especially during the first 2 weeks after propagation or transplantation.
- Waterlogging can favor development of root diseases. Avoid fields that are prone to flooding.
- Avoid injuring roots during cultivation and transplanting as injury predisposes plants to invasion by *Fusarium* and other plant pathogens.
- Use healthy plants for propagation.
- Vesicular-arbuscular mycorrhizal (VAM) fungi are reported to be effective biological control agents for the control of many soil-borne fungal root diseases, including those caused by *Fusarium*, *Pythium*, and *Rhizoctonia* spp. VAM fungi stimulate the activities of organisms that compete with soil-borne pathogens. Incorporating a VAM inoculum into the growth medium may accelerate or increase the colonization of *Rhodiola* roots by VAM fungi and assist in establishing this mutualistic symbiosis.
- Biocontrol agents (*Trichoderma harzianum* Rifai, *Glomus intraradices* Schenck & Sm., *Bacillus subtilis* Ehr., and *Brevibacillus laterosporus* Shilda et al) for the control of root rot pathogens are available commercially for field application and have been shown to be effective against damping-off caused by *Rhizoctonia* and *Fusarium* in other host species (Pal and Gardener 2006). However, validation of the effectiveness of the biocontrol agents for *Rhodiola* production under field conditions is required (Li and Lu 2009).
- No seed treatments are currently registered for use on *Rhodiola* in Canada.

6.3.2 FOLIAR DISEASES

- Ultrafine oil sprays and sulfur are used to treat powdery mildew.
- Practice sanitation by removing infected plant debris at the end of each growing season.
- Since *Rhodiola* plants need 3–5 years to reach the harvest stage, proper spacing should be considered while planting to improve air circulation among plants.
- Grow plants in a sunny location.
- Use drip irrigation in preference to wetting the leaves.
- Crop rotation—*S. sclerotiorum* has a very wide host range and can infect many different broad-leaved crop species. More than 3 years in rotation with cereal crops, corn or grasses, is often required to reduce the pathogen populations in the soil. Therefore, *Rhodiola* crops should not follow canola, sunflower, beans, or other broad-leaved crops.
- Avoid water condensation in greenhouses. Use proper ventilation systems and water plants well before sunset.
- No foliar fungicides are currently registered for *Rhodiola* in Canada.
- Powdery mildew is often less damaging when nitrogen, potassium, and phosphorus fertilizer are correctly proportioned. Control may also be achieved with the use of foliar fungicides applied as soon as the symptoms become visible, where and when registered products are available.

6.3.3 ASTER YELLOWS

- Keeping leafhopper populations low is an important method for controlling phytoplasma diseases because there is no product available, that can directly prevent or cure phytoplasma infections.
- Control and mow down perennial weeds in adjacent fields and field margins, as they may harbor phytoplasmas and/or leafhoppers.
- Grow companion crops, such as yarrow, garlic, and mint, which are repellent to leafhoppers, and apply repellent essential oils such as bergamot and cilantro onto *Rhodiola* plants early in the spring and at least four times during the growing season.
- Apply low-toxicity insecticides, such as malathion or pyrethrin onto the surrounding weeds to control leafhoppers. These products are not registered for direct application to *Rhodiola* in Canada.
- Destroy plants that show typical symptoms of phytoplasma infection.
- Susceptible specialty crops or other perennial crops affected by phytoplasma diseases should not be planted in close vicinity to *Rhodiola*.
- Specialty crops and surrounding fields and headlands should be regularly monitored for leafhoppers using sweep nets, sticky yellow cards and traps.
- Monitoring of leafhoppers and early detection of the symptoms of aster yellows are important for those crops in which hand removal of diseased plants is feasible.

REFERENCES

Ampong-Nyarko, K. 2004. A new *Rhodiola* commercialization project for Alberta. *Alberta Natural Health Agricultural Network Grass Roots Gateway* 1:1–5.

Arthur, J.C. and Cummins, G.B. 1934. *Manual of the Rusts in the United States and Canada.* Hafner Publishing Co., New York, NY.

Bai, Q., Xie, Y., Wang et al. 2012. First report of damping-off of *Rhodiola sachalinensis* caused by *Rhizoctonia solani* AG-4 HG-II in China. *Plant Diseases* 96:142.

Brown, R.P., Gerbarg, P.L. Ramazanov, Z. 2002. *Rhodiola rosea*: A phytomedicinal overview. *Journal of American Botany Council* 56:40–52.

Galambosi, B. 2005. Demand and availability of *Rhodiola rosea* L. raw material. In: R.J. Bogers, L.E. Craker, and D. Lange (Eds.), *Medicinal and Aromatic Plants*, pp. 223–36, Proceedings of the Frontis Workshop on Medicinal and Aromatic Plants. April 17–20, 2005, Waginengen, The Netherlands. http://library.wur.nl/frontis/medicinal_aromatic _plants/16_galambosi.pdf.

Gjærum, H.B. 1989. Rosenrot-rusten, *Puccinia umbilici* gjenfunnet i Sør-Norge. (*Puccinia umbilici* rediscovered on *Sedum rosea* in south Norway.) *Blyttia* 47:189–90.

Hwang, S.F., Ahmed, H.U., Ampong-Nyarko, K., Strelkov, S.E., Howard, R.J., and Turnbull, G.D. 2009a. Causal agents of root rot and the effects of vesicular-arbuscular mycorrhizal fungi in seedlings of *Rhodiola rosea* in Alberta, Canada. *Plant Pathology Journal* 8:120–6.

Hwang, S.F., Feng, J., Strelkov, S.E., Ampong-Nyarko, K., Turnbull, G.D., and Howard, R.J. 2009b. Detection and molecular characterization of an aster yellows phytoplasma in *Rhodiola* in Alberta, Canada. *Journal of Plant Diseases and Protection* 116:145–8.

Jiang, W., Zhong, W., Jiang, M., and Meng, Q. 1996. Root rot disease of the valuable medial herb *Rhodiola sachalinenisi* in Changbai mountain and identification of pathogenicity of the causative agent. *Acta Phytophylacica Sinica* 23:185–6.

Kelly, G.S. 2001. *Rhodiola rosea*: A possible plant adaptogen. *Alternative Medicine Review* 6:293–302.

Li, X., Huang, S., Quan, C., and Shi, T. 2003. Biological characteristics of root rot pathogens on *Rhodiola sachalinensis*. *Journal of Northeastern Forestry University* 31:12–14.

Li, X. and Lu, L. 2009. Prevention and cure of diseases in *Rhodiola sachalinensis*. *Agrochemicals* 48:458–9.

Li, X., Piao, X., and Huang, S. 2005. Spatial distribution patterns and sample techniques for root rot of *Rhodiola sachalinensis*. *Journal of Northeastern Forestry University* 33:40–41.

Liu, J.F. and Cheng, Y.Q. 2011. First report of root rot on *Rhodiola sachalinensis* caused by *Fusarium verticillioides (Gibberella fujikuroi)* in China. *Plant Disease* 95:222.

Meng, Q., Jiang, M., Zhong, W., Cheng, G., and Liu, S. 1994. Controlling the root-rot disease of *Rhodiola sachalinensis* A Bor. with pesticides. *Journal of Shenyang Agricultural University* 25:264–7.

Pal, K.P. and Gardener, B.M. 2006. Biological control of plant pathogens. *The Plant Health Instructor*. http://www.apsnet.org/.

7 Biotechnology of *Rhodiola rosea*

Zsuzsanna György

CONTENTS

7.1 Introduction 173
7.2 *In Vitro* Cultures of *R. rosea* 174
    7.2.1 Explants for *In Vitro* Culture 174
    7.2.2 Micropropagation 175
    7.2.3 Adaptation of Micropropagated Plants 176
    7.2.4 Callus Induction 176
7.3 Biotechnological Methods for Increasing the Pharmaceutically
    Important Metabolite Production in *In Vitro* Cultures of *Rhodiola Rosea* ... 177
    7.3.1 Production of Salidroside in Callus Cultures 178
    7.3.2 Production of Rosavins in Callus Cultures 180
7.4 On the Way to Metabolic Engineering of *Rhodiola rosea* 181
    7.4.1 Biosynthesis of Cinnamyl Alcohol Glycosides 181
    7.4.2 Biosynthesis of Salidroside 182
    7.4.3 Biosynthesis of Tyrosol 182
    7.4.4 Salidroside Formation 183
    7.4.5 Agrobacterium-Mediated Transformation 184
Acknowledgments 184
References 184

7.1 INTRODUCTION

Rhodiola rosea L. (roseroot) has been categorized as an adaptogen and currently is one of the most studied medicinal plants for its enormous pharmaceutical significance and more reputably for its bioactive secondary metabolites. Roseroot is difficult to cultivate and develops very slowly both in its natural environment and in cultivation, justifying development of new methods for the production of its bioactive compounds. Several studies on *in vitro* cell, tissue, and callus culture accompanied by some biotransformation trials have been carried out in the last three decades. To achieve a clear understanding of how to design roseroot *in vitro* culture systems that maximize the content of their specific natural products, the knowledge of biosynthetic pathway of these glycosides is necessary and indispensable. In this review, the results of *in vitro* experiments with roseroot and the first steps concerning the exploration of the biosynthetic pathway of salidroside and cinnamyl alcohol glycosides are summarized.

7.2   *IN VITRO* CULTURES OF *R. ROSEA*

In vitro culture of *R. rosea* has a history of slightly more than 30 years. First studies were done in the former Soviet Union and most of the publications from the initial period are in the Russian language. Most of these report on the effect of different culture media and different plant hormones on the various types of explants, sterilization of different explant types, callus induction, organogenesis, and regeneration. From the beginning of the *in vitro* studies, research was conducted on the production of the most valuable secondary metabolites in *in vitro* cultures as well. The experimental work was mostly organized in a number of research units. The following provides a detailed review of the *in vitro* work on roseroot.

In 1981, Aleksandrova et al. (cited by Furmanowa et al. 1995) patented the method for root regeneration from callus, but did not provide any data on the callus induction and its maintenance. The only information was that Murashige and Skoog (MS) medium was used for root induction from callus supplemented with sucrose (2.5%–3.5%), thiamine, HCl, mesoinositol, NAA (0.8–1.1 mg/L), adenine (0.08–0.013 mg/L), and kinetin (0.01–0.1 mg/L) and 25–28 days were necessary for root formation. First scientific publications about callus induction appeared in the mid-80s (Poletaeva et al. 1984; Baev 1984; Bykov 1986; all cited by Bykov et al. 1999). The explants for callogenesis were sterile-growing 40- to 50-day-old plantlets. For callus induction MS medium was used both under continuous illumination and in the dark.

7.2.1   EXPLANTS FOR *IN VITRO* CULTURE

Leaves of *in vitro* plantlets derived from seed culture are the explants used the most often and with the most success (Tasheva and Kosturkova 2010, 2012). Furmanowa et al. (1995) germinated immature seeds after being washed in running water for 1 hour, dipped into 70% ethanol for 1 minute, soaked in 5% solution of calcium hypochlorite for 10 minutes and finally rinsed three times with sterile water. Ishmuratova (1998) used both seeds and buds successfully for *in vitro* culture establishment. The sterilization protocol for both included 1 minute in 70% ethanol and either 5 minutes in 3% hydrogen peroxide or 7 minutes in aqueous 0.1% mercuric chloride. The sterilization procedure of Martin et al. (2010) included a bath of 3 minutes in 70% ethanol, followed by 20 minutes in 10% chlorinated lime and finally 20 minutes in 10% sodium hypochlorite. Tasheva and Kosturkova (2010) conducted a broad study with 6 schemes for sterilizing not only seeds but also bud, shoot, or rhizome segments. For seed decontamination, 3 minutes in 70% ethanol followed by 15 minutes of 20% (v/v) bleach is recommended by them. Also apical buds with 1 minute in 70% ethanol, 20 minutes in 15% bleach, and 15 minutes in 0.2% mercuric chloride and rhizome buds with 1 minute in 70% ethanol, 17 minutes in 20% bleach were successfully sterilized and further developed. Ghiorghita et al. (2011) reported about difficulties during decontamination of roseroot explants. This fact was previously concluded by Tasheva and Kosturkova (2010) also. Mature rhizome buds, apices, and shoot segments of young plants were treated with short immersion in 5% chloramine-T solution. Only those explants remained viable after the sterilization process, which were previously submitted to low temperature. Based on these reports, a general protocol cannot be determined. The different working groups have contradictory results; what worked for one did not necessarily work for the other. The genotype and the original environment of the plant probably have a major effect on the success of the decontamination.

7.2.2 MICROPROPAGATION

Micropropagation of *R. rosea* is in the focus of research from the 1990s. Furmanowa et al. (1995) gave the first report on micropropagation. In their study, the most effective medium for plant development from shoot tips was the one of Nitsch and Nitsch (NN) supplemented with 0.1 or 1 mg/L kinetin or 0.01 mg/L NAA along with 0.1 mg/L IAA. When NAA in higher concentrations (1 mg/L) was added to the medium, the growth was completely stopped. Ishmuratova (1998) conducted micropropagation of roseroot on MS medium containing 0.2 mg/L BAP and 0.1 mg/L IAA. MS medium containing higher levels of BAP and lower levels of NAA was also found to be the most efficient by Yin et al. (2004).

In 2009, Debnath reported the application of RITA temporary immersion system for the micropropagation of *R. rosea* (Debnath 2009). Thidiazuron (TDZ) and zeatin were used as growth regulators. TDZ at 2–4 µM stimulated shoot induction, but inhibited shoot elongation. It was also observed that the three studied clones differed significantly with respect to the multiplication rate. In the RITA system, 0.5 µM TDZ sustained rapid shoot proliferation, but at higher concentration induced hyperhydricity. These hyperhydric shoots produced normal shoots within 4 weeks when transferred to solid MS medium containing 1–2 µM zeatin. Tasheva and Kosturkova (2010) tested a wide range of plant hormones for inducing organogenesis. The best bud formation was achieved on MS medium supplemented with 2 or 0.2 mg/L zeatin along with 0.2 mg/L IAA in case of stem segments with leafy nodes, whereas multiplication of the developed shoots was the most efficient on MS medium with 1 mg/L zeatin and 0.2 mg/L IAA. Ghiorghita et al. (2011) also studied the possibilities of micropropagation. According to them, the most efficient combination of hormones on MS medium is 0.2 mg/L IAA and 2 mg/L zeatin.

Controversial results have been published about rhizogenesis. According to Ghiorghita et al. (2011), the presence of NAA in the medium leads to the most intense rhizogenesis, but the combination of 0.2 mg/L IAA and 2 mg/L zeatin or 0.5 mg/l NAA and 1 mg/L kinetin is also favorable for root induction. Meanwhile, based on the work of Tasheva and Kosturkova (2010), the presence of NAA induced callogenesis not rhizogenesis. According to them, the most effective root induction was found on medium supplemented with 0.2 mg/L IAA, 2 mg/L IBA, and 1 or 0.4 mg/L giberrelic acid; however, even on simple ½ MS, 58% of the shoots rooted.

Recent studies concerning micropropagation published by Romanian and Bulgarian groups (Ghiorghita et al. 2011; Tasheva and Kosturkova 2010) aim to restore and repopulate the natural habitats of this species, which is endangered in these countries. This strategy might cause a severe genetic drift effect in the natural populations. On the other hand, micropropagated plantlets could be a way to produce uniform material for establishing roseroot plantations.

7.2.3 ADAPTATION OF MICROPROPAGATED PLANTS

The acclimatization of *in vitro* propagated plants is always a critical point. Ishmuratova (1998) used a two-step protocol for acclimatization. Plantlets were transplanted into vermiculite for 2 weeks in 85%–90% relative humidity and then to a 1:1:1 mixture of soil, humus, and vermiculite. Tasheva and Kosturkova (2010) first planted the micropropagated plantlets into a mixture of 1:1:2 perlite, peat, and soil in 90% relative humidity. The survival rate was 57%. After 2–3 months, plants were transferred from the adaptation room to greenhouse with a survival rate of 85%. After 6 months, plants were planted in the field and more than 70% survived the winter. Ghiorghita et al. (2011) accommodated the *in vitro* plantlets in a hydroponic system for a week. This stage was survived by more than 90% of the plantlets. The plantlets were then transferred into soil pots and later planted to their natural habitat to the wild. The first summer was survived by 73.5% of the plants, whereas after the first winter, the percentage of survivors dropped to 57%. Also György and Trócsányi (2012) worked with micropropagated plant material. According to their experience, the key to successful adaptation is the sufficient amount of water. *In vitro* plantlets were planted in 1:2 mixture of perlite and peat and were directly transferred into the greenhouse. After 3 months, the pots were placed to open-air conditions for the winter. The percentage of surviving plants was 70%.

7.2.4 CALLUS INDUCTION

The utilization of callus cultures for the production of the bioactive agents is considered as an alternative way, which is faster and independent from environmental conditions. In case of callogenesis, the results of the different working groups are more consistent than in case of micropropagation. Callus culture has been obtained on MS medium supplemented with different plant hormones. Furmanowa et al. (1995, 1998) concluded that callogenesis is the most effective from leaf explants on BAP along with medium containing NAA or IBA or IAA. The best combination for induction and growth of callus was BAP and IBA. Two strains of callus were described: a deep green and a light cream strain. György et al. (2004, 2005) used also leaf explants and the combination of 1.5 mg/L BAP and 0.5 mg/L NAA. Krajewska-Patan et al. (2007a,b) induced callus from hypocotyls of *in vitro* seedlings on MS medium supplemented with also BAP, NAA, and adenine chloride. Martin et al. (2010) used epicotyls for callus induction and a liquid medium, rather than solid. The most effective combination was found to be 1 mg/L 2,4-D with 1 mg/L IBA or 0.1 or 1 mg/L 2,4-D alone. The main aim of Tasheva and Kosturkova (2010) was the micropropagation of roseroot, but during their thorough experiments, callus-inducing medium composition was also described. Apical buds on MS medium containing 0.2 mg/L BAP and 0.1 mg/L IAA formed a compact, green callus. Leaf explants on medium containing zeatin (2 or 0.2 mg/L zeatin and 0.2 mg/L IAA) produced a poor growing soft callus. Seventy-eight percent of leaf segments cultured on medium containing 2-iP (3 mg/L 2-iP and 0.3 mg/L IAA or NAA) formed compact green callus while 22% formed pale and friable callus. The two previously mentioned strains of roseroot callus were also observed by Ghiorghita et al. (2011).

A compact, green callus was generated on MS medium supplemented with 1 mg/L BAP and 0.5 mg/L 2,4-D from internode fragments, whereas a semi-compact cream color callus developed from leaf fragments inoculated on MS medium with 1 mg/L BAP and 0.5 mg/L 2,4-D.

In 2004, György et al. published an article on the initiation and cultivation of compact callus aggregates (CCA). To establish suspension culture of CCAs, callus from the solid media was gently broken using forceps and was transferred into liquid MS medium containing 0.5 mg/L BAP and 1 mg/L NAA and shaken on a gyratory shaker at 135 rpm. Subcultures were carried out in every 8–10 days by decanting all medium from the flask and adding fresh medium to the cultures. CCA culture was composed of green or light green, spherical, smooth surfaced callus aggregates as described previously by Xu et al. for *Rhodiola sachalinensis*. CCAs were used later in a number of studies detailed later in Sections 7.3.1 and 7.3.2 (György et al. 2005; György 2006; Krajewska-Patan et al. 2007a,b; György and Hohtola 2009).

7.3 BIOTECHNOLOGICAL METHODS FOR INCREASING THE PHARMACEUTICALLY IMPORTANT METABOLITE PRODUCTION IN *IN VITRO* CULTURES OF *RHODIOLA ROSEA*

Tissue culture of *R rosea* has been investigated since the late 1970s. In 1981, Aleksandrova et al. (cited by Furmanowa et al. 1995) patented the method for root regeneration from callus. At that time, the possibility for the production of pharmaceutically active substances of the plant was already being studied. l-Tyrosine, adenine, *p*-fluorophenylalanine, and some other agents were added to the MS medium of *R. rosea* callus cultures. However, none of these agents was found to significantly affect the qualitative or quantitative composition of the target phenol glycosides (Bykov 1986 and Poletaeva et al. 1984 cited by Bykov et al. 1999).

The accumulation of the overall sum of phenolic compounds was found to correlate with the cell growth and could be described by an S-Shaped curve (Bykov et al. 1999). A strain of *R. rosea* was developed for commercial cultivation under the name ZK-1 (Aleksandrova and Danilina 1992 cited by Bykov et al. 1999).

Several pharmacological studies were carried out with the extracts of ZK-1 *R. rosea* callus, and pronounced stimulating and adaptogenic effects were described comparable to those of preparations from natural plant rhizomes (Levina et al. 1987; Krendal 1989; Krendal et al. 1990, all cited by Bykov et al. 1999). The extract was even recommended for clinical trials as an alternative to the extract of intact plants (Levina et al. 1987 cited by Furmanowa et al. 1995). At the same time, extracts of callus were, on the average, only half as active as their natural counterparts with respect to most of the characteristics studied (Krendal et al. 1990 cited by Bykov et al. 1999). The toxicity and mutagenic properties were also studied of this extract (Levina et al. 1987 cited by Bykov et al. 1999). In 1999, Bykov et al. wrote in their review, that by that time the entire complex of preclinical investigation of the tincture of ZK-1 *R. rosea* strain was complete in accordance with the requirements for medicinal preparations to be submitted to the Russian State Pharmacological Committee of the Ministry of Public Health, the given preparation was classified as plant adaptogen with the regular type of action. Subsequent investigations of the authors of this review showed that suspension cultures are even more promising, because their content of triandrin (hydroxylated analogue of rosin) is higher by one order of magnitude than in the callus culture of *R. rosea*.

Poletaeva et al. reported in 1984 (cited by Bykov et al. 1999) that salidroside and tyrosol were extracted from callus cultures of *R. rosea*, but repeated thorough chemical analysis did not confirm these data (Zapesochnaya et al. 1987; Kurkin and Zapesochnaya 1989; Kurkin et al. 1991, all cited by Bykov et al. 1999). Thirteen compounds were isolated from the callus of *R. rosea* and their structures were reported (Kurkin et al. 1992). The main component of suspension cultures was triandrine with 0.19%, while neither salidroside and tyrosol nor rosavin, rosin, and rosarin was found. According to the authors, the *in vitro* culture conditions favor activation of the hydroxylating system of *R. rosea*, and rosin is transformed to triandrine (*p*-hydroxyrosin). Kurkin et al. (1992) suggested that standardization of the extracts from *R. rosea* callus cultures should be based on the determination of the content of phenylpropanoids with the aid of a standard sample of triandrine. While Kurkin et al. (1992) found neither salidroside nor cinnamyl alcohol glycosides in callus, Furmanowa et al. (1995, 1998) reported traces of these glycosides in addition to triandrine and caffeic acid.

7.3.1 PRODUCTION OF SALIDROSIDE IN CALLUS CULTURES

The use of CCAs instead of cell suspensions has been a successful tool in some cases aiming at secondary metabolite production. The application of CCAs has been investigated mostly in connection with production of salidroside by *R. sachalinensis* (Jianfeng et al. 1998a,b; Xu et al. 1999). Culturing CCAs raised the salidroside content sixfold according to Jianfeng et al. (1998b) and threefold according to Xu et al. (1999). The results presented in the study of György et al. (2004) confirm that there is no production of the pharmaceutically important glycosides in callus cultures of *R. rosea*, even in CCA.

Xu et al. (1998a) studied the biotransformation by *R. sachalinensis* cell cultures for producing salidroside. Three possible precursors phenylalanine, tyrosine, and tyrosol were added to the cultures in final concentrations of 0.05, 0.1, 0.5, and 1 mM. All three compounds had adverse effects on the biomass growth in proportion to the concentration (phenylalanine only to a lower extent). One millimolar of tyrosine and tyrosol set it back to one-third of the control. On the other hand, both tyrosine and tyrosol improved the salidroside content. The addition of 1 mM tyrosine gave 1% salidroside and 1 mM tyrosol resulted in 1.44% salidroside, which was nine times higher than in the control. Afterward, Xu et al. (1998b) studied the tyrosol glucosylation in more detail. They determined the activity of the tyrosol glucosyltransferase over the cell growth cycle and found that it was the highest during the exponential growth phase while the intracellular tyrosol accumulated in large amounts at the stationary growth phase. They suggest that this non-synchronization is responsible for the low amounts of salidroside obtained in cell suspensions. Based on this result, 1 mM tyrosol was added at the beginning of the exponential growth phase, which was transformed in 95% after 24 hours. When 0.5, 1, 2, 3, 4, 5 mM tyrosol was added to the cultures, the amount of salidroside formed was proportional to the amount of tyrosol added up to 3 mM. Concentrations higher than 3 mM caused a sharp decrease in the salidroside content. With the repeated addition of 3 mM tyrosol at 24 hour intervals, a 36-fold (516 µmol g^{-1}) increase of the salidroside content was achieved in the cells. Salidroside, however, was not released into the medium. Wu et al. (2003) explored the culturing conditions of *R. sachalinensis* CCAs for a high yield of salidroside. They found that acidic medium and faster shaking speed favorably influenced the salidroside production. Among the several auxins and cytokinins tested, 2,4-D stimulated the salidroside production, but it inhibited the biomass growth. The addition of salicylic acid, phenylalanine, or tyrosine up to 0.5 mM increased the salidroside production slightly, but 4 mM tyrosol boosted the production to 5.77%.

Furmanowa et al. (1999a, 2002) studied the same reaction with *R. rosea* cell cultures. To the cell cultures, 2.5 mM tyrosol was added, of which 50%–67% was transformed after 72 hours and 1.2%–2.3% salidroside yields were obtained. Residual unconverted tyrosol was detected both in the medium and in the cells. The results of György (2006) are similar to those with *R. sachalinensis*. However, up to 3 mM tyrosol did not have such a serious adverse effect as described by Xu et al. (1998b); the growth of biomass was set back by only 10%. Concentration of 2 and 3 mM tyrosol resulted in the highest salidroside yield of 2.62% and 2.72%, respectively. Salidroside was not released into the medium, like in the case of *R. sachalinensis*. After 48 hours, 100% of the tyrosol was transformed when 2 mM tyrosol was added. At this concentration, tyrosol was neither detected in the medium nor in the cells at the end of the experiments.

In 2007, Krajewska-Patan and coworkers published an article on the comparison of biosynthesis of salidroside through biotransformation of tyrosol by roseroot callus cultured on solid media and CCA cultured in liquid media (Krajewska-Patan et al. 2007b). Tyrosol was added to the solid and liquid medium at final concentrations of 5 mM and 2.5 mM (only to the liquid medium), and 7 days after the inoculation, the calli were collected for chemical analysis. Both treatments resulted in the increase of tyrosol and salidroside. The yield of tyrosol was 0.6% in callus on solid medium and 0.8% in liquid cultivated CCAs, while of salidroside was 3.1% on solid medium and 4.3% in liquid medium. Addition of 2.5 mM tyrosol to the CCAs yielded 3.5% of salidroside. An adverse effect on the biomass growth was observed, but described only in case of CCA cultures; 20% set back at 5 mM tyrosol concentration and 9% at 2.5 mM concentration. Similarly to the results of György (2006), neither tyrosol, nor salidroside was released into the medium.

The possibility of elicitation was also studied by Krajewska-Patan et al. (2001, 2002). Yeast extract was added to the media of callus cultures in a concentration of 5 mg/culture. Salidroside content was determined after elicitation of 1–7 days. Salidroside content increased with the duration of elicitation until the 5th day and decreased on the 6th and 7th days. Salidroside content in the elicited callus was 1.5% compared to the 0.8% of the nontreated control. Also the immune-stimulating and antistress properties of the callus extract were studied *in vitro* alongside with plant extracts. Although the salidroside content of the elicited callus was twice as much as the salidroside content of roots, the detected immune-stimulating activity was lower compared to the plant extract.

7.3.2 PRODUCTION OF ROSAVINS IN CALLUS CULTURES

The production of the cinnamyl alcohol glycosides by biotransformation is much less studied, which can be explained by the fact that these compounds are specific for *R. rosea*. According to Furmanowa et al. (1999b), more than 90% of cinnamyl alcohol—when added in 2.5 mM to cell suspensions of roseroot—was transformed into several products, but only rosavin was identified. After 72 hours, 0.03%–1% rosin accumulated in the cells and was not excreted into the medium. The results obtained by György and coworkers are more detailed. In addition to rosin, rosavin and four new products were detected and identified (Tolonen et al. 2004). The optimal cinnamyl alcohol concentration was found to be 2 mM (György et al. 2004), since the resulting rosin concentration was the highest at that concentration. This amount of cinnamyl alcohol did not have an adverse effect on the biomass growth, as seen with higher concentrations, and all cinnamyl alcohol was converted at this concentration, whereas using higher concentrations, residual amounts were detected in the medium. The rosin content was at its highest 3 days after the precursor addition and it decreased if the cells were further cultivated. The maximum achieved rosin contents were between 0.4% and 1.25%. The repeated addition of 2 mM cinnamyl alcohol at 3-day intervals did not improve the rosin production as was demonstrated with salidroside by Xu et al. (1998b); but it did inhibit the production (György et al. 2004). The four new compounds identified (Tolonen et al. 2004) are all closely related to rosin and rosavin. Compound "321" differs from rosin by opening the double bond in the middle of the propyl chain of the aglycone, while "337" differs by an extra hydroxyl group at the C-8 position. Compound "481" has an extra hydroxyl group on the 3rd carbon of the second glucose compared to rosavin, and compound "483" is very similar, but again the double bond in the middle of the propyl chain of the aglycone is opened. The presence of the many closely related products after the biotransformation indicates that either several enzymes take part in the glucosylation of cinnamyl alcohol or at least some of the products form spontaneously.

For increasing the glucosylation of cinnamyl alcohol and tyrosol, a simple trick was applied (György et al. 2005). Since the MS medium contains only sucrose as a sugar source, glucose was added into the medium to be directly used in the glucosylation reaction. This approach was very effective and beneficial for the production of the cinnamyl alcohol glycosides; yields were doubled compared to the control. Rosavin was only produced in the glucose-containing media. However, the salidroside production was not affected at all. This ambiguous effect of sugars on secondary metabolite production was unexpected, but it is not uncommon. Later, Krajewska-Patan et al. (2007a) studied biotransformation of cinnamyl alcohol in callus cultures both on solid and in liquid media. Two types of calli were used in the experiments: callus induced from hypocotyls of seedlings or axillary buds of intact plants. Cinnamyl alcohol was added to the solid and liquid media in final concentrations of 2.5 and 5 mM. The chemical analysis of the samples was performed on the 7th day after the inoculation. The supplementation of 2.5 mM cinnamyl alcohol resulted only in the enhancement of the desired glycosides. This is in line with the findings of György et al. (2004), who suggested the addition of at most 2 mM cinnamyl alcohol; there is no benefit to adding more. The biotransformation was more efficient in the hypocotyl-derived callus but callus from the axillary buds also showed similar tendency. The addition of 2.5 mM cinnamyl alcohol resulted 1056.18 mg/100 g rosin on solid medium and 776.33 mg/100 g in liquid medium. The content of rosavin also increased, although to a lower extent: it was 48.66 mg/100 g on solid medium and 63.60 mg/100 g in liquid cultures. The biotransformation also enhanced the rosarin content, but only in a very low amount (4.89 mg/100 g). The authors mention that the supplementation with cinnamyl alcohol influenced the biosynthesis of tyrosol and salidroside as well. It was emphasized that the control nontreated callus also produced all spectra of glycosides of the intact plant, not like in the works of György et al. (2004, 2005). This difference is probably due to the usage of different genotypes as starting material. Both the work of Krajewska-Patan et al. (2007a) and György et al. (2004, 2005) concluded that with biotransformation of cinnamyl alcohol, mainly rosin can be synthesized, which is the simplest glycoside of cinnamyl alcohol. This is contrary to the results of Furmanowa et al. (1999a,b) where rosavin was found in high amounts (1.1% of the dry weight).

7.4   ON THE WAY TO METABOLIC ENGINEERING OF *RHODIOLA ROSEA*

Salidroside and cinnamyl alcohol glycosides are the most important bioactive compounds of roseroot that have arisen from the phenylpropanoid metabolism via the biosynthesis of aromatic L-amino acids—phenylalanine and tyrosine—in the shikimate pathway (György 2006). A recent and thorough review on the plausible biosynthetic pathway of salidroside and cinnamyl alcohol glycosides is given by Mirmazloum and György (2012).

Phenylalanine ammonia-lyase (PAL) is the first enzyme that can catalyze the phenylalanine deamination and *trans*-cinnamic acid production (Hahlbrock and Scheel 1989), which is the common precursor for the synthesis of both cinnamyl alcohol glycosides and salidroside. PAL activity has been found in all higher plants, and several studies (Cochrane et al. 2004) have been carried out to determine the enzyme characteristics and corresponding gene(s). Ma et al. (2008) reported a cloning and expression pattern of a cDNA encoding a PAL (PALrs1) from *R. sachalinensis*. It turned out that PALrs1 protein, in amino acid level shows about 80% identity with the prior isolated cDNA clones from different investigated dicotyledons. No molecular work has been reported in regard with PAL gene(s) in the case of *R. rosea*.

7.4.1   BIOSYNTHESIS OF CINNAMYL ALCOHOL GLYCOSIDES

In the proposed biosynthesis pathway of cinnamyl alcohol glycosides (György 2006), *trans*-cinnamic acid is converted to cinnamoyl-CoA by the activity of an enzyme, named 4-coumarate-CoA ligase (4CL) via a two-step reaction mechanism that involves the hydrolysis of ATP (Gross and Zenk 1974). 4CL catalyzes the activation of 4-cinnamate and various other cinnamic acid derivatives by forming their corresponding CoA thioesters (Kumar and Ellis 2003). In higher plants, 4CL typically occurs as a gene family consisting of two to three members. The pivotal activity of

4CL is widely studied in lignin biosynthesis pathway, but no gene encoding 4CL activity has been reported from *Rhodiola* species concerning the synthesis of cinnamyl alcohol glycosides.

The reduction of cinnamoyl-CoA to cinnamaldehyde formation is catalyzed by the enzyme cinnamoyl-CoA oxidoreductase (CCR). CCR was first purified from soybean (Wegenmayer et al. 1976). CCR is considered to be the first enzyme committed toward the biosynthesis of monolignols (hydroxycinnamyl alcohols) and shows homology to the flavonoid biosynthetic gene flavonol 4-reductase (Lacombe et al. 1997).

Subsequently, cinnamyl alcohol dehydrogenase (CAD) reduces the cinnamaldehyde to cinnamyl alcohol. CAD is also encoded by a multigene family in *Arabidopsis* (Goujon et al. 2003) and probably in many other species. Conversion of hydroxy-cinnamaldehydes to their corresponding alcohols appears to be catalyzed by a combination of CAD isoforms, some of which have a preference toward one of the available substrates.

The enzyme(s) that take part in the formation of the glycosides of cinnamyl alcohol are not yet described. Rosin is the simplest glycoside of roseroot which is formed when one molecule of glucose attaches to the cinnamyl alcohol. Rosarin can be formed from rosin, by the connection of an arabinose molecule, and rosavin, by the connection of an arabinofuranose molecule. Depending on the sugar type and the site it is attached to, further glycosides may be formed (György 2006).

7.4.2 Biosynthesis of Salidroside

Salidroside is a product of dehydration between the hemiacetal hydroxyl of glucose and the ethanol hydroxyl of tyrosol (4-hydroxyphenylethanol) catalyzed by the activity of glycosyltransferase. Therefore, the biosynthesis of salidroside can be divided into two steps: biosynthesis of tyrosol and the combination of glucose and tyrosol to form salidroside. There are several possible routes for tyrosol formation in plants based on the literature and the KEGG database utilizing both phenylalanine and tyrosine as precursor (Mirmazloum and György 2012).

7.4.3 Biosynthesis of Tyrosol

The most plausible pathway for the tyrosine-derived tyrosol (Ma et al. 2008; György et al. 2009; Zhang et al. 2011) is initiated by the activity of tyrosine decarboxylase (TyrDC) enzyme that can catalyze the formation of tyramine. The enzyme activity and its characteristics have been studied in *Rhodiola* species. A deduced amino acid sequence of *R. sachalinensis* TyrDC showed around 60% identity with those of other TyrDCs of plant origin (Zhang et al. 2011) and 79% identity with studied *R. rosea* TyrDC by György et al. (2009). Tyramine then can be converted to 4-hydroxyphenylacetaldehyde by means of tyramine oxidase (EC: 1.4.3.4) proposed by Landtag et al. (2002).

The role of TyrDC has been supported in roseroot plants by molecular research results. György et al. (2009) have isolated a cDNA encoding for *TyrDC* and found that in plants exhibiting higher salidroside content the expression of this gene was considerably higher than in the low salidroside producer line. Further the expression of the gene was higher in the roots (where salidroside accumulates) than in the leaves. Zhang et al. (2011) reported a cloning and expression pattern of a cDNA encoding TyrDC from *R. sachalinensis*. The biochemical assays of recombinant RsTyrDC and the effects of sense and antisense overexpression of endogenous RsTyrDC in *R. sachalinensis* demonstrated that the RsTyrDC is most likely to have an important function in the initial steps of the salidroside biosynthesis pathway. The functioning role of TyrDC in tyrosol formation and salidroside biosynthesis is also supported by the results of Ma et al. (2007). They proved that tyrosol biosynthesis does not have any direct linkage with p-coumaric acid accumulation in *R. sachalinensis* and these results demonstrate that it is highly unlikely that p-coumaric acid is a precursor for tyrosol biosynthesis. Another supportive result for this hypothesis is in the work of Landtag et al. (2002), where potato was transformed to express parsley TyrDC to study what role tyramine plays in response to *Phytophthora infestans* infection and whether the expression leads to higher tyramine-derived compounds. However, the reaction did not turn out as expected. Instead, it led to the accumulation of another compound, which has not been reported from potato previously, and was identified as tyrosol glycoside, that is salidroside.

7.4.4 SALIDROSIDE FORMATION

The final step in salidroside formation is the connection of a glucose molecule to tyrosol. It is likely that UDP-glycosyltransferase is the enzyme catalyzing the final reaction by means of deploying UDP-Glc (a typical activated sugar in plants) as a glucose donor and tyrosol as its aglycon (Ma et al. 2007). Currently, the Carbohydrate-Active Enzyme database (CAZy, http://www.cazy.org) contains over 12,000 sequences classified in 90 families, encoding glycosyltransferases (GTs) in different organisms. Among the GTs contributing to the many different glycosyl transfer reactions, those glycosylating natural products and small lipophilic molecules belong to the GT1 family. Sugars can be transferred to the OH, SH, NH, and carboxyl groups of the acceptors (Vaistij et al. 2009). Glycosides may also be further glycosylated. Secondary metabolites, such as shikimate derivatives (phenylpropanoids, benzoates, flavonoids), alkaloids, and many amino acid derivatives exist as glycosides in plants. Glycosylation can alter the solubility and transport of the compounds within the cell, stabilizing the product and modulating biodisponibility and storage (Vogt and Jones 2000).

In connection with salidroside formation, Ma et al. (2007) reported an expression pattern of a cDNA encoding a UDP-glycosyltransferase that regulates the conversion of tyrosol to salidroside in *R. sachalinensis*. Overexpression of the endogenous UDP-glycosyltransferase (UGT73B6) increased the salidroside content to double compared with that of untransformed controls in the transgenic *R. sachalinensis* although the lack of tyrosol supply may be a limiting factor for further accumulation of salidroside (Ma et al. 2007).

In their recent research, Yu et al. (2011) studied the overexpression of two endogenous UDP-glycosyltransferases, UGT72B14 and UGT74R1 in hairy root lines of *R. sachalinensis*. UGT72B14 transcripts were mainly detected in roots and exhibited the highest level of activity for salidroside production *in vitro* and *in vivo*. Accordingly, they suggested that UGT72B14 has an important function in this pathway. Compared to UGT74R1, UGT73B6 (Ma et al. 2007) showed a higher level of activity for salidroside production, and this isozyme was highly expressed in the roots, so it is very likely that UGT73B6 can contribute to salidroside synthesis. Compared to the UGT73B6 transgenic plants and transgenic callus systems, the hairy root culture system exhibited the highest level of salidroside production.

7.4.5 AGROBACTERIUM-MEDIATED TRANSFORMATION

To facilitate metabolic engineering of *R. rosea*, Mirmazloum et al. (2012) developed a method for the transformation of roseroot callus. An antibiotic-resistant EHA101 (pTd33) strain of *A. tumefaciens* carrying a *gusA* gene was used in model experiments on roseroot callus transformation on solid and in liquid cultures. The T-DNA of pTd33 binary vector plasmid harbors an *npt ll* gene conferring resistance to kanamycin, and a *gusA* reporter gene encodes the β-glucuronidase enzyme. The callus on solid media was cocultivated with Agrobacterium for 48 hours and in liquid cultures for 20 hours. Subculturing was done on selection media (solid and liquid, respectively) supplemented with kanamycin, claforan (cefotaxim), and carbenicillin for selecting the antibiotic-resistant cells. The green and healthy calli were transferred to fresh medium of the same composition every 2 weeks for further selections. In liquid media, an intermediate culture with claforan and carbenicillin (without kanamycin) had to be conducted for 2 weeks to reduce the stress level for the cells. The GUS test using a triton buffer, verified the transformation of the calluses on the solid media 2 weeks after the transformation. The optimized method in this experiment can be a tool for inserting the gene of interest to the plant genome to increase or improve the quantity or potential quality of these metabolites in bioengineering projects or even for bioreactor culture of *R. rosea* plant.

ACKNOWLEDGMENTS

The author is grateful to Iman Mirmazloum for his help during the preparation of the manuscript. Hungarian scientific research fund (OTKA PD83728), TÁMOP-4.2.1/B-09/1/KMR-2010-0005, and TÁMOP-4.2.2/B-10/1/2010-0023 are acknowledged for financial support.

REFERENCES

Bykov, V.A., Zapesochnaya, G.G., Kurkin, V.A. 1999. Traditional and biotechnological aspects of obtaining medicinal preparations from *Rhodiola rosea* L. (a review). *Pharmaceutical Chemistry Journal* 33(1):29–40.

Cochrane, F.C., Davin, L.B., Lewis, N.G. 2004. The Arabidopsis phenylalanine ammonia lyase gene family: Kinetic characterization of the four PAL isoforms. *Phytochemistry* 65:1557–64.

Debnath, S.C. 2009. Zeatin and TDZ-induced shoot proliferation and use of bioreactor in clonal propagation of medicinal herb, roseroot (*Rhodiola rosea* L). *Journal of Plant Biochemistry and Biotechnology* 18(2):245–8.

Furmanowa, M., Hartwich, M., Alfermann, A.W. 1999a. Salidroside as a product of biotrans-formation by *Rhodiola rosea* cell suspension cultures. *Book of Abstracts, 2000 Years of Natural Products Research*. July 26–30, 1999, Amsterdam, Holland. p. 152.

Furmanowa, M., Hartwich, M., Alfermann, A.W. 2002. Glucosylation of p-tyrosol to salidro-side by *Rhodiola rosea* L. cell cultures. *Herba Polonica* 48(2):71–6.

Furmanowa, M., Hartwich, M., Alfermann, A.W., Kozminski, W. Olejnik, M. 1999b. Rosavin as a product of glycosylation by *Rhodiola rosea* (roseroot) cell cultures. *Plant Cell, Tissue and Organ Culture* 56:105–10.

Furmanowa, M., Oledzka, H., Michalska, M., Sokolnicka, I., Radomska, D. 1995. *Rhodiola rosea* L. (Roseroot): *In vitro* regeneration and the biological activity of roots. *Biotechnology in Agriculture and Forestry Vol. 33, Medicinal and Aromatic Plants* VIII: 412–26.

Furmanowa, M., Skopinska, R.E., Rogala, E., Hartwich, M. 1998. *Rhodiola rosea in vitro* culture: Phytochemical analysis and antioxidant action. *Acta Societatis Botanicorum Polonia* 67(1):69–73.

Ghiorghita, G., Hartan, M., Maftei, D.E., Nicuta, D. 2011. Some considerations regarding the *in vitro* culture of *Rhodiola rosea* L. *Romanian Biotechnological Letters* 16(1):5902–8.

Goujon, T., Sibout, R., Eudes, A., MacKay, J., Jouanin, L. 2003. Genes involved in the biosyn-thesis of lignin precursors in *Arabidopsis thaliana*. *Plant Physiology and Biochemistry* 41:677–87.

Gross, G.G., Zenk, M.H. 1974. Isolation and properties of hydroxycinnamate: CoA ligase from lignifying tissue of *Forsythia*. *European Journal of Biochemistry* 42:453–9.

György, Z. 2006. Glycoside production by *in vitro Rhodiola rosea* cultures. *Acta Universitatis Ouluensis C Technica* 244, Oulu University Press, Oulu.

György, Z., Hohtola, A. 2009. Production of cinnamyl glycosides in compact callus aggregate cultures of *Rhodiola rosea* through biotransformation of cinnamyl alcohol. In: Mohn, J., Saxena P.K. (Eds.). *Protocols for In Vitro Cultures and Secondary Metabolite Analysis of Aromatic and Medicinal Plants*. The Humana Press, Inc., Springer, New York, USA, pp. 305–12.

György, Z., Jaakola, L., Neubauer, P., Hohtola, A. 2009. Isolation and genotype-dependent, organ-specific expression analysis of a *Rhodiola rosea* cDNA encoding tyrosine decar-boxylase. *Journal of Plant Physiology* 166:1581–6.

György, Z., Tolonen, A., Neubauer, P., Hohtola, A. 2005. Enhanced biotransformation capacity of *Rhodiola rosea* callus cultures for glycosid production. *Plant Cell Tissue and Organ Culture* 83:129–35.

György, Z., Tolonen, A., Pakonen, M., Neubauer, P., Hohtola, A. 2004. Enhancement of the production of cinnamyl glycosides in CCA cultures of *Rhodiola rosea* through biotrans-formation of cinnamyl alcohol. *Plant Science* 166(1):229–36.

György, Z., Trócsányi, E. 2012. *Rhodiola rosea* mikroszaporítása. (Micropropagation of rose-root.) XVIII. Növénynemesítési Tudományos Napok, 2012. In: Veisz O. (Eds.). *Book of Abstracts*. MTA, Budapest, p. 82.

Hahlbrock, K., Scheel, D. 1989. Physiology and molecular biology of phenylpropanoid metab-olism. *Annual Review of Plant Physiology and Plant Molecular Biology* 40:347–69.

Ishmuratova, M.M. 1998. Clonal micropropagation of *Rhodiola rosea* L. and *R. iremelica* Boriss. in vitro. *Plant Resources* 34:12–23 (in Russian).

Jianfeng, X., Jian, X., Pusun, F., Zhiguo, S. 1998a. Suspension nodule culture of the Chinese herb, *Rhodiola sachalinensis*, in an air-lift reactor: Kinetics and technical characteristics. *Biotechnology Techniques* 12(1):1–5.

Jianfeng, X., Zhiguo, S., Pusun, F. 1998b. Suspension culture of compact callus aggregate of *Rhodiola sachalinensis* for improved salidroside production. *Enzyme and Microbial Technology* 23:20–7.

Krajewska-Patan, A., Dreger, M., Lowicka, A., Górska-Paukszta, M., Mscisz, A., Mielcarek, S., Baraniak, M., Buchwald, W., Furmanowa, M., Mrozikiewicz, P.M. 2007a. Chemical investigations of biotransformed *Rhodiola rosea* callus tissue. *Herba Polonica* 53(4):77–87.

Krajewska-Patan, A., Furmanowa, M., Dreger, M., Górska-Paukszta, M., Lowicka, A., Mscisz, A., Mielcarek, S., Baraniak, M., Buchwald, W., Mrozikiewicz, P.M. 2007b. Enhancing the biosynthesis of salidroside by biotransformation of p-tyrosol in callus culture of *Rhodiola rosea* L. *Herba Polonica* 53(1):55–64.

Krajewska-Patan, A., Mscisz, A., Kedzia, B., Lutomski, J. 2002. The influence of elicitation on the tissue cultures of roseroot (*Rhodiola rosea* L.). *Herba Polonica* 48(2):77–81.

Krajewska-Patan, A., Mscisz, A., Lutomski, J. 2001. The influence of elicitation on the tissue cultures of roseroot (*Rhodiola rosea* L.). *Book of Abstracts of the International Congress and 49th Annual Meeting of the Society for Medicinal Plant Research*, September 2–6, 2001, Erlangen, Germany.

Kumar, A., Ellis, B.E. 2003. 4-Coumarate: CoA ligase gene family in *Rubus idaeus*: cDNA structures, evolution, and expression. *Plant Molecular Biology* 31:327–40.

Kurkin, V.A., Zapesochnaya, G.G., Dubichev, A.G., Vorontsov, E.D., Aleksandrova, I.V., Panova, R.V. 1992. Phenylpropanoids of callus culture of *Rhodiola rosea*. *Chemistry of Natural Compounds* 27(4):419–25.

Lacombe, E., Hawkins, S., Van Doorsselaere, J., Piquemal, J., Goffner, D., Poeydomenge, O., Boudet, A.M., Grima-Pettenati, J. 1997. Cinnamoyl CoA reductase, the first committed enzyme of the lignin branch biosynthetic pathway: Cloning, expression and phylogenetic relationships. *Plant Journal* 11:429–41.

Landtag, J., Baumert, A., Degenkolb, T., Schmidt, J., Wray, V., Scheel, D., Strack, D., Rosahl, S. 2002. Accumulation of tyrosol glucoside in transgenic potato plants expressing a parsley tyrosine decarboxylase. *Phytochemistry* 60:683–9.

Ma, L.Q., Gao, D.Y., Wang, Y.N., Wang, H.H., Zhang, J.X., Pang, X.B., Hu, T.S., Lu, S.Y., Li, G.F., Ye, H.C., Li, Y.F., Wang, H. 2008. Effects of overexpression of endogenous phenylalanine ammonia-lyase (PALrs1) on accumulation of salidroside in *Rhodiola sachalinensis*. *Plant Biology* 10:323–33.

Ma, L.Q., Liu, B.Y., Gao, D.Y., Pang, X.B., Lu, S.Y., Yu, H.S., Wang, H., Yan, F., Li, Z.Q., Li, Y.F., Ye, H.C. 2007. Molecular cloning and overexpression of a novel UDP-glucosyltransferase elevating salidroside levels in *Rhodiola sachalinensis*. *Plant Cell Reports* 26:989–99.

Martin, J., Pomahacova, B., Dusek, J., Duskova, J. 2010. *In vitro* culture establishment of *Schizandra chinensis* (Turz.) Baill. and *Rhodiola rosea* L., two adaptogenic compounds producing plants. *Journal of Phytology* 2(11):80–7.

Mirmazloum, I., György, Z. 2012. Review of the molecular genetics in higher plants toward salidroside and cinnamyl alcohol glycosides biosynthesis in *Rhodiola rosea* L. *Acta Alimentaria* 41:133–46.

Mirmazloum, I., Forgács, I., Zok, A., György, Z. 2012. Transzgénikus *Rhodiola rosea* kallusz kultúra létrehozása. (Transgenic roseroot callus culture establishment) XVIII. Növénynemesítési Tudományos Napok, 2012. Összefoglalók (Szerk: Veisz O.), MTA, Budapest. p. 90.

Tasheva, K., Kosturkova, G. 2010. Bulgarian golden root *in vitro* cultures for micropropagation and reintroduction. *Central European Journal of Biology* 5(6):853–63.

Tasheva, K., Kosturkova, G. 2012. The role of biotechnology for conservation and biologically active substances production of *Rhodiola rosea*: Endangered medicinal species. *The Scientific World Journal* 2012: Article ID 274942 13p.

Tolonen, A., György, Z., Jalonen, J., Neubauer, P., Hohtola, A. 2004. LC/MS/MS identification of glycosides produced by biotransformation of cinnamyl alcohol in *Rhodiola rosea* compact callus aggregates. *Biomedical Chromatography* 18:550–8.

Vaistij, F.E., Lim, E.K., Edwards, R., Bowles, D.J. 2009. Glycosylation of Secondary metabolites and xenobiotics. In: Osbourn, A.E., Lanzotti, V. (Eds.). *Plant-Derived Natural Products: Synthesis, Function, and Application*, Volume 1. Springer: New York, NY. pp. 209–28.

Vogt, T., Jones, P. 2000. Glycosyltransferases in plant natural product synthesis: Characterization of a supergene family. *Trends in Plant Science* 5:380–6.

Wegenmayer, H., Ebel, J., Grisebach, H. 1976. Enzymic synthesis of lignin precursors: Purification and properties of a cinnamoyl-CoA: NADPH reductase from cell suspension cultures of soybean (*Glycine max* L.). *European Journal of Biochemistry* 65:529–36.

Wu, S., Zu, Y., Wu, M. 2003. High yield production of salidroside in the suspension culture of *Rhodiola sachalinensis*. *Journal of Biotechnology* 106:33–43.

Xu, J.F., Liu, C.B., Han, A.M., Feng, P.S., Su, Z.G. 1998a. Strategies for the improvement of salidroside production in cell suspension cultures of *Rhodiola sachalinensis*. *Plant Cell Reports* 17:288–93.

Xu, J.F., Su, Z.G., Feng, P.S. 1998b. Activity of tyrosol glucosyltransferase and improved salidroside production through biotransformation of tyrosol in *Rhodiola sachalinensis* cell cultures. *Journal of Biotechnology* 61:69–73.

Xu, J.F., Ying, P.Q., Han, A.M., Su, Z.G. 1999. Enhanced salidroside production in liquid-cultivated compact callus aggregates of *Rhodiola sachalinensis*: Manipulation of plant growth regulators and sucrose. *Plant Cell, Tissue and Organ Culture* 55:53–8.

Yin, W.B., Li, W., Du, G.S., Huang, Q.N. 2004. Studies on tissue culture of Tibetan *Rhodiola rosea*. *Acta Botany Boreal-Occident Sinica* 24:1506–10.

Yu, H.S., Ma, Q.L., Zhang, J.X., Shi, G.L., Hu, Y.H., Wang, Y.N. 2011. Characterization of glycosyltransferases responsible for salidroside biosynthesis in *Rhodiola sachalinensis*. *Phytochemistry* 72:862–70.

Zhang, J.X., Ma, L.Q., Yu, H.S., Zhang, H., Wang, H.T., Qin, Y.F., Shi, G.S., Wang, Y.N. 2011. A tyrosine decarboxylase catalyzes the initial reaction of the salidroside biosynthesis pathway in *Rhodiola sachalinensis*. *Plant Cell Reports* 30(8):1443–53.

8 Pharmacological Activities of *Rhodiola rosea*

*Fida Ahmed, Steffany A.L. Bennett,
and John T. Arnason*

CONTENTS

8.1 Introduction 189
8.2 Effects of *R. rosea* on the Central Nervous System 192
    8.2.1 Antidepressant Activity of *R. rosea* 192
    8.2.2 Effects of *R. rosea* on Cognitive Function and
          Mental Performance 193
    8.2.3 Antistress (Adaptogenic) Effects of *R. rosea* 194
    8.2.4 Anxiolytic Activity of *R. rosea* 194
8.3 Antioxidant Effects of *R. rosea* 195
8.4 Effects of *R. rosea* on Physical Endurance 195
8.5 Antidiabetic Effects of *R. rosea* 196
8.6 Other Biological Activities of *R. rosea* 197
8.7 Pharmacokinetics, Bioavailability, and Herb–Drug Interactions 197
8.8 Concluding Remarks 198
References 199

8.1 INTRODUCTION

Rhodiola rosea is classified as an adaptogen, placing it in the same category as *Panax ginseng* C.A. Mey. (Araliaceae), *Eleutherococcus senticosus* (Rupr. & Maxim.) Maxim. (Araliaceae), *Schisandra chinensis* (Turcz.) Baill. (Schisandraceae), and *Withania somnifera* (L.) Dunal (Solanaceae). The term *adaptogen*, initially introduced by Russian pharmacologist N.V. Lazarev in 1947, is used to describe plants, herbal mixtures, and (or) compounds that offer nonspecific resistance against a wide variety of physical, chemical, and biological stressors (Panossian and Wikman 2010). Adaptogens are hypothesized to act as metabolic regulators, restoring the homeostasis of physiological systems altered in disease or under stressful conditions and thus allowing the organism to adapt (Panossian et al. 1999; Panossian 2013). In addition, they are also proposed to have a stimulatory effect on physical and mental capabilities, particularly under stressful situations (Panossian and Wagner 2005).

Apart from their good safety and tolerability profiles, adaptogens are characterized by their pleiotropic modes of action and therapeutic effects upon single (Panossian and Wagner 2005) as well as repeated administration (Panossian et al. 2009a).

Research on the adaptogenic properties of *R. rosea* began in the former Soviet Union in the 1960s but has come to prominence in the "Western" scientific literature only in the last decade or so (Panossian et al. 2010). *Rhodiola rosea* extracts, particularly from the roots and (or) rhizomes, have been tested *in vitro* as well as *in vivo* in many cellular and animal models of human diseases. It has been reported to have broad-spectrum pharmacological activities, including neuroprotective, antidepressant, anxiolytic, antistress, antioxidant, cardioprotective, and fatigue-reducing effects among others. In this chapter, the biological effects of *R. rosea* in preclinical studies are described, as well as the underlying mode of action. Table 8.1 lists selected pharmacological activities of *R. rosea* extracts and the model systems/bioassays that were used. The potential of herb–drug interactions as well as bioavailability and pharmacokinetic studies involving *R. rosea* are also discussed.

TABLE 8.1
Summary of Selected Pharmacological Activities of *Rhodiola rosea*

Biological Activity	Model System/Bioassay	References
Antidepressant	*In vivo*: male Sprague–Dawley rats exposed to chronic stress	Chen et al. (2008a, 2009a)
	In vivo: male Wistar rats; Porsolt behavioral despair assay	Panossian et al. (2008)
	In vivo: female Wistar rats exposed to chronic mild stress paradigm	Mattioli et al. (2009)
Antistress	*In vivo*: male Wistar rats exposed to physical and CRF-induced stress	Mattioli and Perfumi (2007)
	In vivo: male Chinchilla rabbits subjected to immobilization	Panossian et al. (2007)
Antidepressant, adaptogenic, anxiolytic	*In vivo*: female BALB/c mice; forced swim test	Panossian et al. (2009a)
	In vivo: male CD-1 mice Forced swim test (antidepressant) Swimming to exhaustion (adaptogenic) Light dark, open field (anxiety)	Perfumi and Mattioli (2007)
Anxiolytic	*In vivo*: male ICR mice Hole-board, elevated plus-maze (anxiety)	Montiel-Ruiz et al. (2012)
Neuroprotective	*In vitro*: acetylcholinesterase inhibition	Hillhouse et al. (2004), Wang et al. (2007)
	In vitro: monoamine oxidase inhibition	van Diermen et al. (2009)
	In vivo: male SD rats-intracerebroventricular streptozotocin insult; Morris Water Maze-acquisition, probe trials (spatial learning and memory)	Qu et al. (2009)
	In vitro: HCN 1-A cell line exposed to hydrogen peroxide and glutamate	Palumbo et al. (2012)

TABLE 8.1 (*Continued*)
Summary of Selected Pharmacological Activities of *Rhodiola rosea*

Biological Activity	Model System/Bioassay	References
Drug (morphine) addiction recovery	*In vivo*: male CD-1 mice; morphine-induced conditioned place preference	Mattioli and Perfumi (2011), Mattioli et al. (2012)
Antioxidant	*In vitro*: human erythrocytes exposed to hypochlorous acid	De Sanctis et al. (2004)
	In vitro: singlet oxygen, H_2O_2, hypochlorite, ferric reducing, ferrous chelating, and protein thiol protection	Chen et al. (2008b)
	In vitro: xanthine oxidase, lipoxygenase tyrosinase	Chen et al. (2009b)
	In vitro: xanthine oxidase, DPPH free radical scavenging	Horng et al. (2010)
Protection against oxidative stress without antioxidant effects	*In vitro*: human osteosarcoma-derived 143B, human diploid fibroblast IMR-90, and human neuroblastoma IMR-32 cells exposed to UV, paraquat, and H_2O_2	Schriner et al. (2009)
Antihyperglycemic	*In vitro*: α-amylase from ddY strain mouse plasma	Kobayashi et al. (2003)
	In vitro: Porcine pancreatic α-amylase; α-glucosidase Angiotensin I-converting enzyme (ACE) from rabbit lung	Kwon et al. (2006)
	In vivo: C57BL/Ks db/db mice	Sung et al. (2006)
	In vitro: lipase from cardiac blood of ddY mice	Kobayashi et al. (2004)
	In vitro: pRB-deficient mouse embryonic fibroblasts (ME3); 3T3-L1 pre-adipocyte cells	Christensen et al. (2009)
Antifatigue	*In vivo*: Sprague–Dawley rats; exhaustive swimming	Abidov et al. (2003)
	In vivo: white mice, forced swim test	Kurkin et al. (2006)
	In vivo: male Wistar rats; weight-loaded forced swimming test	Lee et al. (2009)
Anticancer	*In vivo*: animals with transplanted tumors, Ehrlich adenocarcinoma, Lewis lung carcinoma	Goldberg et al. (2004)
	In vitro: promyelocytic leukemia cells of the HL-60 line	Majewska et al. (2006)
	In vitro: bladder cancer UMUC3 cells	Liu et al. (2012)
Protection against ischemia/ reperfusion injury	*In vitro*: isolated hearts of male Wistar rats	Afanas'ev et al. (1997)
	In vivo: male Wistar rats	Maslov et al. (2009)
Antiarrhythmia (cardioprotective)	*In vivo*: male Wistar rats exposed to epinephrine induced arrhythmia	Maslov et al. (1998)
Cardioprotective/ antihypoxia	*In vivo*: hypobaric hypoxia (male albino mice) *In vivo*: coronary occlusion (Wistar rats)	Arbuzov et al. (2006)

8.2 EFFECTS OF *R. rosea* ON THE CENTRAL NERVOUS SYSTEM

A number of preclinical and clinical studies have investigated the neurological activities of *R. rosea*. Many of these are in Slavic and Scandinavian languages or in Chinese, and are thus inaccessible for general evaluation. Within English language scientific literature, *Rhodiola rosea* affects the central nervous system (CNS), as evidenced by its antidepressant, neuroprotective, anxiolytic, and antistress properties reviewed below. These effects are purported to be mediated by the action of *R. rosea* extracts and (or) its phytochemicals on the neuroendocrine system, via modulation of key neurotransmitter systems, the hypothalamic–pituitary–adrenal gland (HPA) axis; and the sympatho-adrenal system.

8.2.1 ANTIDEPRESSANT ACTIVITY OF *R. ROSEA*

Depression encompasses a group of several disorders that arise from a complex interaction of genetic, environmental, psychological, and biochemical factors (Saveanu and Nemeroff 2012). It is widely prevalent, particularly in developed countries, and contributes to considerable disability, comorbidity, and mortality (Kessler 2012; Saveanu and Nemeroff 2012). The pathophysiology of depression is complex, involving reduced monoaminergic signaling, dysregulation of the HPA axis, oxidative stress, and increased production of inflammatory cytokines (Chopra et al. 2011). First-line therapies generally include drugs that improve neurotransmission of monoamines, such as selective serotonin and norepinephrine reuptake inhibitors (Connolly and Thase 2012). While these drugs provide temporary attenuation of depressive symptoms, the adverse effects associated with their chronic use have led to the investigation of safer and more effective alternatives. Kessler et al. (2001) reported that complementary and alternative therapies, including the use of herbal remedies are highly preferred adjuvant therapies among patients suffering from mood disorders.

Preclinical studies in rats and mice conducted using several different models of depression have demonstrated an antidepressant activity of *R. rosea*. In animal models of depression induced by chronic mild stress, *R. rosea* extracts have been reported to significantly improve both behavioral and physiological measures of depression including enhanced response to rewarding stimuli, increased body weight gain, and increased exploratory behavior compared to untreated controls (Chen et al. 2008a, 2009a; Mattioli et al. 2009). Treatment with *R. rosea* extracts also reduced immobilization times and increased swimming duration of rodents in the forced swim test, a classical paradigm of depression (Perfumi and Mattioli 2007; Panossian et al. 2008). In humans, Darbinyan et al. (2007) first reported significant improvement of mild to moderate depression symptoms in human subjects upon treatment with SHR-5, a standardized formulation of *R. rosea* in a randomized double-blind placebo-controlled clinical trial.

The antidepressant effects of *R. rosea* seem to be mediated through several neurotransmitter systems. Chen et al. (2009a) demonstrated that treatment with *R. rosea* not only boosted levels of serotonin in rat hippocampi but also induced proliferation of neural stem cells. Similarly, an increase in the diencephalic serotonin levels and the expression of serotonin receptor 1A in *R. rosea*-treated rats alleviated depressive symptoms induced by nicotine withdrawal compared to untreated controls

(Mannucci et al. 2012). In a diet-induced model of obesity, *R. rosea* in combination with *Citrus aurantium* led to increased norepinephrine levels in the hypothalamus and dopamine in the frontal cortex, likely due to the modulatory effects on the central monoaminergic system (Verpeut et al. 2013). *R. rosea* water and methanol extracts strongly inhibited monoamine oxidases A and B *in vitro* (van Diermen et al. 2009). These enzymes play key roles in the metabolism of biogenic amines such as epinephrine, norepinephrine, and dopamine and are thus targets for the treatment of depression and neurodegenerative diseases. The antidepressant effects of *R. rosea* may also be enhanced by its inhibitory actions on oxidative stress (Calcabrini et al. 2010) or its anti-inflammatory effects (Bawa and Khanum 2009).

8.2.2 EFFECTS OF *R. ROSEA* ON COGNITIVE FUNCTION AND MENTAL PERFORMANCE

Rhodiola rosea has traditionally been used to enhance memory and concentration (Panossian et al. 2010). *R. rosea* dietary supplements under the European Food Safety Authority claim that *R. rosea* may be used for "optimal mental and cognitive function" (Panossian et al. 2010). Although several studies show promising indications of the potential of *R. rosea* as a neuroprotective agent, these are mostly *in vitro* studies; further animal and clinical assessments are required to validate these effects.

R. rosea extracts as well as their constituent phytochemicals, have been shown to inhibit the *in vitro* and *in vivo* activity of acetylcholinesterase, an enzyme responsible for the degradation of the neurotransmitter acetylcholine, the levels of which are reduced in neurodegenerative diseases including Alzheimer's disease (AD) (Hillhouse et al. 2004; Wang et al. 2007; Zhang et al. 2013). Salidroside, an important bioactive compound of *R. rosea*, protects cells against amyloid-β toxicity *in vitro* through the induction of antioxidant enzymes and inhibition of reactive oxygen species (ROS) accumulation (Jang et al. 2003; Zhang et al. 2010). *R. rosea* extracts as well as salidroside protect against glutamate excitotoxicity by modulation of intracellular Ca^{2+} ion levels (Cao et al. 2006; Palumbo et al. 2012). In addition, *R. rosea* extracts inhibit important enzymes that regulate inflammatory cascades, including members of the phospholipase A_2 superfamily (Bawa and Khanum 2009), which generate lipid second messenger molecules that can further contribute to AD pathogenesis (Ryan et al. 2009).

In a study by Qu et al. (2009), pretreatment with *R. rosea* significantly improved spatial learning and memory in cognitively impaired rats. *R. rosea* attenuated neuronal injury via reduction of markers of oxidative stress, including malondialdehyde and boosted the levels of glutathione reductase and reduced glutathione in the hippocampus. *R. rosea* improves performance of scopolamine-impaired rats in the passive avoidance task, a fear-motivated learning and memory test (Getova and Mihaylova 2012). In addition to *R. rosea* extracts, salidroside administration improves spatial learning and memory performance of cognitively impaired rats in the Morris water maze task (Zhang et al. 2013). Salidroside acts via multiple mechanisms, including inhibition of expression of inflammatory markers nuclear factor kappa-B (NF-κB), inducible nitric oxide synthase (iNOS), cyclooxygenase-2 (COX-2) and receptor for advanced glycation end-products (RAGE) in the hippocampus (Zhang et al. 2013). Salidroside also reduces apoptosis in neural stem cells obtained from rat hippocampi subjected to streptozotocin insult via its antioxidant effects (Qu et al. 2012).

In humans, clinical trials suggest that *R. rosea* treatment as the standardized formulation SHR-5® (standardized to salidroside content) improves mental performance involving complex cognitive tasks, short-term memory, and concentration under conditions of stress-induced fatigue in healthy adults (Darbinyan et al. 2000). In separate studies, *R. rosea* extract as the proprietary supplement Vigo dana® also significantly ameliorated cognitive deficiencies including forgetfulness, memory loss, and problems in concentration (Fintelmann and Gruenwald 2007). While promising, some of these clinical studies lack rigorous placebo-controlled comparisons, thus limiting the conclusions that could be drawn.

8.2.3 ANTISTRESS (ADAPTOGENIC) EFFECTS OF *R. ROSEA*

Rhodiola rosea exerts its antistress effects via simultaneous action on different arms of the stress response system, including the HPA axis and the sympathetic–adrenal gland axis (Lishmanov et al. 1987; Panossian and Wagner 2005; Mattioli and Perfumi 2007). In a study by Mattioli and Perfumi (2007), *R. rosea* selectively blocked hypophagia induced by prolonged restraint (physical stress) and intracerebroventricular injection of corticotropin-releasing factor (CRF) (physiological stress) in rats. *R. rosea* extract SHR-5® and salidroside significantly reduced circulating levels of phosphorylated stress-activated protein kinase, nitric oxide and cortisol in rabbits subjected to restraint stress compared to placebo controls (Panossian et al. 2007). *R. rosea*, in combination with other adaptogens in a standardized proprietary herbal formulation ADAPT-232®, exerts stress resistance effects via increasing the expression of molecular chaperones, including heat shock protein Hsp 72, and stimulating release of neuropeptide Y, a key mediator of stress (Panossian et al. 2009a; Panossian et al. 2012). Further pre-clinical studies assessing the impacts of *R. rosea* monotherapy directly on glucocorticoid levels are warranted. Clinical trials show that *R. rosea* (SHR-5® formulation) decreases salivary cortisol, improves symptoms of stress-induced fatigue, and enhances mental and physical performance under stressful conditions consistent with HPA modulation (Darbinyan et al 2000; Olsson et al. 2009).

8.2.4 ANXIOLYTIC ACTIVITY OF *R. ROSEA*

Only a handful of studies have assessed the anxiolytic effects of *R. rosea*. Perfumi and Mattioli (2007) tested the anxiolytic effects of *R. rosea* in the light dark and open-field behavioral tests in mice. They reported that mice treated with *R. rosea* spent significantly more time in the exposed, unprotected areas of the maze, indicating fear reduction compared to controls. These results were further supported by a recent study by Montiel-Ruiz et al. (2012), which showed that *R. rosea* treatment decreased the number of head dips in the hole-board test similar to a positive control, clonazepam, a $GABA_A$-benzodiazepine receptor agonist. This study also examined the behavior of mice in the elevated plus-maze test, one of the most commonly used behavioral measures of anxiety in rodents. Although there was a trend toward increased time spent in the open arms of the maze by *R. rosea*–treated mice, the anxiolytic effect was not statistically significant (Montiel-Ruiz et al. 2012). In a recent study on *R. rosea* populations from Québec, Canada, anxiolytic effects were seen in rats in three different behavioural paradigms, the elevated-plus maze, the conditioned emotional response as well as the social interaction test (Cayer et al. 2013). Interestingly, *R. rosea* did not seem to be active at the $GABA_A$-benzodiazepine receptor site; serotonergic involvement or monoamine oxidase inhibition may be responsible for observed activity (Chen et al. 2009a; van Diermen et al. 2009). In terms of clinical studies, Bystritsky et al. (2008) conducted a small, pilot study assessing the effects of a proprietary form of *R. rosea* (Rhodax®) for generalized anxiety disorders and showed a significant decrease in the Hamilton anxiety rating scale. However, due to the small sample size, lack of placebo controls, and the open-label design of this study, the results are inconclusive.

8.3 ANTIOXIDANT EFFECTS OF *R. ROSEA*

The antioxidant effects of *R. rosea* underlie many of its other biological properties. In a comparative study of *R. rosea*, *E. senticosus,* and *Emblica officinalis*, *R. rosea* was the most potent antioxidant in terms of singlet oxygen and hydrogen peroxide scavenging, and iron chelating abilities, (Chen et al. 2008b). Moreover, antioxidant potential was directly proportional to *R. rosea* polyphenol content (Chen et al. 2008b). *R. rosea* administration protects erythrocytes from oxidative stress induced by hypochlorous acid by preventing depletion of reduced glutathione, inactivation of glyceraldehyde-3-phosphate dehydrogenase, and hemolysis (De Sanctis et al. 2004). Similarly, Calcabrini et al. (2010) demonstrated that *R. rosea* protects against several different oxidative stressors in human keratinocytes by improving reduced glutathione levels, improving levels of antioxidant enzymes superoxide dismutase and catalase, and reducing intracellular accumulation of ROS. *R. rosea* also inhibits the activity of xanthine oxidase, tyrosinase, and lipoxygenase, the enzymes that are involved in the generation of ROS and in inflammatory pathways (Chen et al. 2009b; Horng et al. 2010). In rats subjected to amyloid-β_{1-40} challenge, salidroside has been shown to induce protective effects against cognitive deficits via attenuation of levels of lipid peroxidation product malondialdehyde and enhancement of the activity of superoxide dismutase and glutathione peroxidase in the hippocampus (Zhang et al. 2013). Other studies have shown that protection against oxidative stressors can be signalled independently of changes in antioxidant enzymes or through the activation of antioxidant response element (Schriner et al. 2009). In humans, *R. rosea* supplementation in professional rowers significantly increased plasma antioxidant capacity, but did not relieve oxidative stress after exhaustive exercise (Skarpanska-Stejnborn et al. 2009).

8.4 EFFECTS OF *R. ROSEA* ON PHYSICAL ENDURANCE

R. rosea may also improve physical work capacity and endurance, although studies so far show conflicting results. Abidov et al. (2003) showed that oral administration of *R. rosea* in rats increased the duration of exhaustive swimming by stimulating the synthesis of ATP in mitochondria of skeletal muscles. In humans, however, no changes were observed in muscle phosphate levels in trained athletes during or after exercise (Walker et al. 2007). Similarly, *R. rosea* supplementation increased physical activity in mice in the forced swim test (Kurkin et al. 2006). Chronic supplementation in rats showed that *R. rosea* extract not only increased liver glycogen content and upregulated lipogenic enzyme expression (sterol regulatory element binding protein-1 [SREBP-1], fatty acid synthase [FAS]), heat shock protein 70 expression, Bcl-2/Bax ratio, and oxygen content before swimming, but also facilitated recovery by reducing biomarkers of fatigue including blood urea nitrogen and lactate dehydrogenase (Lee et al. 2009). These data are supported by microarray analysis examining the effects of *R. rosea* in an herbal preparation, AdMax™ in cultured human fibroblasts, where expression of genes involved in energy metabolism were affected (Antoshechkin et al. 2008).

Studies on humans have been more variable. In clinical studies in healthy untrained individuals, supplementation with Rhodax (*R. rosea* formulation) significantly decreased inflammatory markers in the blood, including C-reactive protein and creatinine kinase, notably after exhaustive exercise (Abidov et al. 2004; Parisi et al. 2010). Small beneficial changes in the time taken to reach exhaustion and maximal oxygen consumption (VO_2 max) have also been reported under acute (De Bock et al. 2004), but not chronic treatment The acute effect has also been replicated in recreationally active women where *R. rosea* supplementation decreased the perception of effort required during exercise (Noreen et al. 2013). Other studies report that time to exhaustion and VO_2 max are not altered by *R. rosea* (Earnest et al. 2004; Colson et al. 2005; Walker et al. 2007). Taken together, it remains to be determined whether *R. rosea* enhances physical endurance in humans. Certainly, these differences could be explained by the variation in the type of extracts under study, duration of supplementation and monitoring, choice of subjects, levels of physical training of subjects etc. A rigorous series of controlled clinical studies are warranted to address these concerns.

8.5 ANTIDIABETIC EFFECTS OF *R. rosea*

R. rosea extracts have been reported to have inhibitory activities on enzymes important for carbohydrate digestion and therefore, a potential preventive effect on the subsequent rise in blood glucose levels. Several studies have shown that *R. rosea* extracts inhibit pancreatic α-amylase (Kobayashi et al. 2003; Kwon et al. 2006) and α-glucosidase activities (Kwon et al. 2006) and reduce postprandial hyperglycemia (Etxeberria et al. 2012). Yet to be identified compounds in *R. rosea* function as a partial agonist for the peroxisome proliferator-activated receptor γ, a key target of insulin-sensitizing drugs, and increase insulin-stimulated glucose uptake (Christensen et al. 2009). The methanol extracts of *R. rosea* inhibited lipase activity *in vitro* as well as *in vivo*, and oral administration of rhodionin and rhodiosin isolated from *R. rosea* significantly inhibited postprandial rise in triglyceride levels (Kobayashi et al. 2004). In diabetic mice treated with *R. rosea*, fasting blood glucose levels were significantly lower, levels of antioxidant enzymes were higher, and lipid peroxidation in hepatic tissue was lower compared to untreated controls (Sung et al. 2006). In streptozotocin-induced diabetic rats, *R. rosea* did not restore blood glucose or insulin levels, but attenuated cardiac dysfunction via upregulation of peroxisome proliferator-activated receptor δ (Cheng et al. 2012).

8.6 OTHER BIOLOGICAL ACTIVITIES OF *R. rosea*

R. rosea has been shown to possess cardioprotective effects against hypoxic/reperfusion damage (Afanas'ev et al. 1997; Arbuzov et al. 2006; Maslov et al. 2009), probably through its effects on the sympathetic nervous system (Maslov et al. 1998). *R. rosea* inhibits the division of HL-60 leukemia cells by causing apoptosis and necrosis, and inducing cell cycle arrest at the G2/M phase (Majewska et al. 2006). After transplantation of spontaneously metastasizing Lewis lung carcinoma cells into mice, it was shown by Goldberg et al. (2004) that *R. rosea* inhibited metastasis by stimulation of immune system cells and a decrease in levels of glucocorticoids. In patients who underwent chemotherapy for ovarian cancer, administration of AdMax™ led to increased levels of lymphocytes and immunoglobulins, IgG and IgM (Kormosh et al. 2006). The anticancer effects of *R. rosea* may be attributed to its bioactive constituent, salidroside, which exhibits cytotoxic action on multiple cancer cell lines (Hu et al. 2010). Salidroside as well as *R. rosea* extract was found to be selectively cytotoxic to bladder cancer cell line UMUC3 via inhibition of the mammalian target of rapamycin (mTOR) protein and the induction of autophagy (Liu et al. 2012).

8.7 PHARMACOKINETICS, BIOAVAILABILITY, AND HERB–DRUG INTERACTIONS

Pharmacokinetic studies of *R. rosea* extracts are limited. Instead, studies have focused mainly on the elucidation of pharmacokinetic parameters of salidroside and *p*-tyrosol in biological tissue, mainly plasma using chromatographic techniques (HPLC, LC-MS/MS) (Chang et al. 2007; Mao et al. 2007; Yu et al. 2008; Guo et al. 2012). In general, salidroside seems to be rapidly absorbed, within 20–30 minutes, and then eliminated. The oral bioavailability ranges from 51.97% to 98%, probably dependant on the dosage of *R. rosea* used (Chang et al. 2007). However, in a study on the pharmacokinetics of salidroside on *Rhodiola crenulata* administration, the absorption of salidroside was still rapid, but its half-life was prolonged considerably due to the presence of other compounds (Zhang et al. 2008). He et al. (2009) showed that the administration of salidroside at increasing doses saturates the sodium-dependent glucose transporter-1 protein, which was shown to be responsible for the intestinal uptake and absorption of salidroside.

Natural health products can drastically affect the pharmacological action of drugs taken in conjunction with them by influencing their absorption, distribution, bioavailability, or excretion. This can happen via stimulatory or inhibitory effects on cytochrome P450 (CYP) enzymes that are key players in the metabolism of xenobiotics as well as endogenous substances (Na et al. 2011). *R. rosea* extracts and compounds, specifically rosarin, are known to significantly inhibit the activities of CYP3A4 (67%) and CYP19 (83%) (Scott et al. 2006). *R. rosea* extracts also induce CYP1A2 mRNA expression in human colon carcinoma cell lines (Brandin et al. 2007). These findings were further supported by Hellum et al. (2010), who reported potent inhibitory effects of *R. rosea* populations on CYP3A4 activity as well as P-glycoprotein transport, highlighting the possibility of potential drug interactions of *R. rosea*. Interestingly, the variation in the concentrations of salidroside, tyrosol, or the rosavins did not account for differences in inhibitory activity, indicating that other compounds are responsible. In an *in vivo* pharmacokinetic study conducted on rats, the concomitant administration of SHR-5® with theophylline and warfarin, both candidates for CYP metabolism, there were no significant effects on pharmacokinetics or the anticoagulant activity of warfarin (Panossian et al. 2009b). Recently, however, a study by Spanakis et al. (2013) showed that in rabbits, the administration of *R. rosea* and Losartan, a hypertension drug, led to significant increases of Losartan concentrations in plasma, indicating the possibility of herb-drug interactions.

8.8 CONCLUDING REMARKS

There is a wide range of pharmacological activities of *R. rosea*, ranging from effects in the CNS to antibacterial activities, reflecting the heterogeneous traditional use of this herb. Most studies show a positive effect of *R. rosea* extract in the model being studied, as well as indicate its safety, which is encouraging in terms of its development as a multimodal therapeutic. At the same time, however, it is important to remember that many of these studies are conducted *in vitro*, an essential starting point for the elucidation of bioactivity as well as identification of active principles, but requiring further validation by well-designed preclinical and clinical trials. Also, caution should be applied while evaluating results from a wide range of studies using plant material from different sources, different methods of extraction of *R. rosea* resulting in varying concentrations of phytochemicals, different dosage and administration regimes in animals, etc. Several studies report the activities of *R. rosea* in a standardized, but complex mixture of multiple herbs, the individual effects of which are difficult to isolate. In gene expression profiling studies by Panossian et al. (2013), *R. rosea* and its compounds, salidroside, tyrosol and triandrin led to differential regulation of genes associated with nervous system development and function, neurological disease and lipid metabolism in a neuroglial cell model. Not surprisingly, the profile and number of target genes affected were altered due to synergistic and (or) antagonistic interactions when *R. rosea* was administered in a mixture of adaptogens as compared to a single extract. Certainly, more studies using *R. rosea* as a monotherapy are warranted. Tannins, which are present in high concentrations in this plant, often interfere with biological enzyme/protein systems, yet most studies overlook these compounds. The application of an additional extraction step for tannin removal prior to testing would reduce the chance of obtaining false-positives.

Of some concern is the inhibitory/inducing activities of *R. rosea* on CYP enzymes, highlighting the potential ability of this plant to interfere with the metabolism of coadministered drugs. More studies are clearly needed on the bioavailability of *R. rosea* in biological tissue as well as its pharmacokinetics. The bioavailability of phenolic compounds is generally low; therefore, these studies might provide more insight into which compounds or modifications of compounds are actually responsible for the activity.

Thus, there is supporting evidence for the pleiotropic pharmacological activities of *R. rosea*. Future research will need to move towards validation of these activities in more sophisticated animal and human trials as well as determination of pharmacokinetic parameters for safety and efficacy.

REFERENCES

Abidov, M., F. Crendal, S. Grachev, R. Seifulla, and T. Ziegenfuss. 2003. Effect of extracts from *Rhodiola rosea* and *Rhodiola crenulata* (crassulaceae) roots on ATP content in mitochondria of skeletal muscles. *Bulletin of Experimental Biology and Medicine* 136 (6): 585–7.

Abidov, M., S. Grachev, R. D. Seifulla, and T. N. Ziegenfuss. 2004. Extract of *Rhodiola rosea* radix reduces the level of C-reactive protein and creatinine kinase in the blood. *Bulletin of Experimental Biology and Medicine* 138 (1): 63–4.

Afanas'ev, S. A., Y.B. Lishmanov, T. V. Lasukova, and A. V. Naumova. 1997. Effect of *Rhodiola rosea* on the resistance of isolated heart from stressed rats to ischemic and reperfusion damage. *Bulletin of Experimental Biology and Medicine* 123 (5):447–9.

Antoshechkin, A., J. Olalde, M. Antoshechkina, V. Briuzgin, and L. Platinskiy. 2008. Influence of the plant extract complex "AdMax" on global gene expression levels in cultured human fibroblasts. *Journal of Dietary Supplements* 5 (3): 293–304.

Arbuzov, A. G., A. V. Krylatov, L. N. Maslov, V. N. Burkova, and N. V. Naryzhnaya. 2006. Antihypoxic, cardioprotective, and antifibrillation effects of a combined adaptogenic plant preparation. *Bulletin of Experimental Biology and Medicine* 142 (2): 212–5.

Bawa, P. A. S. and F. Khanum. 2009. Anti-inflammatory activity of *Rhodiola rosea*—"A second-generation adaptogen." *Phytotherapy Research* 23 (8):1099–102.

Brandin, H., E. Viitanen, O. Myrberg, and A.–K. Arvidsson. 2007. Effects of herbal medicinal products and food supplements on induction of CYP1A2, CYP3A4 and MDR1 in the human colon carcinoma cell line LS180. *Phytotherapy Research* 21 (3): 239–44.

Bystritsky, A., L. Kerwin, and J. D. Feusner. 2008. A pilot study of *Rhodiola rosea* (Rhodax®) for generalized anxiety disorder (GAD). *Journal of Alternative and Complementary Medicine* 14 (2): 175–80.

Calcabrini, C., R. De Bellis, U. Mancini et al. 2010. *Rhodiola rosea* ability to enrich cellular antioxidant defences of cultured human keratinocytes. *Archives of Dermatological Research* 302 (3): 191–200.

Cao, L. L., G. H. Du, and M. W. Wang. 2006. The effect of salidroside on cell damage induced by glutamate and intracellular free calcium in PC12 cells. *Journal of Asian Natural Products Research* 8 (1–2): 159–65.

Cayer, C., F. Ahmed, V. Filion, A. Saleem, A. Cuerrier, M. Allard, G. Rochefort, Z. Merali, and J. T. Arnason. 2013. Characterization of the anxiolytic activity of Nunavik *Rhodiola rosea* (Crassulaceae). *Planta Medica* 79 (15): 1385–91.

Chang, Y. W., H. T. Yao, S. H. Hsieh, T. J. Lu, and T. K. Yeh. 2007. Quantitative determination of salidroside in rat plasma by on-line solid-phase extraction integrated with high-performance liquid chromatography/electrospray ionization tandem mass spectrometry. *Journal of Chromatography B: Analytical Technologies in the Biomedical and Life Sciences* 857 (1): 164–9.

Chen, C. H., H. C. Chan, Y. T. Chu, H. Y. Ho, P. Y. Chen, T. H. Lee, and C. K. Lee. 2009b. Antioxidant activity of some plant extracts towards xanthine oxidase, lipoxygenase and tyrosinase. *Molecules* 14 (8): 2947–58.

Chen, Q. G., Y. S. Zeng, J. Y. Tang, Y. J. Qin, S. J. Chen, and Z. Q. Zhong. 2008a. Effects of *Rhodiola rosea* on body weight and intake of sucrose and water in depressive rats induced by chronic mild stress. *Journal of Chinese Integrative Medicine* 6 (9): 952–5.

Chen, Q. G., Y. S. Zeng, Z. Q. Qu, J. Y. Tang, Y. J. Qin, P. Chung, R. Wong, and U. Hägg. 2009a. The effects of *Rhodiola rosea* extract on 5-HT level, cell proliferation and quantity of neurons at cerebral hippocampus of depressive rats. *Phytomedicine* 16 (9): 830–8.

Chen, T. S., S. Y. Liou, and Y. L. Chang. 2008b. Antioxidant evaluation of three adaptogen extracts. *American Journal of Chinese Medicine* 36 (6): 1209–17.

Cheng, Y. Z., L. J. Chen, W. J. Lee, M. F. Chen, H. Jung Lin, and J. T. Cheng. 2012. Increase of myocardial performance by *Rhodiola*-ethanol extract in diabetic rats. *Journal of Ethnopharmacology* 144 (2): 234 –9.

Chopra, K., B. Kumar, and A. Kuhad. 2011. Pathobiological targets of depression. *Expert Opinion on Therapeutic Targets* 15 (4): 379–400.

Christensen, K. B., A. Minet, H. Svenstrup et al. 2009. Identification of plant extracts with potential antidiabetic properties: Effect on human peroxisome proliferator-activated receptor (PPAR), adipocyte differentiation and insulin-stimulated glucose uptake. *Phytotherapy Research* 23 (9): 1316–25.

Colson, S. N., F. B. Wyatt, D. L. Johnston, L. D. Autrey, Y. L. FitzGerald, and C. P. Earnest. 2005. *Cordyceps sinensis*- and *Rhodiola rosea*-based supplementation in male cyclists and its effect on muscle tissue oxygen saturation. *Journal of Strength and Conditioning Research* 19 (2): 358–63.

Connolly, K. R. and M. E. Thase. 2012. Emerging drugs for major depressive disorder. *Expert Opinion on Emerging Drugs* 17 (1): 105–26.

Darbinyan, V., A. Kteyan, A. Panossian, E. Gabrielian, G. Wikman, and H. Wagner. 2000. *Rhodiola rosea* in stress induced fatigue—A double blind cross-over study of a standardized extract SHR-5 with a repeated low-dose regimen on the mental performance of healthy physicians during night duty. *Phytomedicine* 7 (5): 365–71.

Darbinyan, V., G. Aslanyan, E. Amroyan, E. Gabrielyan, C. Malmström, and A. Panossian. 2007. Clinical trial of *Rhodiola rosea* L. extract SHR-5 in the treatment of mild to moderate depression. *Nordic Journal of Psychiatry* 61 (5): 343–8.

Darbinyan, V., G. Aslanyan, E. Amroyan, E. Gabrielyan, C. Malmström, and A. Panossian. 2007. Erratum: Clinical trial of *Rhodiola rosea* L. extract SHR-5 in the treatment of mild to moderate depression (Nordic Journal of Psychiatry [2007] 61 (5): 343–8). *Nordic Journal of Psychiatry* 61 (6): 503.

De Bock, K., B. O. Eijnde, M. Ramaekers, and P. Hespel. 2004. Acute *Rhodiola rosea* intake can improve endurance exercise performance. *International Journal of Sport Nutrition and Exercise Metabolism* 14 (3): 298–307.

De Sanctis, R., R. De Bellis, C. Scesa, U. Mancini, L. Cucchiarini, and M. Dachà. 2004. *In vitro* protective effect of *Rhodiola rosea* extract against hypochlorous acid-induced oxidative damage in human erythrocytes. *BioFactors* 20 (3): 147–59.

Earnest, C. P., G. M. Morss, F. Wyatt et al. 2004. Effects of a commercial herbal-based formula on exercise performance in cyclists. *Medicine and Science in Sports and Exercise* 36 (3): 504–9.

Etxeberria, U., A. L. De La Garza, J. Campin, J. A. Martnez, and F. I. Milagro. 2012. Antidiabetic effects of natural plant extracts via inhibition of carbohydrate hydrolysis enzymes with emphasis on pancreatic alpha amylase. *Expert Opinion on Therapeutic Targets* 16 (3): 269–97.

Fintelmann, V. and J. Gruenwald. 2007. Efficacy and tolerability of a *Rhodiola rosea* extract in adults with physical and cognitive deficiencies. *Advances in Therapy* 24 (4): 929–39.

Getova, D.P. and A.S. Mihaylova. 2012. Effects of *Rhodiola rosea* extract on passive avoidance tests in rats. *Central European Journal of Medicine*: 1–6.

Gol'dberg, E. D., E. N. Amosova, E. P. Zueva, T. G. Razina, S. G. Krylova, and D. V. Reikhart. 2004. Effects of extracts from medicinal plants on the development of metastatic process. *Bulletin of Experimental Biology and Medicine* 138 (3): 288–94.

Guo, N., Z. Hu, X. Fan et al. 2012. Simultaneous determination of salidroside and its aglycone metabolite p-tyrosol in rat plasma by liquid chromatography-tandem mass spectrometry. *Molecules* 17 (4): 4733–54.

He, Y. X., X. D. Liu, X. T. Wang, X. Liu, G. J. Wang, and L. Xie. 2009. Sodium-dependent glucose transporter was involved in salidroside absorption in intestine of rats. *Chinese Journal of Natural Medicines* 7 (6): 444–8.

Hellum, B. H., A. Tosse, K. Hoybakk, M. Thomsen, J. Rohloff, and O. Georg Nilsen. 2010. Potent *in vitro* inhibition of CYP3A4 and P-glycoprotein by *Rhodiola rosea*. *Planta Medica* 76 (4): 331–8.

Hillhouse, B. J., D. S. Ming, C. J. French, and G. H. N. Towers. 2004. Acetylcholine esterase inhibitors in *Rhodiola rosea*. *Pharmaceutical Biology* 42 (1): 68–72.

Horng, C. T., I. M. Liu, D. H. Kuo, Y. W. Tsai, and P. Shieh. 2010 Comparison of xanthine oxidase-inhibiting and free radical-scavenging activities between plant adaptogens of *Eleutherococcus senticosus and Rhodiola rosea*. *Drug Development Research* 71 (4): 249–52.

Hu, X., S. Lin, D. Yu, S. Qiu, X. Zhang, and R. Mei. 2010. A preliminary study: The anti-proliferation effect of salidroside on different human cancer cell lines. *Cell Biology and Toxicology* 26 (6): 499–507.

Jang, S. I., H. O. Pae, B. M. Choi et al. 2003. Salidroside from *Rhodiola sachalinensis* protects neuronal PC12 cells against cytotoxicity induced by amyloid-β. *Immunopharmacology and Immunotoxicology* 25 (3): 295–304.

Kessler, R. C. 2012. The costs of depression. *Psychiatric Clinics of North America* 35 (1): 1–14.

Kessler, R. C., J. Soukup, R. B. Davis et al. 2001. The use of complementary and alternative therapies to treat anxiety and depression in the United States. *American Journal of Psychiatry* 158 (2): 289–94.

Kobayashi, K., E. Baba, S. Fushiya et al. 2003. Screening of Mongolian plants for influence on amylase activity in mouse plasma and gastrointestinal tube. *Biological and Pharmaceutical Bulletin* 26 (7): 1045–8.

Kobayashi, K., T. Takahashi, F. Takano et al. 2004. Survey of the influence of mongolian plants on lipase activity in mouse plasma and gastrointestinal tube. *Natural Medicines* 58 (5): 204–8.

Kormosh, N., K. Laktionov, and M. Antoshechkina. 2006. Effect of a combination of extract from several plants on cell-mediated and humoral immunity of patients with advanced ovarian cancer. *Phytotherapy Research* 20 (5): 424–5.

Kurkin, V. A., A. V. Dubishchev, G. G. Zapesochnaya et al. 2006. Effect of phytopreparations containing phenylpropanoids on the physical activity of animals. *Pharmaceutical Chemistry Journal* 40 (3): 149–50.

Kwon, Y. I., H. D. Jang, and K. Shetty. 2006. Evaluation of *Rhodiola crenulata Rhodiola rosea* and for management of type II diabetes and hypertension. *Asia Pacific Journal of Clinical Nutrition* 15 (3): 425–32.

Lee, F. T., T. Y. Kuo, S. Y. Liou, and C. T. Chien. 2009. Chronic *Rhodiola rosea* extract supplementation enforces exhaustive swimming tolerance. *American Journal of Chinese Medicine* 37 (3): 557–72.

Lishmanov, I.B., Z.V. Trifonova, A.N. Tsibin, L.V. Maslova, and L.A. Dementeva. 1987. Plasma beta-endorphin and stress hormones in stress and adaptation. *Bulletin of Experimental Biology and Medicine* 103 (4): 422–4.

Liu, Z., X. Li, A. R. Simoneau, M. Jafari, and X. Zi. 2012. *Rhodiola rosea* extracts and salidroside decrease the growth of bladder cancer cell lines via inhibition of the mTOR pathway and induction of autophagy. *Molecular Carcinogenesis* 51 (3): 257–67.

Majewska, A., G. Hoser, M. Furmanowa et al. 2006. Erratum: Antiproliferative and antimitotic effect, S phase accumulation and induction of apoptosis and necrosis after treatment of extract from *Rhodiola rosea* rhizomes on HL-60 cells (*Journal of Ethnopharmacology* [2005] 103: 43–52). *Journal of Ethnopharmacology* 104 (3): 433.

Majewska, A., H. Grayna, F. Mirosława et al. 2006. Antiproliferative and antimitotic effect, S phase accumulation and induction of apoptosis and necrosis after treatment of extract from *Rhodiola rosea* rhizomes on HL-60 cells. *Journal of Ethnopharmacology* 103 (1): 43–52.

Mannucci, C., M. Navarra, E. Calzavara, A. P. Caputi, and G. Calapai. 2012. Serotonin involvement in *Rhodiola rosea* attenuation of nicotine withdrawal signs in rats. *Phytomedicine* 19 (12): 1117–24.

Mao, Y., X. Zhang, X. Zhang, and G. Lu. 2007. Development of an HPLC method for the determination of salidroside in beagle dog plasma after administration of salidroside injection: Application to a pharmacokinetics study. *Journal of Separation Science* 30 (18): 3218–22.

Maslov, L. N., Y. B. Lishmanov, A. G. Arbuzov et al. 2009. Antiarrhythmic activity of phyto-adaptogens in short-term ischemia-reperfusion of the heart and postinfarction cardio-sclerosis. *Bulletin of Experimental Biology and Medicine* 147 (3): 331–4.

Maslov, L. N., Y. B. Lishmanov, L. A. Maimesculova, and E. A. Krasnov. 1998. A mechanism of antiarrhythmic effect of *Rhodiola rosea*. *Bulletin of Experimental Biology and Medicine* 125 (4): 374–6.

Mattioli, L. and M. Perfumi. 2007. *Rhodiola rosea* L. extract reduces stress- and CRF-induced anorexia in rats. *Journal of Psychopharmacology* 21 (7): 742–50.

Mattioli, L. and M. Perfumi. 2011. Effects of a *Rhodiola rosea* L. extract on acquisition and expression of morphine tolerance and dependence in mice. *Journal of Psychopharmacology* 25 (3): 411–20.

Mattioli, L., C. Funari, and M. Perfumi. 2009. Effects of *Rhodiola rosea* L. extract on behavioural and physiological alterations induced by chronic mild stress in female rats. *Journal of Psychopharmacology* 23 (2): 130–42.

Mattioli, L., F. Titomanlio, and M. Perfumi. 2012. Effects of a *Rhodiola rosea* L. extract on the acquisition, expression, extinction, and reinstatement of morphine-induced conditioned place preference in mice. *Psychopharmacology* 221 (2): 183–193.

Montiel-Ruiz, R. M., J. E. Roa-Coria, S. I. Patiño-Camacho, F. J. Flores-Murrieta, and M. Déciga-Campos. 2012. Neuropharmacological and toxicity evaluations of ethanol extract from *Rhodiola rosea*. *Drug Development Research* 73 (2): 106–13.

Na, D. H., H. Y. Ji, E. J. Park, M. S. Kim, K. H. Liu, and H. S. Lee. 2011. Evaluation of metabolism-mediated herb-drug interactions. *Archives of Pharmacol Research* 34 (11): 1829–42.

Noreen, E. E., J. G. Buckley, S. L. Lewis, J. Brandauer, and K. J. Stuempfle. 2013. The effects of an acute dose of *Rhodiola rosea* on endurance exercise performance. *Journal of Strength and Conditioning Research* 27 (3): 839–47.

Olsson, E. M. G., B. Von Schéele, and A. G. Panossian. 2009. A randomised, double-blind, placebo-controlled, parallel-group study of the standardised extract SHR-5 of the roots of *Rhodiola rosea* in the treatment of subjects with stress-related fatigue. *Planta Medica* 75 (2): 105–12.

Palumbo, D. R., F. Occhiuto, F. Spadaro, and C. Circosta. 2012. *Rhodiola rosea* extract protects human cortical neurons against glutamate and hydrogen peroxide-induced cell death through reduction in the accumulation of intracellular calcium. *Phytotherapy Research* 26 (6): 878–883.

Panossian, A. G. 2013. Adaptogens in mental and behavioral disorders. *Psychiatric Clinics of North America* 36 (1): 49–64.

Panossian, A., A. Hovhannisyan, H. Abrahamyan, E. Gabrielyan, and G. Wikman. 2009b. Pharmacokinetic and pharmacodynamic study of interaction of *Rhodiola rosea* SHR-5 extract with warfarin and theophylline in rats. *Phytotherapy Research* 23 (3): 351–7.

Panossian, A. and G. Wikman. 2010. Effects of adaptogens on the central nervous system and the molecular mechanisms associated with their stress—Protective activity. *Pharmaceuticals* 3 (1): 188–224.

Panossian, A., G. Wikman, and H. Wagner. 1999. Plant adaptogens III. earlier and more recent aspects and concepts on their mode of action. *Phytomedicine* 6 (4): 287–300.

Panossian, A. and H. Wagner. 2005. Stimulating effect of adaptogens: An overview with particular reference to their efficacy following single dose administration. *Phytotherapy Research* 19 (10): 819–38.

Panossian, A., G. Wikman, and J. Sarris. 2010. Rosenroot (*Rhodiola rosea*): Traditional use, chemical composition, pharmacology and clinical efficacy. *Phytomedicine* 17 (7): 481–93.

Panossian, A., G. Wikman, P. Kaur, and A. Asea. 2009a. Adaptogens exert a stress-protective effect by modulation of expression of molecular chaperones. *Phytomedicine* 16 (6–7): 617–22.

Panossian, A., G. Wikman, P. Kaur, and A. Asea. 2012. Adaptogens stimulate neuropeptide Y and Hsp72 expression and release in neuroglia cells. *Frontiers in Neuroscience* (FEB) 6 (6): 1–12.

Panossian, A., R. Hamm, O. Kadioglu, G. Wikman, and T. Efferth. 2013. Synergy and antagonism of active constituents of ADAPT-232 on transcriptional level of metabolic regulation of isolated neuroglial cells. *Frontiers in Neuroscience*.

Panossian, A., N. Nikoyan, N. Ohanyan, A. Hovhannisyan, H. Abrahamyan, E. Gabrielyan, and G. Wikman. 2008. Comparative study of *Rhodiola* preparations on behavioral despair of rats. *Phytomedicine* 15 (1–2): 84–91.

Panossian, A., M. Hambardzumyan, A. Hovhannisyan, and G. Wikman. 2007. The adaptogens *Rhodiola* and *Schizandra* modify the response to immobilization stress in rabbits by suppressing the increase of phosphorylated stress-activated protein kinase, nitric oxide and cortisol. *Drug Target Insights* 2: 39–54.

Parisi, A., E. Tranchita, G. Duranti et al. 2010. Effects of chronic *Rhodiola rosea* supplementation on sport performance and antioxidant capacity in trained male: preliminary results. *Journal of Sports Medicine and Physical Fitness* 50 (1): 57–63.

Perfumi, M. and L. Mattioli. 2007. Adaptogenic and central nervous system effects of single doses of 3% rosavin and 1% salidroside *Rhodiola rosea* L. extract in mice. *Phytotherapy Research* 21 (1): 37–43.

Qu, Z. Q., Y. Zhou, Y. S. Zeng et al. 2012. Protective effects of a *Rhodiola crenulata* extract and salidroside on hippocampal neurogenesis against streptozotocin-induced neural injury in the rat. *PLoS ONE* 7 (1): 1–17.

Qu, Z. Q., Y. Zhou, Y. S. Zeng, Y. Li, and P. Chung. 2009. Pretreatment with *Rhodiola rosea* extract reduces cognitive impairment induced by intracerebroventricular streptozotocin in rats: Implication of anti-oxidative and neuroprotective effects. *Biomedical and Environmental Sciences* 22 (4): 318–26.

Ryan, S. D., S. N. Whitehead, L. A. Swayne et al. 2009. Amyloid-β42 signals tau hyperphosphorylation and compromises neuronal viability by disrupting alkylacylglycerophosphocholine metabolism. *Proceedings of the National Academy of Sciences of the United States of America* 106 (49): 20936–41.

Saveanu, R. V. and C. B. Nemeroff. 2012. Etiology of depression: genetic and environmental factors. *Psychiatric Clinics of North America* 35 (1): 51–71.

Schriner, S. E., A. Avanesian, Y. Liu, H. Luesch, and M. Jafari. 2009. Protection of human cultured cells against oxidative stress by *Rhodiola rosea* without activation of antioxidant defenses. *Free Radical Biology and Medicine* 47 (5): 577–84.

Scott, I. M., R. I. Leduc, A. J. Burt, R. J. Marles, J. T. Arnason, and B. C. Foster. 2006. The inhibition of human cytochrome P450 by ethanol extracts of North American botanicals. *Pharmaceutical Biology* 44 (5): 315–27.

Skarpanska-Stejnborn, A., L. Pilaczynska-Szczesniak, P. Basta, and E. Deskur-Smielecka. 2009. The influence of supplementation with *Rhodiola rosea* L. extract on selected redox parameters in professional rowers. *International Journal of Sport Nutrition and Exercise Metabolism* 19 (2): 186–99.

Spanakis, M., I. S. Vizirianakis, G. Batzias, and I. Niopas. 2013. Pharmacokinetic interaction between losartan and *Rhodiola rosea* in rabbits. *Pharmacology* 91 (1–2): 112–6.

Sung, H. K., H. H. Sun, and Y. C. Se. 2006. Antioxidative effects of cinnamomi cassiae and *Rhodiola rosea* extracts in liver of diabetic mice. *BioFactors* 26 (3): 209–19.

van Diermen, D., A. Marston, J. Bravo, M. Reist, P.-A Carrupt, and K. Hostettmann. 2009. Monoamine oxidase inhibition by *Rhodiola rosea* L. roots. *Journal of Ethnopharmacology* 122 (2): 397–401.

Verpeut, J.L., A.L. Walters, and N.T. Bello. 2013. *Citrus aurantium* and *Rhodiola rosea* in combination reduce visceral white adipose tissue and increase hypothalamic norepinephrine in a rat model of diet-induced obesity. *Nutrition Research* 33 (6): 503–12.

Walker, T. B., S. A. Altobelli, A. Caprihan, and R. A. Robergs. 2007. Failure of *Rhodiola rosea* to alter skeletal muscle phosphate kinetics in trained men. *Metabolism: Clinical and Experimental* 56 (8): 1111–7.

Wang, H., G. Zhou, X. Gao, Y. Wang, and W. Yao. 2007. Acetylcholinesterase inhibitory-active components of *Rhodiola rosea* L. *Food Chemistry* 105 (1): 24–7.

Yu, S., M. Liu, X. Gu, and F. Ding. 2008. Neuroprotective effects of salidroside in the PC12 cell model exposed to hypoglycemia and serum limitation. *Cellular and Molecular Neurobiology* 28 (8): 1067–78.

Zhang, J., X. Chen, P. Wang et al. 2008. LC-MS determination and pharmacokinetic study of salidroside in rat plasma after oral administration of traditional Chinese medicinal preparation *Rhodiola crenulata* extract. *Chromatographia* 67 (9–10): 695–700.

Zhang, J., Y.F. Zhen, Ren Pu Bu Ci, L.G. Song, W.N. Kong, T.M. Shao, X. Li, and X.Q. Chai. 2013. Salidroside attenuates beta amyloid-induced cognitive deficits via modulating oxidative stress and inflammatory mediators in rat hippocampus. *Behavioural Brain Research* 244: 70–81.

Zhang, L., H. Yu, X. Zhao et al. 2010. Neuroprotective effects of salidroside against beta-amyloid-induced oxidative stress in SH-SY5Y human neuroblastoma cells. *Neurochemistry International* 57 (5): 547–55.

9 Evidence-Based Efficacy and Effectiveness of *Rhodiola* SHR-5 Extract in Treating Stress- and Age-Associated Disorders

Alexander Panossian and Georg Wikman

CONTENTS

9.1 Introduction 205
9.2 Data on Efficacy 206
    9.2.1 Stress-Induced Mental and Behavioral Disorders 206
    9.2.2 Experiments in the Nematode *Caenorhabditis elegans*, the Pond
        Snail *Lymnaea stagnalis* and on Isolated Cells 206
    9.2.3 Depression 213
    9.2.4 Experiments in Animals and on Isolated Cells 214
    9.2.5 Aging-Associated Disorders 217
9.3 Pharmacokinetic Properties 220
References 220

9.1 INTRODUCTION

The SHR-5 proprietary extract from *Rhodiola rosea* L. roots and rhizome was developed by the Swedish Herbal Institute. Registered in Sweden as a Natural Remedy (1985) and Traditional Herbal Medicinal Product (2008), it is an active constituent of Arctic Root® film-coated tablets and capsules. It is indicated for use as "an adaptogen, in case of decreased/reduced performance, such as fatigue and sensation/feeling of weakness" based on traditional use only, and not upon clinical trial data. In Denmark, it is used in treating fatigue and during convalescence, whereas in Canada, the health claim is that it "helps to improve mental and physical capacities in case of tiredness and non-specific stress." SHR-5 is distributed as a dietary supplement in the United States, Canada, Denmark, Norway, Iceland, France, Belgium, Italy, Austria, and Germany. It "contributes to optimal mental and cognitive activity," according to the function specified on the European Food Safety Authority's

(EFSA) consolidated list of Article 13 health claims (http://www.efsa.europa.eu
/EFSA/efsa_locale-1178620753812_article13.htm).

The efficacy of SHR-5 extract has been demonstrated in pharmacological studies
on isolated cells (Panossian et al. 2012, 2013), nematodes (Wiegant et al. 2009) and
animals (Panossian et al. 2007, 2008a,b, 2009), as well as in clinical trials in healthy
subjects (Darbinyan et al. 2000; Spasov et al. 2000; Shevtsov et al. 2003) and patients
with mental and behavioral disorders (Wikman and Panossian 2004; Darbinyan
et al. 2007; Olsson et al. 2009). The results of these studies have been summarized in several review articles (Panossian and Wagner 2005; Panossian and Wikman 2005, 2009,
2010; Panossian et al. 2010b,c; Panossian 2013). Collectively, this evidence may be sufficient support for medicinal use of *Rhodiola* in treating mental and behavioral disorders.

In this chapter, we summarize the results of preclinical and clinical studies of
SHR-5 and its active constituents, and discuss its possible indication for use in stress
and aging-associated conditions and diseases.

9.2   DATA ON EFFICACY

Efficacy and effectiveness of SHR-5 in treating stress-induced mental and behavioral disorders as well as aging-associated conditions has been the subject of numerous studies.

9.2.1   STRESS-INDUCED MENTAL AND BEHAVIORAL DISORDERS

Currently, SHR-5 is the only *R. rosea* root extract that has been proven to relieve
symptoms associated with stress, such as fatigue and impaired cognitive functions,
in randomized, double-blind placebo-controlled clinical studies (Darbinyan et al.
2000; Spasov et al. 2000; Shevtsov et al. 2003; Olsson et al. 2009) (see Table 9.1).

Three studies have been conducted in healthy human subjects under various
stressful conditions (Darbinyan et al. 2000; Spasov et al. 2000; Shevtsov et al. 2003).
In the 2009 study of Olsson et al., subjects were patients diagnosed with "fatigue
syndrome" as defined by subdivision F43.8A of the International Classification of
Diseases (ICD) code F43.8 among "Other reactions to severe stress."

In Olsson et al.'s 4-week study of 60 patients, *R. rosea* significantly reduced fatigue,
improved attention and decreased cortisol response to stress upon awakening (Figure 9.1).

R. rosea (SHR-5) demonstrated anti-fatigue effects and improved cognitive functions
during fatigue and under various stressful conditions in a total of 257 healthy adults
(based on a compilation of three studies: Darbinyan et al. 2000; Spasov et al. 2000;
Shevtsov et al. 2003) who had received single and repeated doses of the medication.

9.2.2   EXPERIMENTS IN THE NEMATODE *CAENORHABDITIS ELEGANS*,
THE POND SNAIL *LYMNAEA STAGNALIS* AND ON ISOLATED CELLS

Rhodiola SHR-5 can increase stress resistance of the nematode *C. elegans* (Wiegant
et al. 2009), embryos of the pond snail *L. stagnalis* (Boon-Niermeijer et al. 2000),
and isolated hepatocytes (Wiegant et al. 2008). This has been demonstrated
using various stress models, including heat shock, UV-, menadione-, arsenite- or

TABLE 9.1
Results of Randomized Studies on Humans Involving Effects of *Rhodiola* Preparations on Mental Performance Related to Fatigue

Plant name	Design	Total Subjects (Sample Size of Verum/Control) (Age Range)	Intervention/Control Dosage	Primary Endpoint	Main Results	Frequency of Adverse Effects	Quality Level of Evidence*	Jadad Score (max 5)	References
Rhodiola rosea	PC[a] 2 parallel groups	60 volunteers with stress-induced fatigue (30/30) (20–55 years)	Extract SHR-5 (288 mg twice daily)/placebo for 4 weeks	Symptoms of fatigue, attention, depression, QOL,[b] salivary cortisol	Symptoms of fatigue, attention and salivary cortisol significantly improved compared with control	None	Ib	5	Olsson et al. (2009)
	PC, CO 2 parallel groups	56 healthy subjects (26/30)[c] (24–35 years)	Extract SHR-5 (170 mg once daily)/placebo for 2 weeks	Mental fatigue, perceptive and cognitive functions such as associative thinking, short-term memory, calculation and ability of concentration, and speed of audio-visual perception	Statistically significant improvement in the treatment group (SHR-5) during the first 2-week period	None	Ib	4	Darbinyan et al. (2000)

(Continued)

TABLE 9.1 *(Continued)*

Results of Randomized Studies on Humans Involving Effects of *Rhodiola* Preparations on Mental Performance Related to Fatigue

Plant name	Design	Total Subjects (Sample Size of Verum/ Control) (Age Range)	Intervention/ Control Dosage	Primary Endpoint	Main Results	Frequency of Adverse Effects	Quality Level of Evidence*	Jadad Score (max 5)	References
	PC 2 parallel groups	40 healthy subjects (20/20) (17–19 years)	Extract SHR-5 (50 mg twice daily)/placebo for 20 days	Mental fatigue, physical performance, general well-being	Significant improvement in physical fitness, mental fatigue and neuromotor tests compared with control ($p < 0.01$). General well-being was also significantly ($p < 0.05$) better in the verum group. No significance was seen in the correction of text tests or a neuromuscular tapping test	None	Ib	3	Spasov et al. (2000)
	PC 3 parallel treatment groups	161 healthy subjects, (41/20/40 treated + 20 untreated) (19–21 years)	Extract SHR-5 (single dose of 370 mg or 555 mg)/ placebo	Capacity for mental work	Significant difference in anti-fatigue effects in SHR-5 groups compared with control ($p < 0.001$), while no significant difference between the two dosage groups was observed	One subject in placebo group complained of hyper-salivation lasting 40 minutes after intake	Ib	3	Shevtsov et al. (2003)

Design	Patients	Treatment	Indication	Significant results		Evidence		Reference
PC 3 parallel treatment groups	91 patients with mild and moderate depression (31/30/30) (18–70 years)	Extract SHR-5 (170 or 340 mg twice daily)/ placebo for 6 weeks	Depression in total HAMD and BDI scores	Significant differences in HAMD and BDI scores and scores reflecting levels of insomnia, emotional instability, somatization and self-esteem in SHR-5 groups compared to placebo ($p < 0.001$)	None	Ib	5	Darbinyan et al. (2007)

*According to WHO, FDA, and EMEA: Ia, meta-analyses of randomized and controlled studies; Ib, evidence from at least one randomized study with control; IIa, evidence from at least one well-performed study with control group; IIb, evidence from at least one well-performed quasi-experimental study; III, evidence from well-performed non-experimental descriptive studies as well as comparative studies, correlation studies and case-studies; and IV, evidence from expert committee reports or appraisals and/or clinical experiments by prominent authorities.

[a]CO, crossover; PC, placebo-controlled; M, multicenter.

[b]QOL, quality of life; HAMD, Hamilton Depression Rating Scale; BDI, Beck Depression Inventory; RVI, Rand Vitality Index; HR, heart rate; BP, blood pressure; CDR, Cognitive Drug Research; MMSE, Mini-mental State Examination; ADAS, Alzheimer Disease Assessment Scale; CDRS, Clinical Dementia Rating Scale.

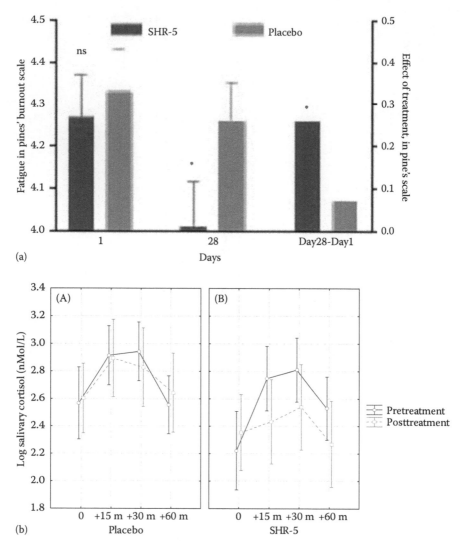

FIGURE 9.1 (a) The level of physical and emotional stress and mental exhaustion, Pines and Aronson Burnout Measure of burnout includes 21 items, evaluated on 7-point frequency scales. (b) Cortisol response to awakening in patients with fatigue syndrome showing pre- and posttreatment logarithm-transformed mean values of salivary cortisol with respect to time after awakening for (A) placebo group ($n = 25$) and (B) group treated with R. rosea extract SHR-5 ($n = 21$) over a period of 28 days. Vertical bars denote standard deviations.

buthionine sulfoximine-induced oxidative stress (Boon-Niermeijer et al. 2000, 2012; Wiegant et al. 2008, 2009) and exposure to high and toxic doses of different environmental pollutants, such as heavy metals, copper, and cadmium (Boon-Niermeijer et al. 2000).

Gene expression profiling was conducted on the human neuroglial cell line, T98G, after treatment with either *Rhodiola* SHR-5 extract, or several of its constituents separately, including salidroside, triandrin, and tyrosol. *Rhodiola* SHR-5 and individual constituents had similar effects on G-protein-coupled receptor (GPCR)-mediated signal transduction through cAMP, phospholipase C, and phosphatidylinositol signaling pathways (Figure 9.2). They may reduce cAMP levels in brain cells by downregulating the adenylate cyclase gene, ADC2Y, and upregulating the phosphodiestherase gene, PDE4D. This activity is essential for energy homeostasis and for switching between catabolic and anabolic states. Downregulation of cAMP by *Rhodiola* may also decrease cAMP-dependent protein kinase A (PKA) activity in various cells. This would inhibit stress-induced catabolic transformations and conserve ATP for many ATP-dependent metabolic transformations. It has also been suggested that cAMP inhibitors can improve working memory, which plays a key role in abstract thinking, planning, organizing, and so on (Panossian et al. 2013).

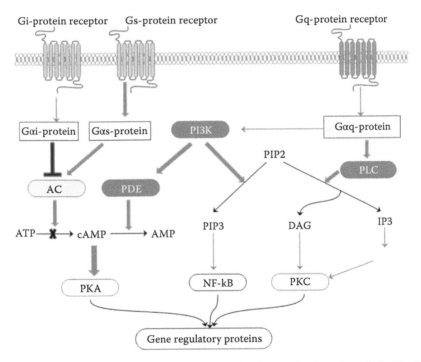

FIGURE 9.2 **(See color insert.)** Hypothetic molecular mechanisms by which *Rhodiola* activate adaptive stress response pathways. Neurons normally receive signals from multiple extracellular stressors that activate adaptive cellular signaling pathways, for example, many neurotransmitters activate GTP-binding protein coupled receptors (GPCR). The receptors in turn activate kinase cascades including those that activate protein kinase C (PKC), protein kinase A (PKA), and phosphatidylinositol-3-kinase (PI3K). Effect of *Rhodiola* on G-protein-coupled receptors pathways: upregulated genes are represented in red, downregulated in blue. The Gs alpha subunit (or Gs protein) stimulates the cAMP-dependent pathway by activating adenylate cyclase. Gi alpha subunit (orGi/G0 or Giprotein) inhibits the production of cAMP from ATP. DAG, diacylglycerol; IP3, inositoltriphosphate; PLC, phospholipase C.

Rhodiola and its constituents unregulated the PLCB1 gene, which encodes phosphoinositide-specific phospholipase C and phosphatidylinositol 3-kinases, key players in the regulation of NF-κB-mediated defense responses. Other common targets of SHR-5, salidroside, triandrin, and tyrosol include genes encoding the ERα estrogen receptor (2.9- to 22.6-fold downregulation), cholesterol ester transfer protein (5.1- to 10.6-fold downregulation), heat shock protein, Hsp70 (3.0- to 45.0-fold upregulation), serpin peptidase inhibitor (neuroserpin), and 5-HT3 serotonin receptor (2.2- to 6.6-fold downregulation). These effects concur with the beneficial effects of *Rhodiola* that have been found in treating stress-induced behavioral and mental conditions and aging-associated disorders, including neurodegeneration, atherosclerosis, and impaired apoptosis (Panossian et al. 2013).

Proposed mechanisms by which *Rhodiola* affects cognitive functions, memory, learning, and attention are

- Deregulation of GPCR, including downregulation of serotonin 5-HT3 GPCR.
- Deregulation of cAMP followed by closure of hyperpolarization-activated channels.
- Upregulation of PI3K, which is required for long-term potentiation of neurons.
- Upregulation of IP3, which plays an important role in the induction of plasticity in cerebellar Purkinje cells.
- Upregulation of the *SERPINI1* gene (serpin peptidase inhibitor, neuroserpin), which plays an important role in synapse development and regulates synaptic plasticity.
- Normalization of cortisol homeostasis (Panossian et al. 2013).

The physiological impact of GPCR is associated with

- Behavior and mood regulation (serotonin, dopamine, GABA, and glutamate receptors in the brain).
- Regulation of immune system activity and inflammation (chemokine, histamine receptors).
- Autonomic nervous system transmission: both the sympathetic and parasympathetic nervous systems are regulated by GPCR pathways, responsible for control of many automatic functions of the body such as blood pressure, heart rate, and digestive processes.
- Growth and metastasis of some types of tumors.
- Sense of smell and vision.
- Homeostasis modulation (e.g., water balance).
- Maintenance of physiological homeostasis during stress.

In a recent study, we hypothesized that adaptation to stress is regulated by increasing or decreasing the total number of GPCRs, and that *Rhodiola* initiates that process via deregulation of the corresponding genes (Panossian et al. 2013) (Figure 9.3).

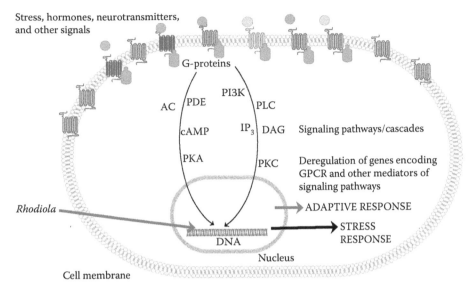

Stress, hormones, neurotransmitters, and other signals

G-proteins

PI3K

AC PDE   PLC

cAMP   IP$_3$ DAG   Signaling pathways/cascades

PKA   PKC   Deregulation of genes encoding
GPCR and other mediators of
signaling pathways

Rhodiola   ADAPTIVE RESPONSE

DNA   STRESS
RESPONSE

Nucleus

Cell membrane

FIGURE 9.3 (**See color insert.**) Evidence suggests that *Rhodiola* SHR-5 extract may initi-ate the activation or suppression of some genes that encode the expression of some GPCRs and key mediators of GPCR-signaling pathways. By reducing the expression of specific GPCRs, SHR-5 decreases cellular sensitivity to stress and increases stress resilience when the individual is exposed to different kinds of stressors, including emotional, physical, heat, chemical, toxic, infectious, malignant, etc.

In stress, activation of serotonin receptors modulates the release of many neu-rotransmitters including glutamate, GABA, dopamine, epinephrine/norepinephrine, and acetylcholine, as well as many hormones including oxytocin, prolactin, vaso-pressin, cortisol, corticotrophin, and substance P. Activation of serotonin receptors also influences various biological and neurological processes such as aggression, anxiety, appetite, cognition, learning, memory, mood (depression), nausea, sleep, thermoregulation, behavioral aging, and longevity. Downregulation of serotonin 5-HT3 GPCR by salidroside and tyrosol may also have an influence on these pro-cesses (Panossian et al. 2013).

9.2.3 DEPRESSION

Darbinyan et al.'s 2007 study is the only randomized, double-blind, placebo-controlled clinical trial to date in which the efficacy of *Rhodiola* for treating mild and moderate depression has been demonstrated. A 6-week DBRPC study of *R. rosea* (SHR-5) in 91 patients with recurrent depression showed significant anti-depressant effect. Two groups were each given a different *R. rosea* dose (340 mg/day and 680 mg/day, 2 and 4 of Arctic Root tablets); the third group received a placebo. Both treatment groups showed significant improvement of depressive symp-toms on the Hamilton Depression Scale (HAMD) and Beck Depression Index (BDI) (Figure 9.4). The herb also reduced insomnia, emotional instability and somatiza-tion, and improved self-esteem.

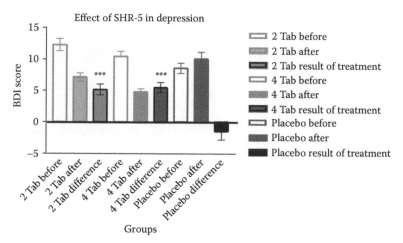

FIGURE 9.4 The effect of treatment with SHR-5 (groups treated with two and four tablets daily) versus placebo as measured by the change in total depression (a-BDI and b-HAMD) scores between the start (day 0—before treatment) and end (day 42—after treatment) of medication. For comparison of improvements between groups, (the differences in scores at days 42 and 0) the significance of the differences is shown as ** = significant with $P < .01$ and *** = very significant with $P < .001$).

9.2.4 EXPERIMENTS IN ANIMALS AND ON ISOLATED CELLS

The purpose of studies in animals was to identify the active principle of the SHR-5 extract, which is required for standardization of herbal preparations and Herbal Medicinal Products to ensure their quality and reproducible activity. Antidepressant effects of SHR-5, its constituents (salidroside, tyrosol, rosavin, rosin, rorarin), and possible metabolites of rosavin (phenylpropanoids cinnamoylalcohol, cinnamaldehyde, and cinnamic acid) were compared with amitriptyline, imiptamine and Hypericum in a Porsolt's behavior despair-forced swimming test of rats (Figure 9.5) (Panossian et al. 2008a). It was found that dose–response correlation is not linear, and that the most active constituents are salidroside and tyrosol. In combination with rosavin, they are more active than any single compound alone, whereas phenylpropanoid metabolites were inactive. Salidroside was found inactive in a MAO-A bioassay (monoamine oxidase A test), where the most active constituents of a *Rhodiola* root extract were found to be cinnamyl alcohol and monoterpene glycoside rosiridin (van Diermen et al. 2009). The content of rosiridin is rather low in *Rhodiola* extract, and is highly unlikely to have any influence on antidepressant effects of *Rhodiola*. Rosiridin has not been detected in *Rhodiola* SHR-5 extract by GS-MS, and apparently has no clinical significance in the treatment of depression. Cinnamyl alcohol was inactive in the Porsolt's depression test in rats (Figure 9.5) (Panossian et al. 2008a). Pathophysiology of depression is associated with multiple impairments in the neuroendocrine system, where many components are involved in addition to catecholamines and monoaminoxidases (MAO-A). Pharmacological activity of some antidepressive drugs is due to MAO-inhibition (unfortunately, all have numerous adverse effects). However,

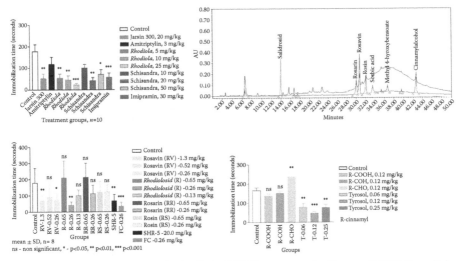

FIGURE 9.5 HPLC-profile of *Rhodiola* SHR-5 extract and efficacy of SHR-5, amitriptyline, imipramine, Hypericum (Jarsin), Schisandra, salidroside, rosavin, rosarin, rosin, tyrosol, cinnamic acid, and phenylpropanoids (cinnamaldehyde and cinnamyl alcohol) in a Porsolt's behavior despair forced swimming test (From Panossian, A., et al., *Phytomedicine* 15, 84–91, 2008a.).

MAO-inhibition is not the only pharmacological target for the treatment of depression. There are other alternatives.

Mechanisms by which SHR-5 and its active constituents act on human emotion and behavior were studied in experiments on isolated cells as well as by analysis of hormones and stress markers in animals. The results of these studies show that the antidepressive effects of SHR-5 and salidroside presumably operate in association with their effects on the following:

- NPY-Hsp70-mediated effects on glucocorticoid receptors, including expression and release of neuropeptide Y (NPY) and stress-activated proteins (Hsp70 and JNK) (Figure 9.6)
  - Activation of expression of neuropeptide Y (Panossian et al. 2012), which is low in depression
  - Activation of expression of Hsp72 (Panossian et al. 2012), which is known to inhibit stress-activated protein kinase JNK (Figure 9.7b), playing an important role in suppressing glucocorticoid receptors and thereby increasing cortisol (Figure 9.7a) during stress and depression (Panossian et al. 2007)
- Downregulation of some of G-protein coupled receptors, particularly receptors of serotonin in isolated neuroglia cells (Panossian et al. 2013)
- Downregulation of estrogen alpha receptors in isolated neuroglia cells (Panossian et al. 2013)

NPY is known to play a role in the pathophysiology of depression. It has been shown that NPY displays antidepressant-like activity in the rat forced swimming

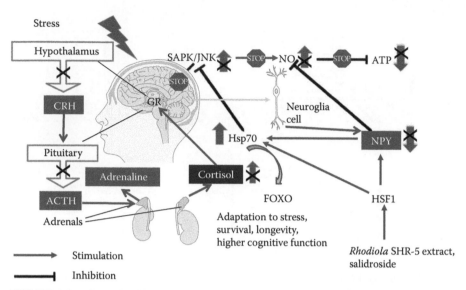

FIGURE 9.6   (See color insert.) Hypothetical neuroendocrine mechanism of stress protection by *Rhodiola* SHR-5 extract and salidroside. Stress induces CRH release from the hypothalamus followed by ACTH release from the pituitary, which simulates release of adrenal hormones and neuropeptide Y (NPY) to mobilize energy resources and cope with the stress. Feedback regulation of overreaction is initiated by cortisol release from the adrenal cortex, followed by binding to glucocorticoid receptors (GR) in the brain. This signal stops the further release of brain hormones and brings the stress-induced increase of cortisol down to normal levels. While brief and mild stress (eustress/challenge) is essential to life, severe stress (distress/overload) is associated with extensive generation of oxygen-free radicals, including nitric oxide (NO), which is known to inhibit ATP formation (energy providing molecules). Stress-activated protein kinases (SAPK/JNK/MAPK) inhibit GR, consequently feedback downregulation is blocked and cortisol content in the blood remains high during fatigue, depression, impaired memory, impaired concentration and other stressful conditions. Adaptogens normalize stress-induced elevated levels of cortisol and other extra- and intracellular mediators of stress response, such as elevated NO, SAPK via upregulation of expression of NPY, heat shock factor (HSF-1) and heat shock proteins Hsp70, which are known to inhibit SAPK. Consequently, NO generation is reduced and ATP production is no longer suppressed. Hsp70 functions intracellularly to enhance anti-apoptotic mechanisms protect proteins against mitochondria-generated oxygen-containing radicals, including nitric oxide, and superoxide anion. The released Hsp70 acts as an endogenous danger signal and plays an important role in immune stimulation. While released NPY plays a crucial role in the HPA axis and maintains energy balance, both NPY and Hsp70 are directly involved in cellular adaptation to stress, increased survival, enhanced longevity and improved cognitive function. Hsp70 inhibits the FOXO transcription factor, playing an important role in adaptation to stress and longevity. These pathways contribute to adaptogenic effects: antifatigue, increased attention and improved cognitive function.

test (Stogner and Holmes 2000; Redrobe et al. 2002). Human studies have revealed a role for NPY in adaptation to stress ("buffering" the harmful effects of stress) (Morgan et al. 2000, 2001; Morales-Medina et al. 2010). There is a plethora of preclinical and clinical evidence suggesting that NPY improves mood and cognitive performance (Fletcher et al. 2010). Higher levels of NPY have been observed

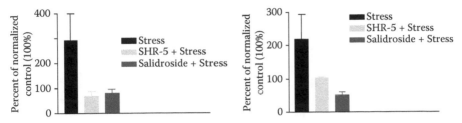

FIGURE 9.7 Effect of repeated administration of SHR-5 and salidroside on the levels of cortisol (a) and p-SAPK/JNK (b) in the blood of immobilized rabbits. (From Panossian, A., et al. *Drug Targets Insights* 1:39–54, 2007.)

in soldiers who either have reduced psychological distress or belong to the elite Special Forces branch (Morgan et al. 2000, 2001). In contrast, decreased levels of NPY were observed in depression and in brain tissues of suicide victims (Morales-Medina et al. 2010).

Rhodiola is an antidepressant that stimulates biosynthesis of NPY in the brain and therefore has an antidepressive effect on humans and animals. Salidroside was found to be active in stimulating NPY-production (Panossian et al. 2012).

Rhodiola downregulates the *ESR1* gene encoding estrogen receptor alpha (ERα) (Panossian et al. 2013), a nuclear receptor, primarily localized in the cytoplasm in complexes with heat-shock-proteins (hsp90, hsp70, and hsp56) and expressed in various tissues (Welshons et al. 1984; Evans 1988). In the human forebrain, ERα are predominantly localized in the amygdala (suggested to be involved in mood and cognition), hypothalamus (involved in learning and memory), and septum (Österlund and Hurd 2001; Yaghmaie et al. 2005; Dahlman-Wright et al. 2006). Affective mood disorders, such as premenstrual syndrome, postnatal depression, and postmenopausal depression are associated with low serum-levels of estrogens.

9.2.5 Aging-Associated Disorders

Rhodiola SHR-5 extract has been shown to significantly increase the mean life span of the nematode *C. elegans*, by about 20% versus control. This increase is dose-dependent: at higher and lower concentrations, less effect was observed, whereas at the highest concentration tested, life span was shortened by 15%–25% (Wiegant et al. 2009) (Figure 9.8a,b). Researchers suggested that this effect may be due to the effect of SHR-5 on nuclear transcription factor DAF-16, which is known to activate transcription of a large number of genes that increase stress resistance and promote longevity. *Rhodiola* SHR-5 can induce translocation of the DAF-16 transcription factor from the cytoplasm into the nucleus (Figure 9.8d), suggesting a reprogramming of transcriptional activities favoring the synthesis of proteins involved in longevity and stress resistance (such as the chaperone HSP-16) (Wiegant et al. 2009).

Recent findings (see Sections 2.1.1.2 and 2.1.2.2) summarized in Table 9.2 below are consistent with the observed beneficial effects of *Rhodiola* in aging-associated disorders, including neurodegeneration, atherosclerosis, and impaired apoptosis.

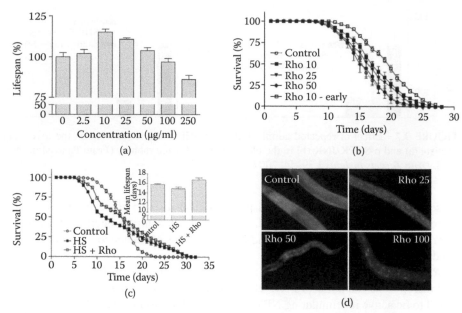

FIGURE 9.8 (a) Effect of different concentrations of *Rhodiola rosea* extract (SHR-5) on longevity of C. elegans and (b) when added later in life (at an age of 7 days) in comparison with non-treated individuals. (c) The protective effect of *Rhodiola* (SHR-5) is shown on the life span of *C. elegans* following exposure to heat shock. (d) *Rhodiola rosea* SHR-5 induces nuclear translocation of DAF16 in C. elegans.

TABLE 9.2
Aging-Associated Disorders and Effects of *Rhodiola* on Genes Involved in Regulation of Aging

Inflammation – atherosclerosis
* Downregulation of *CETP*
* Deregulation of GPCR

Neurodegeneration – impaired cognitive functions (learning, memory, abstract thinking, planning, organizing)
* Downregulation of cAMP
* Downregulation of ESR1
* Upregulation of serpine
* Deregulation of GPCR

Impaired apoptosis – Cancer
* Downregulation of ESR1, OLFM
* Upregulation of IP3, PLC, DAG, PI3K, NFkB
* Deregulation of GPCR

Metabolic disorders and energy metabolism
* Downregulation of cAMP
* Inhibition of ATP metabolism

Rhodiola downregulates the *CETP* gene, which encodes cholesteryl ester transfer protein, a lipid plasma protein that facilitates the transport of cholesterol esters and triglycerides between low-density and high-density lipoproteins (LDL, HDL) (Bruce et al. 1998). Inhibiting *CETP* has been shown to alleviate atherosclerosis and other cardiovascular diseases, as well as metabolic syndrome (Barter et al. 2003).

Since ERs are overexpressed in some type of cancers (Fabian and Kimler 2005; Ascenzi et al. 2006; Deroo and Korach 2006), downregulation of *ESR1* by *Rhodiola* may be effective in the prevention and treatment of aging-related breast, ovarian, colon, prostate, and endometrial cancers.

It has been shown that estrogen signaling via ERα can significantly attenuate an inflammatory neurodegenerative process through its effect on astrocytes. ERα expression is necessary in astrocytes, but not neurons, for neuroprotection in experimental autoimmune encephalomyelitis, the most widely used mouse model of multiple sclerosis, an autoimmune disease characterized by demyelination and axonal degeneration (Spencea et al. 2011). Consequently, it can be expected that the neuroprotective effect of *Rhodiola* (Zhang et al. 2007, 2010; Chen et al. 2009a,b; Bocharov et al. 2010; Fletcher et al. 2010; Li et al. 2011; Palumbo et al. 2012; Shi et al. 2012) must be associated with upregulation ESR1 in glia cells. Surprisingly, the results of a study we conducted are not in line with this hypothesis, because *Rhodiola* downregulate ERα gene expression. How to interpret these contradictory observations—neuroprotection accompanied by downregulation of ERα gene expression? Cell response to adaptogens is stress mimetic. Pretreatment with adaptogen (preconditioning) adapts the cell to stress (Panossian et al. 1999; Wiegant et al. 2009; Boon-Niermeijer et al. 2012). Similarly, it can be speculated that downregulation of ERα gene expression by *Rhodiola* is a signal for glia cells to initiate a feedback regulation of ERα. In a broader sense, this concept is related to inflammation, which is a defense response ("switch on" defense system) to cope with infection. To prevent an overreaction, a feedback mechanism to regulate inflammation is activated, for example, an increase in cortisol and anti-inflammatory cytokines ("switch off" defense system). Mild stress is generally a defense response to activate innate immunity. In this context, *Rhodiola* initiates activation of the innate defense system, including ERα, as one component of stress system.

In addition to the "classical" model of steroid receptor function, where the ligand-bound estrogen receptor regulates transcription of target genes in the nucleus by binding to estrogen response element regulatory sequences in target genes, estrogen-activated membrane-bound CRα initiates rapid signaling PI3K/PLC pathways in astrocytes to indirectly modulate neuronal function and survival (Deroo and Korach 2006; Mhyre et al. 2006). Estrogen treatment attenuates "nonclassical" transcription at estrogen response element sites in glioma cells (Mhyre et al. 2006). *Rhodiola* upregulates *PLCB1*-, *PI3KC2G*-, and c-AMP-related genes, and modulates NO and Stress-activated protein kinases (SAPK). These similarities between estrogens and *Rhodiola* allow us to suggest that their action mechanisms interfere with each other in some way. Although both are neuroprotective, it is still unclear whether they are mimetic and in competition, or are inherently antagonistic.

9.3 PHARMACOKINETIC PROPERTIES

The pharmacokinetics of tyrosol, salidroside, and rosavin—the body's absorption and processing of the active constituents of *R. rosea* SHR-5 extract—have been studied in both rats and human volunteers. Salidroside was found to be quickly and completely absorbed into the blood of rats (bioavailability, 75%–90%), distributed within organs and tissues and rapidly metabolized to tyrosol following oral administration of SHR-5 at doses of 20 and 50 mg/kg. Many of the measured pharmacokinetic parameters of salidroside were significantly different when the pure compound was administered rather than the extract. The basal level of tyrosol in blood plasma of rats increased following administration of SHR-5 as a result of absorption of free tyrosol present in the extract and of biotransformation of salidroside into tyrosol, which occurred within the first 2 hours. The pharmacokinetics and rate of biotransformation of salidroside were essentially the same following single or multiple regimes of administration of SHR-5. Rosavin has low bioavailability (20%–26%) and was quickly eliminated from the blood of rats that had received SHR-5. The results of the study in humans ($n = 16$) showed that after oral administration of two tablets of Rosenroot, healthy volunteers reached a maximum concentration of salidroside (about 3 µM) and rosavin (about 1.2 µM) in blood plasma after 2 hours, where the absorption rate constant was equal in both components. In contrast to salidroside, rosavin was not detected in blood plasma at 0.5 and 1 hours after oral administration of two tablets of Rosenroot. The maximum concentration and elimination half-life of salidroside were two to three times higher than those of rosavin. The $AUC_{0-\infty}$ C_{max} and AUC_{0-t} values of salidroside were 3.5, 2.2, and 3.1 times higher than those of rosavin. Thus, the elimination of salidroside from the blood took 1.8 times longer than elimination of rosavin (Abrahamyan et al. 2004, 2005; Panossian et al. 2010a).

REFERENCES

Abrahamyan, H., Hovhannisyan, A., Gabrielyan, E., Panossian, A. 2004. Pharmacokinetic study of salidroside and rosavin, active principles of *Rhodiola rosea* in rats by high performance capillary electrophoresis system. *Drugs and Medicine* 3:55–61.

Abrahamyan, H., Hovhannisyan, A., Panossian, A., Gabrielyan, E. 2005. The bioavailability of salidroside and rosavin, active principles of *Rhodiola rosea* extract SHR-5 in rats. *Medical Science of Armenia* 45 (1):24–9.

Ascenzi, P., Bocedi, A., Marino, M. 2006. Structure-function relationship of estrogen receptor alpha and beta: Impact on human health. *Molecular Aspects of Medicine* 27: 299–402.

Barter, P.J., Brewer, H.B. Jr., Chapman, M.J., Hennekens, C.H., Rader, D.J., Tall, A.R. 2003. Cholesteryl ester transfer protein: A novel target for raising HDL and inhibiting atherosclerosis. *Arteriosclerosis, Thrombosis, and Vascular Biology* 23:160–7.

Bocharov, E.V., Ivanova-Smolenskaya, I.A., Poleshchuk, V.V., Kucheryanu, V.G., Il'enko, V.A., Bocharova, O.A. 2010. Therapeutic efficacy of the neuroprotective plant adaptogen in neurodegenerative disease (Parkinson's disease as an example). *Bulletin of Experimental Biology and Medicine* 149:682–4.

Boon-Niermeijer, E.K., van den Berg, A., Vorontsova, O.N., Bayda, L.A., Malyshev, I.Y., Wiegant, F.A.C. 2012. Enhancement of adaptive resistance against a variety of chronic stress conditions by plant adaptogens: Protective effects on survival and embryonic development of *Lymnaea stagnalis Adaptive Medicine* 4(4):233–44, 233.

Boon-Niermeijer, E.K., van den Berg, A., Wikman, G., Wiegant, F.A., 2000. Phyto-adaptogens protect against environmental stress-induced death of embryos from the freshwater snail *Lymnaea stagnalis*. *Phytomedicine* 7:389–99.

Bruce, C., Sharp, D.S., Tall, A.R. 1998. Relationship of HDL and coronary heart disease to a common amino acid polymorphism in the cholesteryl ester transfer protein in men with and without hypertriglyceridemia. *Journal of Lipid Research* 39:1071–8.

Chen, Q.G., Zeng, Y.S., Qu, Z.Q, Tang, J.Y., Qin, Y.J., Chung, P., Wong, R., Hägg, U. 2009a. The effects of *Rhodiola rosea* extract on 5-HT level, cell proliferation and quantity of neurons at cerebral hippocampus of depressive rats. *Phytomedicine* 16:830–8.

Chen, X., Zhang, Q., Cheng, Q., Ding, F. 2009b. Protective effect of salidroside against H_2O_2-induced cell apoptosis in primary culture of rat hippocampal neurons. *Molecular and Cellular Biochemistry* 332:85–93.

Dahlman-Wright, K., Cavailles, V., Fuqua, S.A., Jordan, V.C., Katzenellenbogen, J.A., Korach, K.S., Maggi, A., Muramatsu, M., Parker, M.G., Gustafsson, J.A. 2006. International Union of Pharmacology. LXIV. Estrogen receptors. *Pharmacological Review* 58:773–81.

Darbinyan, V., Kteyan, A., Panossian, A., Gabrielian, E., Wikman, G., Wagner, H. 2000. *Rhodiola rosea* in stress-induced fatigue: A double blind cross-over study of a standardized extract SHA-5 with a repeated low-dose regimen on the mental performance of healthy physicians during night duty. *Phytomedicine* 7(5):365–71.

Darbinyan, V., Aslanyan, G., Amroyan, E., Gabrielyan, E., Malmström, C. Panossian, A. 2007. Clinical trial of *Rhodiola rosea* L. extract SHR-5 in the treatment of mild to moderate depression. *Nordic Journal Psychiatry* 61(5):343–8.

Deroo, B.J., Korach, K.S. 2006. Estrogen receptors and human disease. *Journal of Clinical Investigation* 116:561–7.

Evans, R.M. 1988. The steroid and thyroid hormone receptor superfamily. *Science* 240:889–95.

Fabian, C.J., Kimler, B.F. 2005. Selective estrogen-receptor modulators for primary prevention of breast cancer. *Journal of Clinical Oncology* 23;1644–1655.

Fletcher M.A., Rosenthal M., Antoni M., Ironson G., Zeng X.R., Barnes Z., Harvey J.M., Hurwitz B., Levis S., Broderick G., Klimas N.G. 2010. Plasma neuropeptide Y:A biomarker for symptom severity in chronic fatigue syndrome. *Behavioral and Brain Functions* 6:76.

Li, X., Ye, X., Li, X., Sun, X., Liang, Q., Tao, L., Kang, X., Chen, J. 2011. Salidroside protects against MPP(+)-induced apoptosis in PC12 cells by inhibiting the NO pathway. *Brain Research* 1382:9–18.

Mhyre, A.J., Shapiro, R.A., Dorsa, D.M. 2006. Estradiol reduces nonclassical transcription at cyclic adenosine 3,5-monophosphate response elements in glioma cells expressing estrogen receptor alpha. *Endocrinology* 147:1796–1804.

Morales-Medina, J. C., Dumont, Y., Quirion, R. 2010. A possible role of neuropeptide Y in depression and stress. *Brain Research* 1314:194–205.

Morgan, C. A., III, Wang, S., Rasmusson, A., Hazlett, G., Anderson, G., Charney, D.S. 2001. Relationship among plasma cortisol, catecholamines, neuropeptide Y, and human performance during exposure to uncontrollable stress. *Psychosomatic Medicine* 63:412:22.

Morgan, C.A., III, Wang, S., Southwick, S.M., Rasmusson, A., Hazlett, G., Hauger, R.L., Charney, D.S. 2000. Plasma neuropeptide-Y concentrations in humans exposed to military survival training. *Biological Psychiatry* 47:902–9.

Olsson, E.M.G., von Schéele, B., Panossian, A.G. 2009. A randomized double-blind placebo controlled Parallel group study of SHR-5 extract of *Rhodiola rosea* roots as treatment for patients with stress related fatigue. *Planta medica.* 75(2):105–12. Epub 2008 Nov 18.

Österlund, M.K., Hurd, Y.L. 2001. Estrogen receptors in the human forebrain and the relation to neuropsychiatric disorders. *Progress in Neurobiology* 64, 251–67.

Palumbo, D.R., Occhiuto, F., Spadaro, F., Circosta, C. 2012. *Rhodiola rosea* extract protects human cortical neurons against glutamate and hydrogen peroxide-induced cell death through reduction in the accumulation of intracellular calcium. *Phytotherapy Research* 26:878–83.

Panossian, A., Hambartsumyan, M., Hovanissian, A, Gabrielyan, E., Wikman, G., 2007. The adaptogens *Rhodiola* and *Schizandra* modify the response to immobilization stress in rabbits by suppressing the increase of phosphorylated stress-activated protein kinase, nitric oxide and cortisol. *Drug Targets Insights*, 1:39–54.

Panossian, A., Hovhannisyan, A., Abrahamyan, H., Wikman, G. 2010a. Pharmacokinetics of active constituents of *Rhodiola rosea* L. special extract SHR-5. In: V.K. Gupta. (Ed.). *Comprehensive Bioactive Natural Products Vol. 2: Efficacy, Safety & Clinical Evaluation (Part-1)*. Houston, TX: Studium Press LLC. pp. 1–23.

Panossian, A., Nikoyan, N., Ohanyan, N., Hovhannisyan, A., Abrahamyan, H., Gabrielyan, E., Wikman, G. 2008a. Comparative study of *Rhodiola* preparations on behavioural despair of rats. *Phytomedicine* 15(1):84–91.

Panossian, A., Wagner, H. 2005. Stimulating effect of adaptogens: An overview with particular reference to their efficacy following single dose administration. *Phytotherapy Research* 19:819–38.

Panossian, A., Wikman, G. 2005. Effect of adaptogens on the central nervous system. *Arquivos Brasileiros de Fitomedicine Científica* 2:109–31.

Panossian, A., Wikman, G. 2009. Evidence-based efficacy of adaptogens infatigue, and molecular mechanisms related to their stress-protective activity. *Current Clinical Pharmacology* 4(3): 198–219.

Panossian, A., Wikman, G. 2010. Effects of adaptogens on the central nervous system and the molecular mechanisms associated with their stress-protective activity. *Pharmaceuticals* 3:188–224.

Panossian, A., Wikman, G., Andreeva, L., Boykova, A., Nikiforova, D., Timonina, N. 2008b. Adaptogens exert a stress protective effect by modulation of expression of molecular chaperons. *Planta Medica* 74:1018.

Panossian, A., Wikman, G., Kaur, P., Asea, A. 2009. Adaptogens exert a stress-protective effect by modulation of expression of molecular chaperones. *Phytomedicine* 16(6–7): 617–622.

Panossian, A., Wikman, G., Kaur, P., Asea, A. 2010b. Molecular chaperones as mediators of stress protective effect of plant adaptogens. In: Alexzander A. A., Bente K.P. (Eds.). *Heat Shock Proteins and Whole Body Physiology. Heat Shock Proteins*. Volume 5. Heiderberg, London, New York: Springer Dordrecht. pp. 351–64.

Panossian, A., Wikman, G., Kaur, P, Asea, A. 2012. Adaptogens stimulate neuropeptide Y and Hsp72 expression and release in neuroglia cells. *Frontiers in Neuroscience* 6:6.

Panossian. A., Wikman, G., Sarris, J. 2010c. Rosenroot (*Rhodiola rosea*): Traditionaluse, chemical composition, pharmacology and clinical efficacy. *Phytomedicine* 17(7):481–93. http://www.ncbi.nlm.nih.gov/pubmed/20378318.

Panossian A., Wikman G., Wagner H. 1999. Plant adaptogens. III. Earlier and more recent aspects and concepts on their mode of action. *Phytomedicine* 6(4):287–300. Review. PMID 10589450.

Panossian, A.G. 2013. Adaptogens in mental and behavioral disorders. *Psychiatric Clinics of North America* 36(1):49–64.

Panossian, A.G., Hamm, R., Kadioglu, O., Wikman, G., Efferth, T. 2013. Synergy and antagonism of active constituents of ADAPT-232 on transcriptional level of metabolic regulation of isolated neuroglial cells. *Frontiers in Neuroscience* 7: 16.

Redrobe, J.P., Dumont, Y., Fournier, A., Quirion, R. 2002. The neuropeptide Y (NPY) Y1 receptor subtype mediates NPY-induced antidepressant-like activity in the mouse forced swimming test. *Neuropsychopharmacology* 26:615–24.

Shevtsov, V.A., Zholus, B.I., Shervarly, V.I., Vol'skij, V.B., Korovin, Y.P., Khristich, M.P., Roslyakova, N.A., Wikman, G. 2003. A randomized trial of two different doses of a SHR-5 *Rhodiola rosea* extract versus placebo and control of capacity for mental work. *Phytomedicine*10:95–105.

Shi, T.Y., Feng, S.F., Xing, J.H., Wu, Y.M., Li, X.Q., Zhang, N., Tian, Z., Liu, S.B., Zhao, M.G. 2012. Neuroprotective effects of Salidroside and its analogue tyrosol galactoside against focal cerebral ischemia *in vivo* and H_2O_2-induced neurotoxicity *in vitro*. *Neurotoxicity Research* 21:358–67.

Spasov, A.A., Wikman, G.K., Mandrikov, V.B., Mironova, I.A., Neumoin, V.V. 2000. A double blind placebo controlled pilot study of the stimulating effect of *Rhodiola rosea* SHR-5 extract on the physical and mental work capacity of students during a stressful examination period with a repeated low-dose regimen. *Phytomedicine* 7(2):85–89.

Spencea, R.D., Hambyb, M.E., Umedaa, E., Itoha, N., Dua, S., Wisdoma, A.J., Caoa, Y., Bondarb, G., Lama, J., Aob, Y., Sandovala, F., Surianya, S., Sofroniew, M.V., Voskuhla, R.R. 2011. Neuroprotection mediated through estrogen receptor-α in astrocytes. *Proceedings of the National Academy of Sciences of the United States of America* 108:8867–72. http://www.pnas.org/cgi/doi/10.1073/pnas.1103833108.

Stogner, K. A., Holmes, P. V. 2000. Neuropeptide-Y exerts antidepressant-like effects in the forced swim test in rats. *European Journal of Pharmacology* 387: R9–10.

van Diermen, D., Marston, A., Bravo, J., Reist, M., Carrupt, P.A., Hostettmann, K. 2009. Monoamine oxidase inhibition by *Rhodiola rosea* L. roots. *Journal of Ethnopharmacology* 122:397–401.

Welshons, W.V., Lieberman, M.E., Gorski, J. 1984. Nuclear localization of unoccupied oestrogen receptors. *Nature* 307:747–9.

Wiegant, F.A.C., Limandjaja, G., de Poot, S.A.H., Bayda, L.A., Vorontsova, O.N., Zenina, T.A., Langelaar Makkinje, M., Post, J.A., Wikman, G. 2008. Plant adaptogens activate cellular adaptive mechanisms by causing mild damage. In: Lukyanova, L., Takeda, N., Singal, P.K. (Eds.). *Adaptation Biology and Medicine: Health Potentials*. Volume 5. New Delhi: Narosa Publishers. pp. 319–32.

Wiegant, F. A. C., Surinova, S., Ytsma, E., Langelaar-Makkinje, M., Wikman, G., Post, J. A. 2009. Plant adaptogens increase life span and stress resistance in *C. elegans*. *Biogerontology* 10(1):27–42.

Wikman G., Panossian A. 2004. Medical herbal extract Carpediol for the treating of depression. US Patent 20040131708.

Yaghmaie, F., Saeed, O., Garan, S.A., Freitag, W., Timiras, P.S., Sternberg, H. 2005. Caloric restriction reduces cell loss and maintains estrogen receptor-alpha immunoreactivity in the pre-optic hypothalamus of female B6D2F1 mice. *Neuro Endocrinology Letters* 26:197–203.

Zhang, L., Yu, H., Sun, Y., Lin, X., Chen, B., Tan, C., Cao, G., Wang, Z. 2007. Protective effects of salidroside on hydrogen peroxide-induced apoptosis in SH-SY5Y human neuroblastoma cells. *European Journal of Pharmacology* 564:18-25.

Zhang, L., Yu, H., Zhao, X., Lin, X., Tan, C., Cao, G., Wang, Z. 2010. Neuroprotective effects of salidroside against beta-amyloid-induced oxidative stress in SH-SY5Y human neuroblastoma cells. *Neurochemistry International* 57:547–55.

10 Rhodiola rosea in Psychiatric and Medical Practice

Patricia L. Gerbarg, Petra A. Illig, and Richard P. Brown

CONTENTS

10.1 Introduction 226
10.2 Psychiatric and Medical Conditions 227
    10.2.1 Depression 227
    10.2.2 Stress, Anxiety, and PTSD 228
        10.2.2.1 Case Presentation...229
    10.2.3 Cognitive Function, Memory, and Attention Deficit Disorder 230
        10.2.3.1 Neuroprotection 230
        10.2.3.2 Attention Deficit Disorder... 231
        10.2.3.3 Cognitive Function in Dementia, Post-Stroke, and
                Traumatic Brain Injury 231
    10.2.4 Stress and Fatigue..232
    10.2.5 Sexual Function, Fertility, and Menopause 232
    10.2.6 Infections and Immune Functions 234
    10.2.7 Cancer, Chemotherapy, and Radiation 235
        10.2.7.1 Cancer-Related Stress and Fatigue 235
        10.2.7.2 Adaptogens Augment Chemotherapy While Protecting
                Liver and Bone Marrow..235
10.3 Aerospace Medicine 236
    10.3.1 Acute Mountain Sickness and Hypoxia...236
    10.3.2 Jet Lag...237
    10.3.3 Aircrew Stress and Fatigue...237
    10.3.4 Space Exploration 238
10.4 Treatment Guidelines..239
    10.4.1 Dosages...239
    10.4.2 Side Effects and Medication Interactions of *R. rosea* 239
        10.4.2.1 Activation...239
        10.4.2.2 Hormonal 240
        10.4.2.3 Herb–Drug Interactions 240

10.5 Counteracting Side Effects of Prescription Medications...........................241
    10.5.1 Asthenia, Fatigue, Somnolence 242
    10.5.2 Weight Gain, Insulin Resistance, and Hyperlipidemia...................242
    10.5.3 Impairment of Cognition and Memory or
           Word-Finding Problems...243
    10.5.4 Sexual Dysfunction and Hormonal Changes...................................243
    10.5.5 Neurological Side Effects 243
10.6 More Future Directions 244
    10.6.1 Sports Medicine...244
    10.6.2 Psychiatry: Stress, Anxiety, PTSD, and Mass Disasters 244
    10.6.3 Stress and Fatigue—Home, School, Workplace..............................245
    10.6.4 Dermatology 245
    10.6.5 Dentistry 245
    10.6.6 Veterinary Medicine..245
10.7 Quality and Potency of *R. rosea* Brands 246
Acknowledgments..246
    Conflicts of Interest: 246
References..247

10.1 INTRODUCTION

Rhodia radix [some call it Rhodida] grows in Macedonia being like to Costus, but lighter, & uneven, making a scent in ye bruising, like that of Roses. It is of good use for ye agrieved with headache, being bruised & layed on with a little Rosaceum, & applied moist to ye forehead, & ye temples.

Dioscorides, AD 77

In AD 77, the famous physician, Dioscorides, who served with the Roman army in diverse regions of the Empire, included *R. rosea* in *De Materia Medica*, the first pharmacopeia and the standard for the practice of medicine for over 1500 years (Gunther 1968). However, in searching for earlier sources, we found that the first recorded medicinal use of *R. rosea* root extract appeared in an epic poem, the story of Jason and the Argonauts, dating back to the thirteenth century BC Bronze Age Greece. *R. rosea* grows in the Caucasus Mountains of the Republic of Georgia, a land once called Colchis, an area known for artifacts left by Greek trading expeditions 3000 years ago. According to *The Argonautica*, Hecate, the goddess of medicine, told the sorceress, Medea, to climb high in the mountains and to unearth the flesh-colored root of a plant with twin stalks measuring a cubit (15 inches) high, bearing flowers the color of yellow saffron, an accurate description of *R. rosea*. Medea drew sap from the roots of that herb to create the Charm of Prometheus, a potion that imbued Jason with the strength to vanquish fire-breathing bulls. The potion's energizing, strengthening effect on Jason was akin to what Dr. Brown observed using the fresh cut root of wild growing plants, which are even more potent than most cultivated products (Brown and Gerbarg 2004).

In central Asia, *R. rosea* was prescribed for cold and flu. Mongolian physicians used it to treat tuberculosis and cancer. Vikings used the adaptogen to enhance physical strength, endurance, and wound healing. Chinese medicines included *R. rosea* and other *Rhodiola* species. In 1725, the Swedish botanist, Carl Linneas, recommended it as an astringent and for treatment of hernias, leucorrhoea (vaginal discharge), hysteria, and headaches. German doctors found benefits of *R. rosea* as a stimulant and in the treatments of pain, headache, scurvy, hemorrhoids, and inflammation. In traditional folk medicines, *R. rosea* was used as a tonic to increase physical endurance, work productivity, and longevity, as well as to treat fatigue, depression, anemia, impotence, gastrointestinal ailments, infections, and nervous system disorders. For hundreds of years, folk medicine has used *R. rosea* as a tonic (strengthening) agent to improve resistance to physical and mental stress, infection, cancer, high altitudes, extreme cold, and mental decline, as well as to enhance sexual function and fertility.

In discussing the clinical uses of *R. rosea*, we will draw from both the scientific literature and patient observations. Dr. Gerbarg and Dr. Brown have prescribed *R. rosea* for over 20 years to more than 500 patients. Dr. Illig is an aerospace medicine expert. This chapter will focus on current psychiatric and medical applications of *R. rosea*, clinical and research evidence, and future directions for its use in medical care.

10.2 PSYCHIATRIC AND MEDICAL CONDITIONS

10.2.1 DEPRESSION

R. rosea can be effective as a solo treatment for mild to moderate outpatient depression or as an adjunctive to prescription antidepressants for moderate to severe depression. The first study of *R. rosea* for depression was done in the former Soviet Union in 1987. In this study, 128 patients with mixed types of depression were given 150 mg three times daily of *R. rosea* or placebo. Two-thirds of those on *R. rosea* improved significantly (Brichenko and Skorokhova 1987). Although this study does not fulfill today's methodological standards, it pointed the way for further research.

In a randomized double-blind, placebo-controlled 6-week parallel study of 89 adults, a standardized extract (SHR-5) of *R. rosea* rhizomes was used (Darbinyan et al. 2007). Adults between 18 and 70 years of age who met Diagnostic and Statistical Manual of the American Psychiatric Association, 1994 (DSM-IV) criteria for mild to moderate depression and whose initial Hamilton Depression Scale (HAMD) scores were between 12 and 31 were randomly assigned to one of the three groups. Group A was given moderate doses, SHR-5 (340 mg/day); Group B SHR-5 680 mg/day; and Group C placebo tablets. After 6 weeks, mean HAMD scores dropped significantly, showing a marked decrease in symptoms of depression, in Groups A (from 24.52 to 15.97) and B (from 23.79 to 16.72), but not in C (from 24.7 to 23.4). No serious adverse effects were reported in any group. SHR-5 showed modest but significant effectiveness in treating patients with mild to moderate depression when administered over 6 weeks in doses of 340 or 680 mg/day.

Although there has been only one Russian study of *R. rosea* in combination with a tricyclic antidepressant, authors Gerbarg and Brown routinely use it as an adjunct to medication because it increases mental and physical energy, which are often low in depression. Furthermore, it improves mood and stress tolerance. Prescription anti-depressants tend to alleviate negative mood states, but often leave residual symptoms of low motivation, lack of interest, difficulty getting out of bed, lack of enthusiasm, and a less-than-positive mood (McClintock et al. 2011). Anecdotally, many patients report that antidepressant augmentation with *R. rosea*, increases interest, motivation, a sense of well-being, impetus to get up and do things, and feelings of enjoyment. *R. rosea* can be particularly beneficial for menopausal women with depression (see Section 10.2.5.).

Possible mechanisms of action include elevation of neurotransmitters including serotonin, norepinephrine, and dopamine (Brown et al. 2002; Kurkin and Zapesochnaya 1986). In addition, *R. rosea* has been shown to increase transport of tryptophan and 5-HT (serotonin) into brain cells (Brown et al. 2002). Furthermore, depression can be the result of prolonged stress and exhaustion of the stress-response system. *R. rosea* may help to replenish and maintain cellular energy supplies (Panossian and Wikman 2009).

10.2.2 Stress, Anxiety, and PTSD

Stress is a major cause of anxiety, and extreme stress can lead to post traumatic stress disorder (PTSD), a condition that can occur following the experience or exposure to a terrifying or life-threatening event. It is characterized by persistent distressing symptoms of increased arousal, re-experiencing, and avoidance. Numerous studies have shown that *R. rosea* improves resistance to stress in normal subjects (Shevtsov et al. 2003; Spasov et al. 2000a, b). Patients often report that the herb reduces anxiety symptoms (Panossian 2013).

In a small, open pilot study, 10 adults with DSM-IV diagnosis of generalized anxiety disorder (GAD) from the UCLA Anxiety Disorders Program received 340 mg per day of *R. rosea* for 10 weeks. Individuals treated with *R. rosea* showed significant decreases in anxiety symptoms, Hamilton Anxiety Rating Scale (HARS) scores at endpoint ($t = 3.27$, $p = .01$). Adverse events were mild or moderate, most commonly, dizziness or dry mouth. Two of the 10 highly anxious subjects became more anxious when given *R. rosea* (Bystritsky et al. 2008). Such a reaction is consistent with the herb's stimulative effects. Small sample size is a limitation.

R. rosea has a unique dual effect: it calms the emotions while stimulating the intellect (Baranov 1994a; Brown et al. 2002). These benefits have been attributed in part to increased levels of neurotransmitters—notably serotonin, norepinephrine, and dopamine—in the brainstem, cerebral cortex, and hypothalamus (Petkov et al. 1986, 1990; Stancheva and Mosharoff 1987). An additional mechanism may involve restoring the balance of sympathetic and parasympathetic activity. Severe or chronic stress entails increased sympathetic activity, often accompanied by reduced parasympathetic activity, causing imbalance in the stress–response system. Prescription medications such as anxiolytics and antidepressants have been shown to lower sympathetic tone, but not to improve the underactive parasympathetic system as reflected in low heart rate variability. *R. rosea* is the only extract that has been shown to elevate parasympathetic activity while also strengthening the sympathetic system (Baranov 1994a). This may contribute to shifting the autonomic stress–response system toward balance, thereby decreasing the excess reactions to internal and external cues.

Although *R. rosea* is energizing, it is also calming. In patients with PTSD, damage to areas involved in emotion and memory processing and disruption of neurotransmitter function has been attributed to excess levels of cortisol and glutamate during traumatic events. The hippocampus is critical for emotion processing and memory. Animal studies indicate that extracts of *R. rosea* not only protect hippocampal neurons from damage *in vitro* but also enhance neurogenesis following stress-induced injury *in vivo* (Cao et al. 2006). Hippocampal neurons are particularly vulnerable to damage by excess levels of the excitatory neurotransmitter, glutamate, oxidative stress, and cortisol. Some of the main active constituents of *R. rosea* are the salidrosides. In cultured rat hippocampal neurons, salidroside protected against free radical damage by hydrogen peroxide (H_2O_2) (Chen et al. 2009) and from the effects of glutamate (Chen et al. 2008). Furthermore, studies of rats exposed to mild chronic stress showed impaired neural stem cell proliferation in the hippocampus and reduced levels of serotonin (5-HT). After 3 weeks of *R. rosea* feeding, the levels of neuronal stem cell proliferation and serotonin returned to normal in rats exposed to chronic mild stress, indicating enhanced cellular repair (Chen et al. 2009). *R. rosea* extract protected human cortical neurons against glutamate and hydrogen peroxide-induced cell death through reduction in the accumulation of intracellular calcium (Palumbo et al. 2012).

Evidence suggests that *R. rosea* helps to restore more normal function to areas of damage associated with PTSD by increasing mitochondrial energy production, cellular repair (Panossian and Wikman 2009), and neurotransmitter activity while preventing further damage. Two of the main diagnostic criteria for PTSD are increased arousal and re-experiencing (American Psychiatric Association 1994). Increased arousal is characterized by elevated sympathetic and reduced parasympathetic tone, difficulty sleeping, irritability, outbursts of anger, difficulty concentrating, hypervigilance, and exaggerated startle response. Re-experiencing includes distressing symptoms of intrusive recollections, recurrent distressing dreams, sense of reliving, illusions, flashbacks, intense distress when exposed to triggers, and physiological reactivity to internal and external cues that resemble or symbolize the traumatic event. In addition, *R. rosea* may protect vulnerable areas from further damage during the retraumatization of reexperiencing, for example, during flashbacks, by preventing surges of excess cortisol and by protecting neurons from oxidative damage by glutamate. The following case illustrates how *R. rosea* helped one woman with PTSD.

10.2.2.1 Case Presentation

Angela, a 30-year-old computer programmer, suffered from PTSD due to emotional abuse and neglect from both parents, and the prolonged illness and death of her beloved grandmother when she was 7 years old. Although sexual abuse was suspected, she had no specific recall of any event. Internal and external cues triggered memories, anxiety, and physiological and psychological symptoms of hyperarousal such as elevated heart rate, rapid breathing, feeling a "knot in the stomach," sweating, and a foreboding sense of doom. She developed a host of behaviors to avoid causing others to become angry, reprimand, or punish her. As a result, she was extremely timid, self-effacing, compliant, and non-confrontational. Despite being highly competent and productive at work, Angela was seen as weak and was passed over for promotions. She was maintained on very low doses of a selective serotonin reuptake inhibitor (SSRI) and an anxiolytic as needed because she was exquisitely sensitive to medication side effects. She was initially given a trial of *R. rosea* at 50 mg/day to relieve fatigue due to chronic overwork. In addition to improving her energy level, when the dose was increased to 100 mg/day, the herb reduced her overall level of fear reactions and avoidance behaviors, as well as increasing her ability to focus and get things done. Consequently, she began to speak up at work, set limits on inappropriate demands for her time, demonstrate her leadership qualities, improve her evaluation scores, win the respect of her supervisors, and obtain the bonuses and promotions she deserved.

10.2.3 COGNITIVE FUNCTION, MEMORY, AND ATTENTION DEFICIT DISORDER

10.2.3.1 Neuroprotection

Through a plethora of effects, the bioactive compounds in *R. rosea* roots could preserve, enhance, and restore brain function. For example, by balancing the stress–response system, *R. rosea* prevents excess release of stress hormones such as cortisol, which can damage brain cells (Baranov 1994a). Animal studies showed that it also increases permeability of the blood–brain barrier to dopamine and 5-hydroxytryptophan precursors (Stancheva and Mosharrof 1987). Furthermore, the herb reversed the blockade of acetylcholine in nicotinic pathways from the limbic system to the cerebral cortex (Hill-House et al. 2004). *R. rosea* extracts contain many antioxidants that protect cell membranes, mitochondria, and DNA from free radical damage (Bolshakova et al. 1997; Furmanowa et al. 1998; Hill-House et al. 2004). By providing more energy to cellular repair systems and protecting against oxidative damage, *R. rosea* could prevent inflammation and the cumulative damage found in aging, degenerative changes in the brain, and disease progression (Panossian 2013).

The neuroprotective effect has been hypothetically attributed to the antioxidant synergistically active components that may be responsible for stabilizing cellular Ca^{2+} homeostasis. An *in vitro* study found that *R. rosea* protected against the oxidative stress of hydrogen peroxide (H_2O_2) and glutamate (GLU)-induced cell apoptosis (cell death) in a human cortical neuron cell line (HCN-A) maintained in culture (Palumbo et al. 2012). In rats *in vivo*, *R. rosea* prevented cerebral ischemic injury induced by a 2-hour middle cerebral artery occlusion and a 24-hour reperfusion (Shi et al. 2012).

In healthy subjects, *R. rosea* enhances attention, concentration, memory, learning, and intellectual performance (Shevtsov et al. 2003; Spasov et al. 2000a). In a double-blind, randomized, placebo-controlled single-dose effect study, a combination of *R. rosea*, *Eleutherococcus senticosus*, and *Schizandra chinensis* improved attention, speed, and accuracy on cognitively demanding tasks (Aslanyan et al. 2010). Through the antioxidant protection, mitochondrial enhancement, anticholinesterase activity, and elevation of neurotransmitter levels described above, *R. rosea* may help to reverse—or even prevent—some of the effects of aging, trauma, and illness on the brain (Furmanowa et al. 1998; Kurkin and Zapesochnaya 1986). In many of the authors' (Gerbarg and Brown) patients, the herb helps reduce symptoms of age-associated memory impairment, age-associated cognitive decline, Parkinson's disease, Alzheimer's disease, attention deficit disorder, and brain injury due to trauma, stroke, vascular insufficiency, infection, and chemotherapy. Middle-aged and elderly patients report having increased energy, clarity of thought, and memory. In addition, increasing cellular energy and stimulation of the reticular activating system may explain how *R. rosea* enables patients with Parkinson's disease to increase their spontaneous activity level, engage in activities, and respond more to conversation and social interaction (Bocharov et al. 2010). Moreover, it is possible that some benefits derive from the herb's effect on dopamine and other neurotransmitters. It may also inhibit COMT (carboxy-*o*-methyl transferase), similar to the Parkinson's drug COMTAN (Blum et al. 2007). Further research may clarify the role of these mechanisms.

10.2.3.2 Attention Deficit Disorder

Several studies have shown that *R. rosea* improves alertness and attention. Although no studies have been published on the effects of *R. rosea* in attention deficit disorder (ADHD), the known neurophysiological effects suggest it could be beneficial, while anecdotal clinical benefits are promising. ADHD is characterized by inattention, hyperactivity, and impulsivity. These symptoms have been attributed to underactivity of dopamine and norepinephrine transmission such that areas of the brain responsible for attention and inhibition of inappropriate behaviors are sluggish or underactive (Brown and Gerbarg 2012). Stimulation of the dopamine and norepinephrine systems by *R. rosea* is one possible mechanism to explain the improvements observed in clinical practice with ADHD patients. Compared to prescription stimulants, *R. rosea* does not cause as many problematic side effects. Moreover, ADHD is often associated with imbalance of the stress–response system, an additional indication for *R. rosea*. In practice, Gerbarg and Brown observe that *R. rosea* can be a useful adjunctive treatment for ADHD. Anecdotally, their patients report improved mental focus and productivity.

10.2.3.3 Cognitive Function in Dementia, Post-Stroke, and Traumatic Brain Injury

For cognitive function, *R. rosea* has been effective when combined with other adaptogenic herbs. Many studies using combinations of *R. rosea*, *E. senticosus*, and *S. chinensis* show improvements in mental focus, accuracy, cognitive function, and memory in healthy subjects (Aslanyan et al. 2010; Brown et al. 2009). In patients with brain injury, *R. rosea* has a mild cognitive stimulation effect while emotionally calming (Brown and Gerbarg 2011) Authors (Gerbarg and Brown) found that in three post-stroke and 12 traumatic brain injury cases, *R. rosea* enhanced the action of piracetam and aniracetam, improving language recovery. Piracetam increases nerve cell membrane fluidity, activates brain waves, and improves blood flow in speech areas of the brain (De Deyn et al. 1997; Kessler et al. 2000).

10.2.4 Stress and Fatigue

R. rosea has been shown to reduce mental and physical fatigue, particularly under stress (Baranov 1994a; Spasov et al. 2000b). These effects are further enhanced by combining *R. rosea* with other adaptogens and medicinal herbs.

A combination of *R. rosea*, *E. senticosus*, and *S. chinensis* called ADAPT-232 (Swedish Herbal Institute) was recently tested in a double-blind, randomized, placebo-controlled single-dose effect study in 40 healthy, but psychologically stressed, tired women aged 20–68 years. Compared to the placebo group, within 2 hours of consuming 270 mg ADAPT-232, subjects gained significant improvements in attention, speed, and accuracy during stressful cognitive tasks (Aslanyan et al. 2010). See Chapter 9 by Panossian and Wikman in this volume for molecular mechanisms involved in adaptogen amelioration of the effects of stress and fatigue.

In clinical practice, Gerberg and Brown have found that in over 200 patients fatigue was rapidly relieved by *R. rosea*, regardless of the underlying cause. Usually, energy enhancement occurs within 1–7 days, depending on how many dose increases are needed for response. Accordingly, Gerberg and Brown routinely use *R. rosea* for fatigue due to external stress, aging, physical exertion, emotional strain, medical illness (e.g., infection, cardiovascular disease, cancer, HIV), medication side effects, and jet lag. In some cases, *R. rosea* was taken continuously without loss of effect for over 10 years. In patients who experience a loss of effect over time, it may be necessary to switch to a more potent brand or to discontinue use for 1–3 weeks to restore the effect.

10.2.5 Sexual Function, Fertility, and Menopause

Although there have been no controlled studies of *R. rosea* as an aphrodisiac, it has been used for centuries to enhance sexual function in both men and women. Sexual desire and response are affected by age, hormonal status, physical condition, medical illness, prescription medications, stress, fatigue, mood, and other emotional issues. Studies of sexual function tend to show a high placebo response rate. Nevertheless, Russian physicians routinely prescribe *R. rosea* to enhance libido, pleasure, and performance. In clinical practice, authors Gerberg and Brown note that patients often report greater sexual interest, desire, and responsiveness with *R. rosea*. The increase in sexual desire and response may be due in part to improvements in mood, energy, and interest, but other *R. rosea* effects, such as hormonal, may play a role as well. Additional mechanisms could involve dopamine stimulation and effects on nitric oxide production (Panossian 2013).

A review of studies of phosphodiesterase-5 inhibitors (PDE5I), such as sildenafil (Viagra), in men with erectile dysfunction found that 11%–44% of patients were nonresponders to PDE5I monotherapy (Dhir et al. 2011) with 20%–25% placebo response. Furthermore, some patients have contraindications to PDE5I treatment, for example, concurrent use of nitrates. Others cannot tolerate side effects, such as headaches, sinus congestion, dizziness, and visual and hearing changes. In comparison, a small open study of 35 men with erectile dysfunction and/or premature ejaculation, found that 100 mg *R. rosea* daily for 3 months improved sexual performance in 74% of the subjects (Saratikov and Krasnov 1987).

It is interesting to note that this study documented an increase in 17-ketosteroids compared to normal and an increase in liposomes (lecithin globules in semen) (Saratikov and Krasnov 1987). 17-ketosteroids are a measure of both adrenal function and fertility. Lecithin protects sperm from damage by excess heat, cold, chemical toxins, or radiation (Saratikov and Krasnov 1987). Storage of sperm in lecithin by animal breeders results in increased rates of successful insemination and reduction of abnormal sperm forms (Aires et al. 2003). The quality of human sperm has been declining, probably due to exposure to pollutants as well as lifestyle and dietary changes (Aitken and Sawyer 2003). Research is needed to determine whether adaptogens such as *R. rosea* could help to improve sperm quality and fertility in men.

In a study in the former Soviet Union, *R. rosea* was given either by mouth or by intramuscular injection for 10–14 days to 40 women of whom seven had no menstrual cycles due to primary amenorrhea (total absence of menses) and 33 had secondary amenorrhea (previously had menses) (Gerasimova 1970; Saratikov and Krasnov 1987). In some cases, the treatment was repeated several times. Normal periods were successfully restored in 25 of the women. Among that group, 11 subsequently became pregnant. Although the report of this study lacks sufficient details to fully interpret the results, it does suggest hormonal effects. This is consistent with our clinical observations that five peri-menopausal and menopausal women who had not menstruated for 6 to 12 months responded to treatment with *R. rosea* by resumption of regular monthly menstrual cycles. Two of Dr. Brown's patients in their early 40s who had failed to become pregnant after 2 years of *in vitro* fertilization and synthetic hormone treatments stopped all hormonal medications and started taking 300 mg/day of *R. rosea*. Three months later, both women became pregnant, carried their babies to term and bore healthy children. In modern society, many women delay childbearing beyond age 35 only to discover that they are unable to conceive when they try to do so. The possibility that *R. rosea* might help women maintain their fertility is a worthy of future research.

The decline in estrogen with menopause can lead to symptoms of estrogen deficiency including vaginal dryness, dyspareunia (pain with intercourse), loss of sexual pleasure, and sexual avoidance. In five peri-menopausal patients treated by authors Gerberg and Brown, *R. rosea* helped ameliorate symptoms of estrogen deficiency, such as vaginal dryness.

These observations raised the question of whether *R. rosea*, like hormone replacement therapy, might increase the risk of cancer through estrogenic effects. To evaluate this possible risk, Dr. Patricia Eagon, professor at the University of Pittsburg Medical Center, tested *R. rosea* for estrogen receptor binding, estrogenic activity, and carcinogenic effects. Her studies of orally administered *R. rosea* (brand Rosavin, Ameriden International) to ovariectomized rats showed that although *R. rosea* bound to estrogen receptors, it did not activate them or cause increases in circulating levels of estradiol. In addition, *in vitro* tests in cultures of estrogen-sensitive and non-estrogen-sensitive breast cancer cells showed that it did not increase cancer cell proliferation (Eagon et al. 2004). Authors Gerberg and Brown suggest that *R. rosea* root extract may constitute a natural selective estrogen receptor modulator.

Current research by pharmaceutical companies is focused on developing selective estrogen receptor modulators (SERMS) to reduce the risk of bone fractures and cardiovascular events in menopausal women without increasing the risk of breast, uterine, or ovarian cancer (Blizzard 2008). An ideal SERM would have estrogen receptor agonistic effects on bone, the cardiovascular system, and central nervous system, but anti-estrogenic or no effects on healthy breast and endometrial tissues (Hendrix and McNeely 2001). Tamoxifen was the first SERM used to reduce the risk of breast cancer, but caused increased risks of endometrial cancer and blood clots (Cuzick et al. 2013). Raloxifene is an estrogen receptor agonist for bone and lipid metabolism and an estrogen antagonist for breast tissue. Bazedoxifene is an estrogen agonist in endometrial and breast tissues (Cuzick et al. 2013). It reduced bone loss, the incidence of vertebral fractures, serum total cholesterol and low-density lipoproteins, but elevated triglycerides (Gatti et al. 2013). Adverse events included headache, infection, arthralgia, pain, hot flashes, back pain, abdominal pain, and deep vein thrombosis (Sobeieraj 2011). The combination of a SERM with estrogen is being investigated in the quest for the best balance of selective estrogen receptor agonism and antagonism (Kawate and Takayangi 2011).

R. rosea has fewer side effects than synthetic SERMS. Studies in menopausal women are warranted to determine whether the combination of bioactive compounds in its roots provides a more favorable balance of estrogen agonist/antagonist effects on target tissues than those that have been created artifcially thus far.

10.2.6 Infections and Immune Functions

Doctors in Scandinavia, Russia, Mongolia, and China have used *R. rosea* for centuries to treat infections, including colds, flu, pneumonia, tuberculosis, lung disease, wounds, dental infections, and intestinal worms. Animal studies showed improved cellular and humoral immune response, following *R. rosea* injections and dietary salidroside derived from *R. rosea* (Guan et al. 2011; Lu et al. 2013).

In a controlled study of 200 patients hospitalized for severe acute dental infections that had spread to the mouth, face, and neck, those who were given *R. rosea* as supplementation to standard antibiotics and surgical treatment recovered more rapidly with reduced inflammation and accelerated normalization of cellular immune response (Yarmenko 1998).

A double-blind, placebo-controlled, randomized, pilot (phase III) study evaluated the effect of Chisan (ADAPT-232) (combination extracts of *R. rosea*, *S. chinensis*, and *E. senticosus*) on patients with acute nonspecific pneumonia. Sixty patients (men and women 18–65 years old) were administered standard treatment with cephazoline, bromhexine, and theophylline. In addition, 30 patients were given Chisan, while the other 30 were given placebo for 10–15 days. Mean duration of antibiotic treatment for recovery from the acute phase was 2 days shorter in patients treated with Chisan compared with the placebo group. The Chisan group scored higher on all Quality of Life domains (physical, psychological, social, and ecological) than patients in the control group. Adjuvant therapy with ADAPT-232 also increased mental performance during rehabilitation (Narimanian et al. 2005).

Stress, fatigue, and depression can impair immune function. Prolonged elevation of epinephrine and norepinephrine can suppress immune defense (Dünser and Hasibeder 2009; Wong et al. 2012). Antibacterial effects of *R. rosea* have been demonstrated. For example, *R. rosea* (roseroot) inhibited the growth of all *Nisseria gonorrhoeae* (Ng) including strains that had become resistant to most antimicrobial agents (Cybulska et al. 2011). In addition to antibacterial effects, *R. rosea* may support immune function through antistress, antifatigue, and antidepressant effects. Given the increasing spread of antibiotic-resistant bacterial strains, augmentation with *R. rosea* could be a valuable addition to the treatments of many infectious diseases (Cybulska et al. 2011).

10.2.7 CANCER, CHEMOTHERAPY, AND RADIATION

10.2.7.1 Cancer-Related Stress and Fatigue

The most common complaint of cancer patients is fatigue. The authors (Gerbarg and Brown) find that in some patients *R. rosea* is very beneficial for fatigue due to cancer and/or chemotherapy. The anti-fatigue properties may be due to increased cellular production of high energy molecules, ATP, and creatine phosphate as well as dopamine and nicotinic effects (Abidov et al. 2003, 2004). *R. rosea* has already been found to improve muscle strength, energy, and fatigue in animal studies, normal subjects, and patients with other illnesses (Hung et al. 2011; Olsson et al. 2009; Perfumi and Mattioli 2007).

Patients with cancer experience severe, life-threatening stress from their illness, worries about the results of each test, anxiety about the future, and the many distressing physical symptoms related to cancer and its treatment. Some develop anxiety, depression, and PTSD. Cancer patients need to mobilize all of their resources to endure and comply with treatments. *R. rosea* could be used to enhance their stress resilience, mood, sleep, and adaptation (Zubeldia et al. 2010).

10.2.7.2 Adaptogens Augment Chemotherapy While Protecting Liver and Bone Marrow

When chemotherapy drugs damage stem cells in the liver and bone marrow, patients become more susceptible to infections and other medical sequelae. If blood cell counts fall too low, interruption of chemotherapy may be necessary. Studies of human cancers (Lewis lung carcinoma, Ehrlich's adenocarcinoma, Pliss lymphosarcoma, NK/Ly tumor, and melanoma B16) transplanted into mice, demonstrated significant antitumor and antimetastatic activity of *R. rosea* (Dementéva and Iaremenko 1987; Razina et al. 2000; Udintsev and Shakhov 1989, 1990, 1991; Udintsev et al. 1992). Moreover, in these studies, *R. rosea* protected liver and bone marrow cells from chemotoxic effects of adriamycin and cyclophosphamide while augmenting the effectiveness of chemotherapy agents in destroying cancer cells and reducing metastases.

In vitro R. rosea extract and one of its bioactive components, salidroside, inhibited the growth of bladder cancer cell lines with a minimal effect on nonmalignant bladder epithelial cells TEU-2 (Liu et al. 2012). In 12 patients with superficial bladder carcinoma, *R. rosea* extract improved urothelial leukocyte integrines and T-cell immunity with a trend toward reduction in frequency of relapse (Bocharova et al. 1995). Also, a 95% ethanol extract of *R. rosea* proved cytototoxic against prostate cancer cells (Ming et al. 2005).

In a study of 28 women with stage III–IV ovarian cancer treated with cisplatin and cyclophosphamide, those who took AdMax 270 mg/day (combination of root extracts from *Leuzea carthamoides*, *R. rosea*, *E. senticosus* and fruits of *S. chinensis*) for 4 weeks following chemotherapy had increased T-cell subclasses (CD3, CD4, CD5, and CD8), IgG, and IgM compared with those who did not take AdMax (Kormosh et al. 2006).

Although there are no published studies on the use of *R. rosea* for protection from radiation exposure in cancer patients, *in vitro* and *in vivo* animal studies have demonstrated reduced radiation effects in animals given *R. rosea* and related subspecies. At high altitudes where *R. rosea* grows naturally, the plants are exposed to higher radiation levels. This may account for the development of radiation-resistant compounds.

In vitro cell culture studies on 20% and 40% alcohol extracts of *R. rosea* showed high antimutagenic activity against numerous mutagens, reaching a 90% level in some cases (Duhan et al. 1999). The ability to protect DNA from mutation contributes to the anticarcinogenic effects of *R. rosea*.

10.3 AEROSPACE MEDICINE

10.3.1 Acute Mountain Sickness and Hypoxia

Acute mountain sickness (AMS), also called high altitude sickness or simply altitude sickness, is a constellation of unpleasant symptoms including headache, dizziness, sleep disturbance, and stomach ache that can occur at high altitudes in unacclimatized people and that usually remit after a few days of rest and appropriate treatment. AMS generally occurs at elevations above 8000 ft (2.438 m) above sea level especially if this altitude has been achieved in a short period. In susceptible individuals, including people who are not physically fit, AMS can occur between 5000 and 8000 ft. It is the result of complex and some yet unknown physiological responses to the reduced atmospheric pressure and, thereby, the lowered oxygen levels (hypoxia) associated with increasing altitude. There is no known screening tool that can identify individuals susceptible to AMS. If it does not resolve or if higher altitude is gained, it can progress to high altitude pulmonary edema (HAPE) or high altitude cerebral edema (HACE), which are potentially fatal.

R. rosea has been used for centuries alone or in combination with other herbal preparations in the traditional medicines of China, Tibet, the Republic of Georgia, and the Soviet Union for the prevention and treatment of AMS as well as the effects of prolonged exposure to low oxygen conditions. There is evidence to support this application (Baranov 1994a), but more research is necessary to identify the mechanisms. For example, a randomized controlled study of 24 healthy young men found that after 1 year of living at high altitude (5380 m above sea level), those given *R. rosea* for 24 days had significantly increased waking oxygen saturation and mean oxygen saturation ($p < .01$). Furthermore, they showed significantly decreased times of oxygen desaturation $\geq 4\%$ per hour as well as the percentage of time at oxygen saturation below 80% ($p < .01$) (Ha et al. 2002). *R. rosea* may protect the brain from the adverse effects of low oxygen conditions. Studies show that *R. rosea* extracts, including salidroside, can reduce ischemic injury to the brain and heart in models demonstrating hypoxic damage (Pogorelyĭ and Makarova 2002; Shi et al. 2012; Yu et al. 2008). In one study of 15 young adults, pretreatment with *R. rosea* did not prevent a decline in arterial capillary blood oxygen (PcO_2) after 30- or 60-minute hypoxic exposure, but it did reduce oxidative stress markers such as lipid peroxidase, indicating another possible mechanisms—reduction of oxygen free radical damage during hypoxia (Wing et al. 2003). Increased production of nitric oxide (NO) may also be involved (Panossian 2013). NO is necessary for high altitude acclimation. Tonizid, a complex of extracts from plant adaptogens, including *R. rosea*, *Aralia mandshurica*, *Panax ginseng*, and *E. senticosus*, significantly reduced the area of brain and heart damage in animal studies of ischemia-reperfusion (Arbuzov et al. 2006; Lishmanov et al. 2008). The traditional recommendation is to start taking *R. rosea* at least 1 week before high-altitude exposure, throughout the period of exposure, and for 1 week afterward.

Depending on the type of flight profile (e.g., high-altitude, long-distance polar flights), aircrews may be exposed to elevated radiation compared to terrestrial workers. Others who regularly fly at elevated altitudes in unpressurized aircraft may be exposed to chronic or intermittent hypoxia, for example, high-altitude rescue, flight seeing, or mountain charter flight crews. Depending on the country, pilots in unpressurized aircraft are permitted to fly up to 12,500–13,000 ft above sea level before oxygen supplementation requirements begin. *R. rosea* may be an effective countermeasure to potential adverse effects of high-altitude exposure. Author (Brown) has enhanced the effectiveness of *R. rosea* in the prevention and treatment of AMS by augmentation with *Ginkgo biloba* and another high-altitude adaptogen, Maca (*Lepidium meyenii*), in 15 cases (Gonzales 2012; Leadbetter et al. 2009; Moraga t al. 2007).

10.3.2 JET LAG

Travelers rapidly crossing three or more time zones will experience some form of circadian desynchronization, commonly called jet lag. It is exacerbated by stress, overeating, dehydration, increasing age, traveling east, sleep deprivation, and alcohol consumption. These negative effects can be minimized by breaking up a long flight into shorter flights, taking transit breaks, and increased exposure to sunlight on arrival. Depending on the individual, it can take a full day for each 3- to 4-hour time zone change to fully recover from jet lag. Recovery time increases with age. Besides fatigue and general discomfort, there are potential problems associated with dosage adjustments of certain medications such as insulin, sleep aides, anticonvulsants, and anticoagulants (Illig 2009). *R. rosea* can be helpful in counteracting jet lag in two ways. It can be used as a daytime stimulant as needed or it can be timed to improve alertness and relieve combat fatigue. It can also be used throughout the jet lag recovery period as one would in times of unusual stress.

10.3.3 AIRCREW STRESS AND FATIGUE

Because of the flight environment, commercial aircrews are often acutely and chronically stressed by imbalances in circadian cycles, nutrition, and immune systems. Chronic stress promotes elevated cortisol output, which can be harmful to the body. Because herbal supplements are generally not prohibited in civil aviation and *R. rosea* is quite safe, it may be useful for aircrews and air traffic controllers to help counteract stress and fatigue. Although not a substitute for rest and good stress management, judicious use of appropriate herbal medicinals or food supplements may be considered for use to safely combat fatigue and circadian desynchronization. Whether passenger or crew member, *R. rosea* may be considered as a boost to general stress resilience during travel.

10.3.4 SPACE EXPLORATION

It has been reported that *R. rosea* was tested as part of the ADAPT formula (*R. rosea, E. senticosus,* and *S. chinensis*) in Russian cosmonauts (Vastag 2007). Results obtained in Soviet research laboratories in the 1980s and 1990s were compared with data obtained on the MIR space station by Polyokov, the physician cosmonaut. The findings were published in the Baranov Reports (1994a, b) submitted to the Department of Defense of the Former Soviet Union. Polyokov also presented data at a 1994 conference in Gutenberg (Polyakov 1996).

Baranov also performed extensive terrestrial tests for the Russian Institute of Medical and Biological Problems (IMBP) to demonstrate the effects of ADAPT on the cardiovascular systems as well as on functional mental capacities. The goals were to investigate the adaptogenic effects of this formula on aspects of human physiology and psychology that would be particularly important in space flight. Physical loads were created with cycle ergometry. It was found that a 7-day course of ADAPT increased the working capacity by 28%, and that general adaptogenic effects (balancing of the sympathetic and parasympathetic nervous systems) were documented (Baranov 1994b). Mental capacity studies were performed using computer-based tests to simulate metal work load tasks (Baranov 1994a). Not surprisingly, very positive effects were noted: significant reduction in mistakes made, improved performance parameters, better abstract thinking, and reduced fatigue (Baranov 1994b).

Another study of particular importance regarding long-duration space flight was performed by testing the effects of a single dose of ADAPT on concentration, eye–hand coordination, and short-term memory during a 90-day isolation experiment (Bogatova et al. 1997). Again, a significant reduction in mistakes was found 4 hours after taking a dose of ADAPT, as well as an improvement in the accuracy of compensatory tracking.

Together, these and other Soviet-Russian studies illustrate the potential usefulness of *R. rosea* as a countermeasure to the stressors of space travel. Short-duration suborbital flights for space adventure tourists and exploratory flights to near-earth asteroids or Earth's moon entail considerable stress. However, longer missions—such as those aboard orbital platforms or missions to other planets within our solar system—would involve long, tedious missions fraught with monotonous tasks that require unwavering concentration. Considering that *R. rosea* enhances physical and mental performance (concentration and accuracy) under conditions of stress, it could be an important addition to the armamentarium needed to survive the rigors of space travel.

Space flight, whether of short or long duration, inherently comes with a host of environmental stressors to human physiology and psychology. In general, astronauts, cosmonauts, and now taikonauts, have been highly trained professionals. In the next generation of the commercial space industry, astronauts will likely be less fit and less trained than those in the original government-sponsored space programs. These individuals will be exposed in various amounts and intensities to micro and zero gravity, variations in oxygen pressure, acceleration and deceleration forces, vibration, radiation, space motion sickness, nutritional and metabolic changes, dehydration, and unknown immune function disturbances, among others. It would be useful to study the efficacy of *R. rosea* on organisms in space during various flight profiles and durations.

10.4 TREATMENT GUIDELINES

10.4.1 DOSAGES

R. rosea is best absorbed when taken on an empty stomach 20 min before breakfast and/ or lunch or 2 hours after a meal. Starting with 100–150 mg/day, doses are increased by 100–150 mg every 3–7 days to a maximum of 600 mg/day. If, for example, a patient's depression responds well to the herb, then it may be continued until the physician determines that an attempt could be made to reduce the dose while monitoring the patient for symptom reoccurrence. If tolerated, the entire dose can be given in the morning to increase compliance. In some cases, the single morning dose has a better effect. Some individuals may be sensitive to stimulative effects and may react with anxiety, jitteriness, or agitation. In such cases, it is better to give a fraction of the usual dose and then increase slowly in small increments as tolerated. Elderly, medically ill or anxious patients should start by taking one-fourth to one-half of a capsule (37–75 mg) per day dissolved in tea or juice and increased slowly. Response takes between 1 and 12 weeks.

For mild to moderate depression, a usual starting dose is 100–150 mg during the first 2 days and then an increase of one capsule every 3–7 days to a maximum of 600 mg/day. As an adjunctive treatment for depression, *R. rosea* is usually effective in doses of 200–600 mg/day. As a solo treatment, higher doses up to 750 mg/day may be needed. *R. rosea* can be used for as long as necessary as a maintenance treatment to prevent remission.

10.4.2 SIDE EFFECTS AND MEDICATION INTERACTIONS OF *R. rosea*

10.4.2.1 Activation

R. rosea has an energizing or mildly stimulating effect. Unlike prescription stimulants (e.g., amphetamines), *R. rosea* does not cause addiction, habituation, or withdrawal symptoms. Individuals who are sensitive to stimulants such as caffeine may initially feel more anxious, agitated, jittery, or "wired" on *R. rosea*. Patients should reduce their intake of caffeine when using this herb because the stimulating effects can be additive. Some people who begin taking *R. rosea* feel that it has no effect until they are under severe stress, when it becomes evident that they are able to function much better than expected. Highly anxious patients may not tolerate higher doses because the activating effects sometimes exacerbate anxiety. This can be corrected by combining *R. rosea* with other more calming adaptogens.

R. rosea should be taken in the early part of the day to avoid interference with sleep. Some people report vivid dreams during the first 2 weeks. Occasionally, the herb may cause mild nausea. This can usually be managed by taking two ginger capsules 20 minutes before the *R. rosea* dose or by drinking ginger tea. Occasionally, patients report headaches with *R. rosea*. Rarely, an individual may feel sleepy after taking *R. rosea*, an effect seen at high doses in animal studies. This may be related to the individual's slow metabolism. The solution is to reduce the dose of *R. rosea* and take it at bedtime.

10.4.2.2   Hormonal

It is possible that *R. rosea* may increase fertility in women of childbearing age, a common usage in folk medicines of people living at high altitudes in the Caucuses (Brown and Gerbarg 2004). Theoretically, women who rely on birth control pills may be at greater risk for pregnancy. Although no cases of unwanted pregnancy have been reported, it is advisable to use an additional barrier method of birth control.

In older Soviet studies, *R. rosea* was reported to exacerbate a condition in patients described as "volatile or euphoric" (Saratikov and Krasnov 1987). These were probably patients with bipolar disorder, and there may be some risk of inducing agitation or mania in bipolar patients as can occur with other antidepressants and stimulants. However, in the authors' experience, *R. rosea* can be quite helpful for depression in bipolar disorder patients on mood stabilizers whose mood swings are primarily the depressed type with occasional mild hypomanic symptoms. Highly anxious patients may not tolerate higher doses because the activating effects sometimes exacerbate anxiety. This can be corrected by combining *R. rosea* with other more calming adaptogens.

10.4.2.3   Herb–Drug Interactions

R. rosea has no reported adverse interactions with drugs; however, caffeine and other stimulants can have addictive effects. In practice, patients are advised to reduce or discontinue their use of caffeine when taking *R. rosea* to prevent overstimulation.

One *in vitro* method of screening substances for potential interactions with medications is to test effects on cytochrome CYP450 and other enzyme systems that metabolize drugs. If the substance strongly inhibits any of the critical enzymes, then it could reduce the metabolism of drugs that are also metabolized by the same enzyme, thereby raising the serum level of the drug. If the substance significantly induces a CYP enzyme, it may increase the drug's metabolism and lower the serum level. Conversely, medications that inhibit or induce CYP enzymes could increase or decrease levels of medicinal herbal compounds. However, *in vitro* testing does not predict or confirm that a substance will interact adversely with prescription medications when tested in live animals. For most medications, these effects are not clinically relevant. However, there can be adverse effects if serum levels change in a small number of medications, particularly anticoagulants such as warfarin, theophylline, chemotherapy agents, and others. *In vitro* studies use test medications such as theophylline that are known to be substrates of CYP3A4 to evaluate the ability of other substances to alter its metabolism. Substances that alter P-glycoprotein (P-gp) activity can potentially affect the absorption of certain medications such as digoxin. In addition, interactions can result when both an herb and a drug have similar or opposite effects, for example, the addictive effects of two stimulatory agents.

An *in vitro* study found that 95% ethanol extracts from *R. rosea* plants grown in different areas of Norway showed inhibition of CYP3A4 and P-gp (Hellum et al. 2010). There was no correlation between the concentrations of the six marker constituents (which were presumed to be active) with these inhibitory effects. The component(s) responsible for *in vitro* inhibition of CYP3A4 and P-gp have not been identified. This study raised the possibility that *R. rosea* could affect levels of medications, such as warfarin, digoxin, and others. However, as with other herbs, inhibitory activity *in vitro* does not necessarily translate to inhibitory activity *in vivo* or in human studies as the following study shows.

In an *in vivo* study of rats treated with theophylline, the addition of *R. rosea* (SHR-5) did not significantly affect the pharmacokinetics of theophylline, a substrate of CYP3A4. In addition, the simultaneous administration of SHR-5 and warfarin did not significantly affect the pharmacokinetics or anticoagulant activity of warfarin. This study indicates that interactions of *R. rosea* with these co-administered drugs, which are commonly used for CYP substrate testing, are likely to be negligible in humans (Panossian et al. 2009).

R. rosea can have a dose-related mild effect on platelets. Because of genetic variability, some people are more susceptible to this effect and may experience increased bruising at doses above 600 mg/day. For most people, this does not occur unless they exceed the maximum recommended therapeutic dose of 900 mg/day. There have been no reports of bleeding problems related to *R. rosea* use.

The authors have used *R. rosea* in patients taking many different medications without adverse effects. Dr. Brown has treated two patients on warfarin with *R. rosea* 150 mg/day and 300 mg/day with no change in measured International Normalizing Ratio (INR). Without more information from *in vivo* and human studies, it cannot be assumed that *R. rosea* would interact adversely with medications. Nevertheless, the checking of serum levels (e.g., digoxin) and INR (e.g., warfarin) would resolve any lingering concerns.

10.5 COUNTERACTING SIDE EFFECTS OF PRESCRIPTION MEDICATIONS

Intolerance of medication side effects is a common cause of discontinuation. Dose reduction or switching medications is the first approach. Unfortunately for many patients, switching or reducing dosage does not prevent side effects. In some cases, *R. rosea* can alleviate or eliminate the following distressing drug side effects: fatigue, weight gain, insulin resistance, hyperlipidemia, impairment of cognition and memory, sexual dysfunction, and neurological side effects.

10.5.1 ASTHENIA, FATIGUE, SOMNOLENCE

Many drugs cause fatigue or somnolence that can be mistaken for symptoms of depression. Common offenders include psychotropics (anxiolytics, antidepressants, antipsychotics); antihypertensives, such as terazosin (Hytrin), catapres (Clonidine), or methyldopa (Aldomet); antihistamines; anticonvulsants; opiate pain relievers; beta blockers; chemotherapy; drugs to reduce symptoms of benign prostatic hypertrophy, for example, terazosin (Hytrin) or tamsulosin (Flomax); and chemotherapy agents. *R. rosea* alone or in combination with *E. senticosus*, and *P. ginseng* can increase alertness and daytime energy (mental and physical) without causing addiction.

10.5.2 WEIGHT GAIN, INSULIN RESISTANCE, AND HYPERLIPIDEMIA

Increased weight is a side effect of numerous medications, including SSRI antidepressants, mood stabilizers (lithium and valproate), and antipsychotics. Prescription medications can lead to insulin resistance, significant weight gain, and increased risks for diabetes, hyperlipidemia, and cardiovascular disease. Certain medications cause weight gain despite a careful diet and regular exercise. If possible, the physician should try switching to a drug that is less likely to cause weight gain. However, if no substitute can be found, then it becomes necessary to engage the patient in an exercise and weight-loss program. Supplements such as *R. rosea* and *Rhododendron caucasicum* may accelerate weight loss. Taken at the beginning of a meal, *Rhododendron caucasicum* blocks absorption of about 20% of fat from food. Studies suggest that *R. rosea* accelerates fat burning by stimulating lipolysis and by reducing perilipins, proteins that coat lipid droplets in adipocytes protecting them from lipolysis, the splitting of triglycerides into glycerol and free fatty acids to be used as fuel in metabolism. Perilipin, a regulator of lipid storage, is elevated in obese individuals (Adamchuk 1969; Adamchuk and Salnik 1971).

In a double-blind placebo-controlled study of 273 obese men and women with body mass index of 29–34 kg/m², half were given one tablet of Rhodalean-400 (200 mg *R. rosea* + 200 mg *Rhododendron caucasicum*) three times a day for 20 weeks, whereas the other subjects were given placebo. Both groups were required to walk 20 minutes after lunch and dinner and to limit daily caloric intake to 1800 cal/day. Those given Rodalean-400 had a mean weight loss of 9.3 ± 1.4 kg (20.5 ± 3.1 pounds) compared to 1.2 ± 1.6 kg (2.6 ± 3.5 pounds) in those given placebo. Postprandial cortisol levels were 17% lower in the Rhodalean-400 group than in the placebo group. Furthermore, those given Rhodalean-400 had lower levels of perilipins, and there were no side effects (Abidoff and Nelubov 1997). *R. rosea* affects a norepinephrine-sensitive lipase in fat cells after 2 months of daily use, leading to marked increases in metabolism and fat burning for hours after modest exercise, such as walking. Rhodalean-200, a combination of 100 mg *R. rosea* plus 100 mg *Rhododendron caucasicum* taken three times a day, accelerated weight loss in a double-blind placebo-controlled study of 45 women who had given birth within the previous year and who were no longer lactating. On average, the women were 42 pounds above their ideal weight. They were required to maintain a calorie intake of 1750–1850 cal/day by reducing consumption of carbohydrates and fat.

Those given Rhodalean-200 lost 5%–6% of their body weight (8–10 pounds) after 6 weeks compared to the placebo group loss of 0.4%–0.7% (0.7–1.1 pounds). On Rhodalean-200, 11.5% of the weight loss occurred from the waist area. There were no adverse effects (Abidoff 1997). Although there have been no studies of *R. rosea* and *Rhododendron caucasicum* to counteract weight gain secondary to prescription medications, in clinical practice, Gerberg and Brown notice modest benefits, particularly at increased doses of *R. rosea* up to 600 mg/day. Weight loss often takes 6–8 weeks in patients on psychotropics and requires at least 20 minutes of walking once or twice every day, as shown in the study above. Patients need to be informed not to expect the metabolic shift to become evident for 6–8 weeks. Daily exercise for at least 20 minutes and some caloric restriction are necessary for weight loss to occur.

10.5.3 IMPAIRMENT OF COGNITION AND MEMORY OR WORD-FINDING PROBLEMS

Many patients report cognitive impairment, decreased memory, and/or word-finding problems while being treated with antidepressants, anxiolytics, mood stabilizers, antipsychotics, and chemotherapy agents, and other pharmaceuticals. Cognitive and memory problems often improve when *R. rosea* is given as solo treatment or in combination with other adaptogens, aniracetam, and/or cholinesterase inhibitors such as Huperzine-A and Aricept in these suggested doses (Brown et al. 2009):

1. *R. rosea* 450–750 mg/day
2. ADAPT-232: 2–4 tablets/day
3. Aniracetam 750 mg b.i.d.
4. Huperzine-A 200–400 mg b.i.d.
5. Aricept 5–10 mg/day

10.5.4 SEXUAL DYSFUNCTION AND HORMONAL CHANGES

Just as *R. rosea* can improve sexual dysfunction from many causes (see Section 10.2.5), it can alleviate sexual dysfunction secondary to medication, for example, antidepressants or antihypertensives. In clinical practice, libido may be enhanced with *R. rosea* alone or in combination with Maca (*Lepidium meyenii*), Muira Puama, dehydroepiandrosterone (DHEA), and/or 7-keto dehydroepiandrosterone (Brown et al. 2009).

10.5.5 NEUROLOGICAL SIDE EFFECTS

Antipsychotic medications and SSRIs used in large doses over many years sometimes affect the dopaminergic nerves in the basal ganglia, the same that are damaged in Parkinson's disease. When these nerves are injured or destroyed, patients may develop "Parkinsonian" symptoms, including stiffness, tremors, bradykinesia (slowed movements), loss of facial expression, and others. Anticholinergic medications are used to relieve these side effects of antipsychotic medication; however, they sometimes fail to help. In schizophrenic patients whose anticholinergic medications had failed to relieve

Parkinson symptoms, *R. rosea* was found to be of benefit (Saratikov and Krasnov 1987). Similarly, in clinical practice, authors Gerbarg and Brown find that in some patients the herb can reduce extrapyramidal symptoms when they are caused by primary Parkinsonism or when they are secondary to psychotropic medications.

10.6 MORE FUTURE DIRECTIONS

The potential benefits of *R. rosea* in psychiatric and medical disorders as well as for physical and cognitive enhancement have been reviewed. Additional research is needed in all of these areas to extend the currently available data, to validate preliminary findings, to refine the dosage and timing for specific conditions, to identify the components that exert specific biological activities, and to better understand the many mechanisms of action. Furthermore, it is possible to change the concentrations of bioactive compounds by changing the proportions of nutrients used in cultivation. Increased knowledge regarding the impact of soil quality, climate, and processing methods is needed to assure quality and potency of products. Reliable methods for testing the purity and strength of commercial preparations may require both genomic testing of roots and high-power liquid chromatography of plant extracts.

10.6.1 SPORTS MEDICINE

The Soviet Union began using adaptogens in their cosmonauts, military personnel, and elite athletes in the early 1970s. For over 30 years, extensive research conducted under the Ministry of Defense was hidden in classified documents (Vastag 2007). Following the collapse of the former Soviet Union, these documents were declassified and eventually translated (Brown et al. 2009). During the past 10 years, the usefulness of *R. rosea* for physical and mental performance enhancement has gained recognition by the athletic community and it has become increasingly popular worldwide as a sports supplement. More research in this area is needed to better guide consumers and trainers. Unfortunately, some manufacturers use excess quantities of *R. rosea* and caffeine in sport drinks. If an athlete consumes too many of these overloaded preparations, the excess caffeine can cause stimulatory effects such as increased heart rate, jitteriness, or anxiety, which could interfere with performance. Moreover, if consumption exceeds 900 mg/day of *R. rosea*, there could be an effect on platelet aggregation with the possibility of increased bruising or bleeding. Used in recommended doses, *R. rosea* improves strength, endurance, and accuracy reduces lactic acid accumulation and shortens recovery time (Brown et al. 2009).

10.6.2 PSYCHIATRY: STRESS, ANXIETY, PTSD, AND MASS DISASTERS

R. rosea could be used to treat mild depression or to augment the effects of antidepressants. It may also reduce sexual side effects of antidepressants. Additional research would also be worthwhile in the uses of *R. rosea* for treatment of stress-related conditions, anxiety disorders, and victims of mass disasters. Survivors experience acute trauma during disasters and prolonged stress from loss, economic decline, displacement, and poor health. Compared to standard treatments, *R. rosea* would be less expensive, have fewer side effects, and cause no habituation or addiction (Brown et al. 2009; Brown et al. 2013). The development of guidelines for dosages to be used in children of different ages and sizes would enable pediatricians and pediatric psychiatrists to use the herb safely.

10.6.3 STRESS AND FATIGUE—HOME, SCHOOL, WORKPLACE

Stress affects the quality of relationships at home and at work. It also impacts performance at school and at work. Excess stress adversely affects parenting and care giving both at home and at work. Parents and professional caregivers can experience fatigue, burnout, irritability, and loss of motivation. When these tendencies become chronic, the individual is at greater risk for substance abuse, poor performance, poor judgment, increased errors, and accidents. *R. rosea* can be a helpful addition to other interventions aimed at reducing the effects of stress and improving stress resiliency in any setting. It may be particularly helpful in situations that involve long hours of work requiring high levels of attention, risk of injury, night shifts, or intense pressure such as the weeks leading up to school exams or standardized tests. The beneficial effects of *R. rosea* are most evident under conditions that involve stress.

10.6.4 DERMATOLOGY

Russian women have been using topical applications of *R. rosea* for years for its purported antiaging properties. This concept has been spreading in the North American and European beauty industry, as there may be significant commercial potential in skin care products. Recent findings of beneficial effects on sensitive skin warrant further research (Dieamant et al. 2008). Because *R. rosea* absorbs ultraviolet light, it could provide some skin protection. However, manufacturers may need to remove tannins that could cause browning or leathering of skin over time.

10.6.5 DENTISTRY

Traditional uses for *R. rosea* include the treatment of infections. Antibacterial effects of *R. rosea* have been documented (see Section 10.2.6). *R. rosea* tablets have been used to treat "trench mouth," a condition characterized by bleeding swollen gums and bacterial infections that has been associated with severe stress. Commercial toothpaste containing *R. rosea* is already being marketed in Eastern Europe and the former Soviet Union.

10.6.6 VETERINARY MEDICINE

R. rosea preparations are being used to help animals in stressful situations, whether physical or psychological. Further veterinary medicine research is needed, particularly regarding animals used in competition, such as racing or steeplechase, and aging animals. *R. sacra* (a related subspecies) has been shown to increase the quality of frozen sperm used to artificially inseminate livestock (Zhao et al. 2009).

Living at high altitudes can reduce fertility in animals, including livestock. Future studies of *R. rosea* in animals at high altitudes could discover benefits for improving fertility and increasing the size of herds.

10.7   QUALITY AND POTENCY OF *R. rosea* BRANDS

The roots of the *R. rosea* plant contain hundreds of medicinal compounds. No one has been able to identify what combination of these is responsible for its many effects. The time and place of harvest, the age of the plants, the proportion of nutrients in the soil, climate, and altitude affect the concentration of bioactive compounds in the roots. Moreover, the methods and temperatures during drying and extraction (e.g., alcohol, water, or supercritical carbon dioxide extraction) affect which compounds are preserved and which may be volatilized. Considerable variation was found in an analysis of constituents from ethanol extracts of six *R. rosea* clones grown in different areas in Norway: salidrosides, tyrosol, rosavin and rosarin/rosin/cinnamyl-(6'-*o*-β-xylopyranosyl-*o*-β-glucopyranoside, and cinnamic alcohol (Hellum et al. 2010). These differences in bioactive compound profiles that result from the different growing conditions and extraction processes may explain some of the heterogeneity found among studies of *R. rosea* effects. It is important to use only those brands that have been proven to be clinically effective and that run test routinely.

Unfortunately, *R. rosea* can be adulterated with less expensive plants. Although it is important that labels indicate the product has met standardization requirements with at least 3% rosavins and 1% salidrosides, this does not guarantee the quality. Salidrosides are found in other plants. Rosavins are unique to *R. rosea*, but these could be synthesized and added to camouflage adulterants. It may become necessary to compare the high-pressure liquid chromatographic profile of extracts with a standardized profile that documents the concentrations of dozens of key constituents. Genomic testing would also be useful once the genetic profiles of potent plants become known. Given the current lack of reliable, available information on product constituents, it is wise to use brands that have proven to be effective in research studies and that are recommended by experienced clinicians (Brown et al. 2009).

ACKNOWLEDGMENTS

Dr. Brown and Dr. Gerberg acknowledge with deep appreciation the late Dr. Zakir Ramazanov for his assistance in obtaining research documents on *R. rosea* and for his generosity in sharing his extensive knowledge of this herb.

CONFLICTS OF INTEREST:

Dr. Richard P. Brown and Dr. Patricia L. Gerberg have no financial or other conflicts of interest regarding the material in this chapter.

REFERENCES

Abidoff, M. T. 1997. Synergistic effect of *Rhodiola rosea* and *Rhododendron caucasicum* herbal supplement on weight loss in healthy female volunteers: Placebo controlled clinical study (Report No. Grant 77-1997), Moscow, Russia.

Abidoff, M. T. and Nelubov, M. 1997. Russian anti-stress herbal supplement promotes weight loss, reduces plasma perilipins and cortisol levels in obese patients: Double-blind placebo controlled clinical study. *Stress and Weight Management at Russian Perestroika/ Healthy Diet* June 1-3, 1997, Caucasian Republic of Dagestan, Russia.

Abidov, M., Crendal, F., Grachev, S., Seifulla, R., and Ziegenfuss, T. 2003. Effect of extracts from *Rhodiola rosea* and *Rhodiola crenulata* (Crassulaceae) roots on ATP content in mitochondria of skeletal muscles. *Bull Exp Biol Med* 136(6): 585–7.

Abidov, M., Grachev, S., Seifulla, R. D., and Ziegenfuss, T. N. 2004. Extract of *Rhodiola rosea* radix reduces the level of C-reactive protein and creatinine kinase in the blood. *Bull Exp Biol Med* 138(1): 63–4.

Adamchuk, L. B. 1969. Effects of *Rhodiola* on the process of energetic recovery of rat under intense muscular workload. Unpublished doctoral dissertation, Tomsk, Russia, Tomsk State University and Medical Institute.

Adamchuk, V. and Salnik, B. U. 1971. Effect of *Rhodiola rosea* extract and piridrol on metabolism of rats under high muscular load. *Proc Inst Cytol Russ Acad Sci* 89–92.

Aires, V. A., Hinsch, K. D., Mueller-Schloesser, F., Bogner, K., Mueller-Schloesser, S., and Hinsch, E. 2003. *In vitro* and in vivo comparison of egg yolk-based and soybean lecithin-based extenders for cryopreservation of bovine semen. *Theriogenology*, 60(2): 269–79.

Aitken, R. J. and Sawyer, D. 2003. The human spermatozoon—not waving but drowning. *Adv Exp Med Biol* 518: 85–98.

American Psychiatric Association. 1994. *Diagnostic and Statistical Manual of Mental Disorders*, 4th ed. Washington, DC: American Psychiatric Association.

Arbuzov, A. G., Krylatov, A. V., Maslov, L. N., Burkova, V. N., and Naryzhnaya, N. V. 2006. Antihypoxic, cardioprotective, and antifibrillation effects of a combined adaptogenic plant preparation. *Bull Exp Biol Med* 142(2): 212–5.

Aslanyan, G., Amroyan, E., Gabrielyan, E., Nylander, M., Wikman, G., and Panossian, A. 2010. Double-blind, placebo-controlled, randomized study of single-dose effects of ADAPT-232 on cognitive functions. *Phytomedicine* 17(7): 494–9.

Baranov, V. B. 1994a. Experimental trials of herbal adaptogen effect on the quality of operation activity, mental and professional work capacity. Contract 93-11-615. Moscow: Russian Federation Ministry of Health Institute of Medical and Biological Problems (IMBP).

Baranov, V. B. 1994b. The response of cardio-vascular system to dosed physical load under the effect of herbal adaptogens. Contract 93-11-615. Moscow: Russian Federation Ministry of Health Institute of Medical and Biological Problems (IMBP).

Blizzard, T. A. 2008. Selective estrogen receptor modulator medicinal chemistry at Merck: A review, *Curr Top Med Chem*, 8(9): 792–812.

Blum, K., Chen, T. J., Meshkin, B., et al. 2007. Manipulation of catechol-O-methyl-transferase (COMT) activity to influence the attenuation of substance seeking behavior, a subtype of Reward Deficiency Syndrome (RDS), is dependent upon gene polymorphisms: A hypothesis. *Med Hypotheses* 69(5): 1054–60.

Bocharov, E. V., Ivanova-Smolenskaya, I. A., Poleshch, V. V., et al. 2010. Therapeutic efficacy of the neuroprotective plant adaptogen in neuro degenerative disease (Parkinson's disease as an example). *Bull Exp Biol Med*, 149(6): 682–4.

Bocharova, O. A., Matveev, B. P., Baryshnikov, A. I., et al. 1995. The effect of a *Rhodiola rosea* extract on the incidence of recurrences of a superficial bladder cancer. *Urol Nefrol* (Mosk) (2): 46–7.

Bogatova, R. I., Shlykova, L. V., Salnitsky, G., et al. 1997. Evaluation of the effect of a single dose of a phytoadaptogen on the working capacity of human subjects during prolonged isolation. *Aerospace Environ Med* 31(4): 51–4.

Bolshakova, I. V., Lozovskaia, E. L., and Sapezhinskii, I. I. 1997. Antioxidant properties of a series of extracts from medicinal plants. *Biofizika* 42(2): 480–3.

Brichenko, V. S. and Skorokhova, T. F. 1987. Herbal adaptogens in rehabilitation of patients with depression, clinical and organisational aspects of early manifestations of nervous and mental diseases (p. 15). Soviet Union: Barnaul.

Brown, R. P. and Gerbarg, P. L. 2004. *The Rhodiola Revolution*. New York: Rodale Press.

Brown, R. P. and Gerbarg, P. L. 2011. Integrative treatments in brain injury. In *Neuropsychiatry of Traumatic Brain Injury*, 3rd ed., eds. J. M. Silver, S. C. Yudofsky, T. W. McAllister. Washington, DC: American Psychiatric Press, pp. 599–622.

Brown, R. P. and Gerbarg, P. L. 2012. *Non-drug Treatments for ADHD. New Options for Kids, Adults, and Clinician*. New York: W.W. Norton & Company.

Brown, R. P., Gerbarg, P. L., and Muench, F. 2013. Breathing practices for treatment of psychiatric and stress-related medical conditions. *Psychiatr Clin North Am* 36(1):121–40.

Brown, R. P., Gerbarg, P. L., and Muskin, P. R. 2009. *How to Use Herbs, Nutrients, and Yoga in Mental Health Care*. New York: W.W. Norton.

Brown, R. P., Gerbarg, P. L., and Ramazanov, Z. 2002. A phytomedical review of *Rhodiola rosea*. *Herbalgram* 56:40–62.

Bystritsky, A., Kerwin, L., and Feusner, J. D. 2008. A pilot study of *Rhodiola rosea* (Rhodax) for generalized anxiety disorder (GAD). *J Altern Complement Med* Mar 14(2): 175–80.

Cao, L. L., Du, G. H., and Wang, M. W. 2006. The effect of salidroside on cell damage induced by glutamate and intracellular free calcium in PC12 cells. *J Asian Nat Prod Res* 8(1-2), 159–65.

Chen, X., Liu, J., Gu, X., and Ding, F. 2008. Salidroside attenuates glutamate-induced apoptotic cell death in primary cultured hippocampal neurons of rats. *Brain Res* 1238: 189–98.

Chen, X., Zhang, Q., Cheng, Q., and Ding F. 2009. Protective effect of salidroside against H_2O_2-induced cell apoptosis in primary culture of rat hippocampal neurons. *Mol Cell Biochem*, 332(1-2): 85–93.

Cuzick J., Sestak I., Bonanni B., Costantino J.P., Cummings S., DeCensi A., Dowsett M., Forbes J.F., Ford L., LaCroix A.Z., Mershon J., Mitlak B.H., Powles T., Veronesi U., Vogel V., Wickerham D.L., and SERM Chemoprevention of Breast Cancer Overview Group. Selective oestrogen receptor modulators in prevention of breast cancer: An updated meta-analysis of individual patient data. *Lancet* 381: 1827–34.

Cybulska, P., Thakur, S. D., Foster, B. C., Scott, I. M., Leduc, R. I., Arnason, J. T., and Dillon, J. A. 2011. Extracts of Canadian first nations medicinal plants, used as natural products, inhibit neisseria gonorrhoeae isolates with different antibiotic resistance profiles. *Sex Transm Dis* 38(7): 667–71.

Darbinyan, V. G., Aslanyan, G, Amroyan, E., et al. 2007. Clinical trial of *Rhodiola rosea* L. extract SHR-5 in the treatment of mild to moderate depression. *Nord J Psychiatry* 61(5):343–8.

De Deyn, P. P., Reuck, J. D., Deberdt, W., et al. 1997. Treatment of acute ischemic stroke with piracetam. Members of the Piracetamin Acute Stroke Study (PASS) Group. *Stroke* 28(12): 2347–52.

Dementeva, L. A. and Iaremenko, K. V. 1987. Effect of *Rhodiola* extract on the tumor process in an experiment. *Vopr Onkol* 33(7):57–60.

Dhir, R. R., Lin, H.-C., Canfield, S. E., and Wang, R. 2011. Combination therapy for erectile dysfunction: An update review. *Asian J Andrology* 13: 382–90.

Dieamant Gde, C., Velazquez Pereda Mdel, C., Eberlin, S., Nogueira, C., Werka, R. M., and Queiroz, M. L. 2008. Neuroimmunomodulatory compound for sensitive skin care: *In vitro* and clinical assessment. *J Cosmet Dermatol* 7(2): 112–9.

Duhan, O. M., Baryliak, I. R., Nester, T. I., et al. 1999. The antimutagenic activity of biomass extracts from the cultured cells of medicinal plants in the Ames test. *Ukrainian Tsitl Genet* 33(6): 19–25.

Dünser, M. W. and Hasibeder, W. R. 2009. Sympathetic over stimulation during critical illness: Adverse effects of adrenergic stress. *J Intensive Care Med* 24(5): 293–316.

Eagon, P. K., Elm, M. S., Gerbarg, P. L., Brown, R. P., Check, J. J., Diorio, G. J., and Houghton, F. Jr. 2004. Evaluation of the medicinal botanical *Rhodiola rosea* for estrogenicity [Abstract]. *Proceedings of the American Association for Cancer Research Annual Meeting* 45, 663.

Furmanowa, M., Skopinska-Rozewska, E., Rogala, E., and Malgorzata, H. 1998. *Rhodiola rosea in vitro* culture—Phytochemical analysis and antioxidant action. *Acta Societis Botanicorum Poloniae* 76(1): 69–73.

Gatti, D., Rossini, M., Sblendorio, I., and Lello, S. 2013. Pharmacokinetic evaluation of bazedoxifene for the treatment of osteoporosis. *Expert Opin Drug Metab Toxicol* 9(7): 883–92.

Gerasimova, H. D. 1970. Effect of *Rhodiola rosea* extract on ovarian functional activity. *Proc. of Scientific Conference on Endocrinology and Gynecology*. Sverdlovsk, Russia. 46–8.

Gonzales, G. F. 2012. Ethnobiology and Ethnopharmacology of *Lepidium meyenii* (Maca), a Plant from the Peruvian Highlands. *Evid Based Complement Alternat Med* 2012:193496.

Guan, S., He, J., Guo, W., Wei, J., Lu, J., and Deng, X. 2011. Adjuvant effects of salidroside from *Rhodiola rosea* L. on the immuneresponses to ovalbumin in mice. *Immunopharmacol Immunotoxicol* 33(4): 738–43.

Gunther, R. T. 1968. trans., *The Greek Herbal of Dioscorides*. Vol. 4, Sec. 45, Rhodiola Radis, Sedum Rhodiola. London: Hafner Publishing Company..

Ha, Z., Zhu, Y., Zhang X., et al. 2002. The effect of *Rhodiola* and acetazolamide on the sleep architecture and blood oxygen saturation in men living at high altitude (in Chinese). *Zhonghua* 25(9): 527–30.

Hellum, B. H., Tosse, A., Hoybakk, K., et al. 2010. Potent *in vitro* inhibition of CYP3A4 and P-glycoprotein by *Rhodiola rosea*. *Planta Med.* 76(4): 331–8.

Hendrix, S. L. and McNeeley, S. G. 2001. Effect of selective estrogen receptor modulators on reproductive tissues other than endometrium. *Ann NY Acad Sci* 949: 243–50.

Hill-House, B. J, Ming, D. S, French, C. J., and Towers, N. G. H. 2004. Acetylcholine esterase inhibitors in *Rhodiola rosea*. *Pharm Biol*, 42(1): 68–72.

Hung, S. K., Perry, R., and Ernst, E. 2011. The effectiveness and efficacy of *Rhodiola rosea* L.: A systematic review of randomized clinical trials. *Phytomedicine* 18(4): 235–44.

Illig, P. I. 2009. Passenger health. In *Principles and Practice of Aviation Medicine*. eds C. C. Christiansen. J. Drager, J Kriebel. Singapore: World Scientific. pp. 667–708.

Kawate, H. and Takayanagi, R. 2011. Efficacy and safety of bazedoxifene for postmenopausal osteoporosis. *Clin Interv Aging* 6:151–60.

Kessler, J., Thiel, A., Karbe, H. et al. 2000. Piracetam improves activated blood flow and facilitates rehabilitation of poststroke aphasic patients. *Stroke* 31(9): 2112–6.

Kormosh, N., Laktionov, K., and Antoshechkina, M. 2006. Effect of a combination of extract from several plants on cell-mediated and humoral immunity of patients with advanced ovarian cancer. *Phytother Res* 20(5): 424–5.

Kurkin, V. A. and Zapesochnaya, G. G. (1986). Khimicheskiy sostav i farmakologicheskiye svoystva rasteniy roda *Rhodiola*. Obzor. [Chemical composition and pharmacological properties of *Rhodiola rosea*]. *Khim-Farm Zh Chemical and Pharmaceutical Journal Moscow* 20(10): 1231–44.

Leadbetter, G., Keyes, L. E., Maakestad, K. M., Olson, S., Tissotvan Patot, M. C., and Hackett, P. H. 2009. Ginkgo biloba does—and does not—prevent acute mountain sickness. *Wilderness Environ Med* 20(1): 66–71.

Lishmanov Iu, B., Maslov, L. N., Arbuzov, A. G., Krylatov, A. V., Platonov, A. A., Burkova, V. N., and Kaiumova, E. A. 2008. Cardioprotective, inotropic, and anti-arrhythmia properties of a complex adaptogen "Tonizid" [Russian]. *Eksp Klin Farmakol* 71(3): 15–22.

Liu, Z., Li, X., Simoneau, A. R. et al. 2012. *Rhodiola rosea* extracts and salidroside decrease the growth of bladder cancer cell lines via inhibition of the mTOR pathway and induction of autophagy. *Mol Carcinog Mar* 51(3): 257–67.

Lu, L., Yuan, J., and Zhang, S. 2013. Rejuvenating activity of salidroside (SDS): Dietary intake of SDS enhances the immune response of aged rats. *Age* (Dordr), 35(3): 637–46.

McClintock, S. M., Husain, M. M., Wisniewski, S. R., Nierenberg, A. A., Stewart, J. W., Trivedi, M., Cook, I., Morris, D., Warden, D., and Rush, A. J. 2011. Residual symptoms in depressed outpatients who respond by 50% but do not remit to antidepressant medication. *J Clin Psychopharmacol* 31(2): 180–6.

Ming, D. S., Hillhouse, B. J., Guns, E. S., Eberding, A., Xie, S., Vimalanathan, S., and Towers, G. H. 2005. Bioactive compounds from *Rhodiola rosea* (Crassulaceae). *Phytother Res* 19(9): 740–3.

Moraga, F. A., Flores, A., Serra, J., Esnaola, C., and Barriento, C. 2007. *Ginkgo biloba* decreases acute mountain sickness in people ascending to high altitude at Ollagüe (3696 m) in northern Chile. *Wilderness Environ Med* 18(4): 251–7.

Narimanian, M., Badalyan, M., Panosyan, V. et al. 2005. Impact of Chisan (ADAPT_232) on the quality-of-life and its efficacy as an adjuvant in the treatment of acute non-specific pneumonia. *Phytomedicine* 12(10): 723–9.

Olsson, E. M, von Schéele, B., and Panossian, A. G. 2009. Arandomised, double-blind, placebo-controlled, parallel-group study of the standardised extracts hr-5 of the roots of *Rhodiola rosea* in the treatment of subjects with stress-related fatigue. *Planta Med*; 75(2): 105–12.

Palumbo, D. R., Occhiuto, F., Spadaro, F. et al. 2012. *Rhodiola rosea* extract protects human cortical neurons against glutamate and hydrogen peroxide-induced cell death through reduction in the accumulation of intracellular calcium. *Phytother Res* 26(6): 878–83.

Panossian, A., Hovhannisyan, A., Abrahamyan, H. et al. 2009. Pharmacokinetic and pharmacodynamic study of interaction of *Rhodiola rosea* SHR-5 extract with warfarin and theophylline in rats. *Phytother Res Mar* 23(3): 351–7.

Panossian, A. and Wikman, G. 2009. Evidence-based efficacy of adaptogens in fatigue, and molecular mechanism related to their stress-protective activity. *Curr Clin Pharmacol* 4(3): 198–219.

Panossian, A. G. 2013. Adaptogens in mental and behavioral disorders. *Psychiatr Clin North Am* 36(1): 49–64.

Perfumi, M. and Mattioli, L. 2007. Adaptogenic and central nervous system effects of single doses of 3% rosavin and 1% salidroside *Rhodiola rosea* L.extract in mice. *Phytother Res* 21(1): 37–43.

Petkov, V. D., Stancheva, S. L., Tocuschieva, L. et al. 1990. Changes in brain biogenic monoamines induced by the nootropic drugs adafenoxate and meclofenoxate and by citicholine (experiments on rats). *Gen Pharmacol* 21:71–5.

Petkov, V. D., Yonkov, D., Mosharoff, A. et al. 1986. Effects of alcohol aqueous extract from *Rhodiola rosea* L. roots on learning and memory. *Acta Physiol Pharmacol Bulg* 12:3–16.

Pogorelyĭ, V. E. and Makarova, L. M. 2002. *Rhodiola rosea* extract for prophylaxis of ischemic cerebral circulation disorder. *Eksp Klin Farmakol* 65(4): 19–22.

Polyakov, V. V. 1996. The use of a new phytoadaptogem under conditions of space flight. Presented at Gothenburg Seminar, Nov. 4-5, Gothenburg, Sweden.

Razina, T. G., Zueva, E. P., Amosova, E. N. et al. 2000. Medicinal plant preparations used as adjuvant therapeutics in experimental oncology. *Eksp Klin Farmakol* 63(5): 59–61.

Saratikov, A. S. and Krasnov, E. A. 1987. Clinical studies of *Rhodiola*. In *Rhodiola rosea is a Valuable Medicinal Plant (Golden root)*, eds. A. S. Saratikov and E. A. Krasnov. Tomsk, Russia: Tomsk State University Press.

Shevtsov, V. A., Zholus, I., Shervarly, V. I. et al. 2003. A randomized trial of two different doses of a SHR-5 *Rhodiola rosea* extract versus placebo and control of capacity for mental work. *Phytomedicine* 10(2-3): 95–105.

Shi, T. Y., Feng, S. F., Xing, J. H. et al. 2012. Neuroprotective effects of Salidrosides and its analogue tryrosol galactoside against focal cerebral ischemia in vivo and H_2O_2-induced neurotoxicity *in vitro*. *Neurotox Res* 21(4): 358–67.

Sobeieraj, D. M. 2011. Bazedoxifene: An investigational selective estrogen receptor modulator for the treatment and prevention of osteoporosis in postmenopausal women. Formulary. http://formularyjournal.modernmedicine.com/formulary-journal/news/clinical/clinical-pharmacology/bazedoxifene-investigational-selective-estroge. Posted May 11, 2011. Viewed 6-06-13.

Spasov, A. A., Mandrikov, V. B., Miranova, I.A. et al. 2000a. The effect of the preparation rodakson on the psychophysiological and physical adaptation of students to an academic load. *Eksp Klin Farmakol* 63(1): 76–8.

Spasov, A. A., Wikman, G. K., Mandrikov, V. B. et al.2000b. A double-blind placebo-controlled pilot study of the stimulating and adaptogenic effect of *Rhodiola rosea* SHR-5 extract on the fatigue of students caused by stress during an examination period with a repeated low-dose regimen. *Phytomedicine* 7(2): 85–9.

Stancheva S. L. and Mosharoff, A. 1987. Effect of the extract of *Rhodiola rosea* L. on the content of the brain biogenic monoamines. *Proc Bulg Acad Sci Med* 40:85–87.

Udintsev, S. N. Fomina, T. I., and Razina, T. G. 1992. An experimental model of metastatic liver involvement by using Ehrlich's ascitic cancer. *Vopr Onkol* 38(6): 723–6.

Udintsev, S. N. and Shakhov, V. P. 1989. Decrease in the growth rate of Ehrlich's tumor and Pliss' lymphosarcoma with partial hepatectomy. *Vopr Onkol* 35(9): 1072–5.

Udintsev, S. N. and Shakhov, V. P. 1990. Changes in clonogenic properties of bone marrow and transplantable mice tumor cells during combined use of cyclophosphane and biological response modifiers of adaptogenic origin. *Eksp Onkol* 12(6): 55–6.

Udintsev, S. N. and Schakhov, V. P. 1991. Decrease of cyclophosphamide haematotoxicity by *Rhodiola rosea* root extract in mice with Ehrlich and Lewis transplantable tumors. *Eur J Cancer* 27(9): 1182.

Vastag, B. 2007.Warming to a Cold War herb: Soviet secret finds its way west. *Science News* 172 (12):184–9.

Wing, S. L., Askew, E. W., Luetkemeier M. J. et al. 2003. Lack of effect of *Rhodiola* or oxygenated water supplementation on hypoxemia and oxidative stress. *Wilderness Environ Med* 14(1): 9–16.

Wong, D. L., Tai, T. C., Wong-Faull, D. C., Claycomb, R., Meloni, E. G., Myers, K. M., Carlezon, W. A. Jr., and Kvetnansky, R. 2012. Epinephrine: A short- and long-term regulator of stress and development of illness: A potential new role for epinephrine in stress. *Cell Mol Neurobiol* 32(5): 737–48.

Yarmenko, A. 1998. Complex treatment of severe infectious diseases [dissertation] I. P. Pavlov, St. Petersburg State Medical University.

Yu, S., Liu, M., Gu, X., and Ding, F. 2008. Neuroprotective effects of salidroside in the PC12 cell model exposed to hypoglycemia and serum limitation. *Cell Mol Neurobiol.* 28(8): 1067–78.

Zubeldia, J. M., Nabi, H. A., Jiménez del Río, M., and Genovese, J. 2010. Exploring new applications for *Rhodiola rosea*: Can we improve the quality of life of patients with short-term hypothyroidism induced by hormone withdrawal? *J Med Food* 13(6): 1287–92.

11 Toxicology and Safety of *Rhodiola rosea*

Hugh Semple and Brandie Bugiak

CONTENTS

11.1 Introduction 253
11.2 *In Vitro* Studies 253
11.3 *In Vivo* Animal Toxicity Studies 255
11.4 Clinical Trials 256
11.5 Case Report 257
11.6 Pharmacokinetics 257
      11.6.1 Pharmacokinetics of *Rhodiola* Active Constituents 257
      11.6.2 Pharmacokinetic/Pharmacodynamic Interactions 259
11.7 Critical Assessment of Current Information 260
      11.7.1 Assessment of *R. rosea* Safety Based on Current Evidence 260
      11.7.2 Gaps and Topics for Further Study 260
11.8 Conclusion 261
References 261

11.1 INTRODUCTION

Although *Rhodiola rosea* is touted to be one of the safest herbs in popular literature, only a smattering of studies has been conducted, which can be used to support or refute the claim. In recent years, as the popularity of *R. rosea* has surged, the safety and potential risks for this herb to be involved in toxic interactions with drugs are beginning to receive more attention. Many gaps in understanding still remain, however. In the following review, the currently available information on *R. rosea* safety and toxicity is summarized and areas for follow-up are identified.

11.2 *IN VITRO* STUDIES

The review literature tends to extol the excellent safety profile of *R. rosea* and long history of safe use, although this herb has not undergone the kind of profiling for potential drug interactions that constitute an important aspect of its toxicity assessment. As its use increases in Europe and North America, screening for membrane transporter and metabolic enzyme inhibition has become an urgent need. We have found only three reports of *in vitro* screening, and all of these point to potentially dangerous interactions between *R. rosea* and drugs or endogenous compounds.

In a screening study on ethanol (55%) extracts of 10 North American botanicals (Scott et al. 2006), the effects on three isoenzymes of human cytochrome P450 (CYP) important in endogenous compound and drug metabolism were evaluated. CYP3A4, mainly found in the liver, was studied because it is an important inducible CYP that is involved in the metabolism of a broad range of pharmaceuticals. CYP19 was included because it is responsible for the conversion of testosterone to estrogen, and CYP2C19 is another important enzyme in the metabolism of xenobiotics. *R. rosea* extract was not tested against CYP2C19. Two *R. rosea* extracts from different accessions contained quite different concentrations of the two marker compounds, salidroside (417 vs. 1460 μg/mL) and rosarin (115 vs. 1443 μg/mL). In an enzyme inhibition assay, the *Rhodiola* extracts significantly inhibited CYP3A4 and CYP19 by 67% and 83%, respectively. Of the 10 botanical species, *R. rosea* extract was ranked as the most potent enzyme inhibitor, and the degree of inhibition was positively related to the concentration of rosarin. It was concluded that *R. rosea* has the potential to influence the bioavailability and pharmacokinetics of a wide variety of drugs.

A second, more detailed *in vitro* study on the inhibition of CYP3A4 and also the membrane efflux transporter P-glycoprotein (P-gp), which has a number of drugs as substrates, was conducted on ethanolic extracts from six clones of *R. rosea* from different regions of Norway (Hellum et al. 2010). The percent inhibition of testosterone (a CYP3A4 model substrate) metabolism was measured at extract concentrations from 0.1 to 10 μg/mL, and IC_{50} and IC_{25} values were calculated. IC_{50} values ranged from 1.7 to 3.1 μg/mL and IC_{25} values ranged from 1.2 to 2.0 μg/mL. These values were noted to be considerably lower than those reported for other common herbs, including St. John's Wort and Ginkgo Biloba, and were comparable to *in vitro* IC_{50} values for drugs such as ketoconazole, erythromycin, and fluoxetine that are known to increase plasma levels of other CYP3A4 substrates. Thus, *R. rosea* was deemed a potent inhibitor of CYP3A4. Inhibition of P-gp by *R. rosea* was measured on Caco-2 cell membranes using digoxin as a model transporter substrate. IC_{50} values for the six clones ranged from 16.7 to 51.7 μg/mL. For context, these were in the lower range of values reported for other herbs and comparable to those of drugs such as omeprazole and quinidine (3–25 μg/mL), both of which increase digoxin bioavailability in man, causing significant interactions. The degree of inhibition of either CYP3A4 or P-gp could not be related to the concentrations of any of the constituents measured (including cinnamyl alcohol, salidroside, tyrosol, rosavin, and rosarin/ rosin/cinnamyl-(6'-o-β-xylopyranosyl)-O-β-glucopyranoside), leading the authors to conclude that other constituents not presumed to be biologically active could be responsible for the inhibition. Another explanation not mentioned would be the possible additive or synergistic inhibitory effect of two or more of the constituents. The similarity of inhibition levels among the different clones led to a secondary conclusion that the geographical origin of the plants was not an important factor.

In an investigation of a possible mechanism to explain the influence of *R. rosea* on mood disorders, an *in vitro* assay measuring monoamine oxidase (MAO) inhibition was conducted on 12 isolated compounds, and dichloromethane, methanol, and water extracts of *R. rosea* roots (van Diermen et al. 2009). MAO A and MAO B inhibition were studied. Both methanol and water extracts produced marked MAO inhibition. At an extract concentration of 100 µg/mL, MAO A was inhibited by 92.5% and 84.3% and MAO B was inhibited by 81.8% and 88.9%, respectively, for the methanol and water extracts. These activities were attributed to the presence of isolated fractions of rosaridin, rhodioloside B and C isomers, cinnamyl alcohol, triandrin, and epigallocatechin gallate dimer, which could act additively or possibly synergistically. It was concluded that *R. rosea* root may influence serotonin and nor-epinephrine in nerve terminals, conferring antidepressant activity through MAO A activity and influence the progress of other neurodegenerative diseases through MAO B inhibition. The toxicological implications of MAO inhibition were not mentioned, however, interactions between MAO-inhibiting antidepressants can lead to elevated serotonin levels and a potentially life-threatening toxic syndrome. The results of this study open the possibility that *R. rosea* also has the potential to interact with MAO inhibitors causing serotonin-related toxicity.

The significance of *in vitro* screening studies can only be determined through follow-up *in vivo* studies in animal and human because many factors influencing toxicity are not accounted for. Nevertheless, the three studies described above indicate the potential for a wide variety of interactions between *R. rosea* products and commonly prescribed drugs.

11.3 *IN VIVO* ANIMAL TOXICITY STUDIES

As part of a project to develop *R. rosea* as a new crop in Alberta, Canada, initial general toxicology studies were undertaken on an alcoholic extract of Alberta-grown *R. rosea* (Semple 2010). The extract used in the studies contained approximately 2.7% rosavins (total of rosavin, rosarin, and rosin). Three studies were conducted on Sprague–Dawley rats following Organisation for Economic Cooperation and Development (OECD) guidelines: an acute oral toxicity study (males only), a 7-day repeated dose study (males only), and a 28-day repeated dose study (males and females). In the acute toxicity study, the highest of three dosage levels tested was a single oral (gavage) dose of 1000 mg/kg of *R. rosea* extract. In the 7- and 28-day repeated oral dose studies, the highest of three dosage levels was 50 mg/kg via oral gavage once daily, a severalfold multiple of dosage levels commonly used in humans (200–600 mg or ~3–9 mg/kg daily). Animals were observed for signs of toxicity for the duration of the studies. At the end of each of the studies, blood samples were collected for hematology, clinical chemistry, and coagulation. All animals were sacrificed and underwent a gross postmortem examination for macroscopic abnormalities and tissues were collected for histological examination for evidence of toxicity. In these three studies, none of body weight, food consumption, hematological, clinical, pathological, and coagulation parameters, gross or histopathological findings in *R. rosea*-treated animals were different from those in normal animals. It was concluded that the Alberta *R. rosea* extract was not toxic at repeated doses up to 50 mg/kg. Even though this dose exceeded the usual dose in humans (up to ~10 mg/kg when a dose of 600 mg/day is used) by severalfold, the limit of safety was not reached, and follow-up studies at higher doses would be of value.

There is one report (Brown et al. 2002) of a Russian review (Kurkin and Zapeso-chnaya 1985) giving an LD_{50} value of 28.6 mL/kg or approximately 3360 mg/kg. This is well above the normally used OECD limit dose of 2000 mg/kg used to establish the safety of relatively nontoxic substances and is, therefore, supportive of a high degree of safety for *R. rosea*. It is difficult to substantiate the quality of this data as the original paper and Russian review were unavailable to us.

Although the above studies are the only ones reported for *R. rosea* extracts, a geno-toxicity study on a major *R. rosea* active ingredient (usually present in standardized extracts at a minimum of 1%; Natural Standards 2008), salidroside, has been reported (Zhu et al. 2010). Salidroside was evaluated via a bacterial reverse mutation (Ames) assay, an *in vitro* mammalian chromosome aberration test, and an *in vivo* mouse micro-nucleus assay in which doses of up to 1500 mg/kg of salidroside were administered daily over 3 days. All studies were conducted under OECD guidelines. The results indicated that salidroside was not genotoxic under the conditions of any of the three assays.

11.4 CLINICAL TRIALS

There have been several human clinical trials for *R. rosea* examining effects of the whole plant or the active ingredients. A handful of groups have published systematic reviews of these trials with each group focusing on different aspects of the studies (Brown et al. 2002; Blomkvist et al. 2009; Panossian et al. 2010b; Hung et al. 2011; Ulbricht et al. 2011; Ishaque et al. 2012). The efficacy of *R. rosea* and the safety of its use for the treatment of various afflictions, as well as the validity of results based on statistical analyses, have been called into question by the reviewing groups. Fifteen clinical trials were evaluated among these review groups; five of these were reviewed by all groups. The studies had exposure periods ranging from a single dose (Shevtsov et al. 2003) to a 12-week study (Fintelmann and Gruenwald 2007) with doses from 60 mg of active substances per day (Abidov et al. 2004) to 1000 mg in one exposure (Walker et al. 2007).

Blomkvist et al. (2009) chose seven clinical trials to review. The review focused on the validity of trial conclusions based on the statistical analysis methods used. The authors decided that six of the seven trials were of poor scientific quality with unsupportable conclusions.

Brown et al. (2002) approached their review by evaluating the traditional uses of *R. rosea* and its phytochemistry while highlighting ongoing studies and medical applications. Brown and colleagues discussed 17 human studies, including clinical, preclinical, and open studies. Despite presenting a large amount of information and positive responses within the studies, the authors maintain that larger controlled studies are warranted to expand the current knowledge regarding the safety and toxicity.

Hung et al. (2011) published a review of 11 clinical trials and focused on the apparent effectiveness or efficacy of *R. rosea*. The authors concluded that replicating the clinical trials to obtain similar results in a strict and regimented protocol would be beneficial for the field.

Ishaque et al. (2012) chose to focus on the efficacy and safety of the use of *R. rosea* for the treatment of mental and physical fatigue in their review of 10 clinical trials. The authors acknowledged that *R. rosea* may be beneficial for alleviating mental fatigue and enhancing physical performance; however, they also noted that the evidence appears contradictory and inconclusive. The review concluded that an accurate assessment of *R. rosea* efficacy was limited due to methodological flaws.

A review compiled by Panossian et al. (2010b) examined seven clinical trials in the course of summarizing data accumulated relating to chemical composition, pharmacological activity, and both traditional and official medicinal uses of *R. rosea*. Although this review group recognizes the obstacles presented in other summaries, they advocate the use of *R. rosea* for the treatment of fatigue and believe that there is promising evidence to support the use for mood and cognition.

Ulbricht et al. (2011) compiled information from a large number of studies, including eight clinical trials, to examine the use of *R. rosea* on a number of conditions. The authors, like many of review groups, resolved that although there are a number of trials published with encouraging results, larger, well-designed randomized clinical trials are needed before any conclusions can be drawn.

On the basis of the above-mentioned systematic reviews, it appears that a total of approximately 700 subjects were included in clinical trials. Adverse effects were reported by six individuals from the 700 included in various studies. Of these six, three had been a part of the placebo groups (Shevtsov et al. 2003; De Bock et al. 2004), two had received *R. rosea* and complained of minor effects (headache and insomnia) (De Bock et al. 2004), and one from an unknown treatment group reported an unspecified illness (Wing et al. 2003). The consensus from the groups reviewing the clinical trials was that *R. rosea* appears relatively safe for use in healthy individuals who are not taking any other medication. *R. rosea* remains not recommended for use by pregnant or lactating women or by juvenile populations due to the lack of information.

11.5   CASE REPORT

We found only one case report of a toxic drug interaction with *Rhodiola* (McGovern and McDonnell 2010). A 26-year-old patient on the selective serotonin reuptake inhibitor (SSRI), escitalopram, developed supraventricular tachycardia 3 days after commencing *Rhodiola*. Because *Rhodiola* potently inhibits CYP3A4, one of the enzymes involved in escitalopram metabolism, and also because *Rhodiola* inhibits MAO A, it was reasoned that levels of serotonin, a monoamine neurotransmitter, may have become elevated to excessive levels, producing a toxic "serotonin syndrome." Although the patient did not exhibit the full spectrum of symptoms of serotonin syndrome, the authors highlighted the importance of unexpected interactions between drugs and herbal medications, and recommended increased vigilance around the possibilities of herb–drug interactions.

11.6   PHARMACOKINETICS

11.6.1   Pharmacokinetics of *Rhodiola* Active Constituents

The pharmacokinetics of the active constituents, salidroside and rosavin, have been investigated in rats (Abrahamyan et al. 2005; Li et al. 2006; Panossian et al. 2010a) and humans (Panossian et al. 2010a).

In rats, *Rhodiola* extract was given at oral doses of 20 or 50 mg/kg of SHR-5 extract containing 10 mg/g of salidroside and 20 mg/g of rosavin (Panossian et al. 2010a). These doses were deemed by the authors to be therapeutically equivalent to human doses of one or two tablets of a product containing 144 mg of SHR-5, 3.63 mg of salidroside, and 4.2 mg of rosavin per tablet. Salidroside was completely absorbed. The extract was also administered intravenously at a dose of 20 mg/kg and salidroside was administered at a dose of 0.570 mg/kg, equivalent to that in 50 mg/kg of SHR-5 extract. At these doses, peak concentrations of salidroside occurred at 1 hour after oral administration and after a rapid decline over the next hour, declined to undetectable levels (<100 ng/mL) by 5–6 hours post-dosing. C_{max} values were 400 and 700 ng/mL and half-lives were 3.6 and 4.5 hours, respectively, for the 20 and 50 mg/kg doses. Kinetic parameters were unchanged after five daily repeated 20 or 50 mg/kg doses. When salidroside was administered alone, the T_{max} was 1.44 hours, the C_{max} was 334 ng/mL, the half-life was 2.6 hours, and the $AUC_{0-\infty}$ was about 80% of that for the 50 mg/kg oral dose of SHR-5. Therefore, salidroside in the *Rhodiola* product was more rapidly absorbed and more slowly eliminated than pure salidroside. It was concluded that other compounds in the extract impeded the metabolism of salidroside; however the authors did not include information on the matrix of the SHR extract solution, which could potentially be involved in any interactions. The kinetics of the salidroside metabolite tyrosol supported this con-clusion. These data were in contrast to those of Li et al. (2006) who administered 10 mg/kg of salidroside either alone or in a water extract of *R. rosea*. In this study, T_{max} was 50 minutes for both preparations, and half-life values were similar, between 35 and 40 minutes; however, C_{max} of salidroside alone, at 4.43 µg/mL, was over 70% higher than that of salidroside in extract, at 2.59 µg/mL. The discrepancies between the two studies may relate to the lower assay sensitivity of Li et al., which allowed measurement of the distribution phase, and not the elimination phase, of the concen-tration–time curves, and the differences between the extracts used (different regions of origin, water vs. alcoholic). In addition, subject differences could also contribute to the discrepancies; however, these are difficult to evaluate. While both studies employed male Wistar rats, Panossian et al. (2010a) did not specify the weight and age of the animals used in their study and Li et al. (2006) did not specify the ages. The animal numbers and sampling were different between the two studies. The doses used by Li et al. were also manyfold higher than those used by Panossian et al. The salient information from both studies is that salidroside concentrations in rat plasma after doses that are therapeutically equivalent to those in humans can be expected to not exceed 1 µg/mL and to decline rapidly over a few hours. No accu-mulation of salidroside is expected after multiple dosing; however, this does not rule out accumulation of the aglycone (or other constituents when the plant extract is administered).

In the study by Panossian et al. (2010a), rosavin kinetics in rats were evaluated similarly to those of salidroside. After the 20 and 50 mg/kg oral doses of SHR-5, T_{max} occurred at 1 hour, C_{max} values were 250 and 579 ng/mL, respectively, and concen-trations declined to below detectable levels (100 ng/mL) after 1.5 hours. Comparison with data from an intravenous dose yielded bioavailability estimates of 20% and 26% for the two doses, respectively. C_{max} after 5 days of repeated daily dosing of

50 mg/kg of SHR-5 was 620 mg/kg, similar to that after a single dose. Therefore, rosavin does not appear to accumulate on multiple dosing.

In humans, administration of two tablets of a product containing SHR-5 *R. rosea* extract produced a dose of 9.34 mg of salidroside and 7.74 mg of rosavin (Panossian ct al. 2010a). Both ingredients reached T_{max} at 2 hours, and C_{max} values were 948 and 446 ng/mL, respectively. The 2.5-hour half-life of salidroside was longer than that of rosavin at 1.4 hours. Thus, *Rhodiola* ingredients in extract are absorbed and eliminated more slowly in humans than in rats, but the same pattern of lower bio-availability and more rapid elimination of rosavin than of salidroside is the same in both species. At high therapeutic doses of *Rhodiola* extract in humans, plasma concentrations of neither ingredient, on average, exceeded 1 µg/mL, and detectable concentrations were not present beyond 6 hours.

11.6.2 Pharmacokinetic/Pharmacodynamic Interactions

One study has been conducted with the purpose of elucidating the interaction between *Rhodiola* extract and two drugs that undergo therapeutic monitoring due to their narrow margins of safety (Panossian et al. 2009). Both test drugs, theophylline (used to treat chronic obstructive pulmonary disease and asthma) and warfarin (an anticoagulant used to prevent thrombosis and thromboembolism), are CYP 1A2 substrates and warfarin is also metabolized by CYP 2C19. In this study, water or a *R. rosea* extract (SHR-5) was administered at 50 mg/kg daily to male Wistar rats for 3 days, followed an hour after the final dose by either 192 mg/kg of theophylline or 2 mg/kg of warfarin. Blood samples were taken at appropriate intervals and analyzed for drug, and the pharmacokinetic parameters of the test drugs were calculated. Although the pharmacokinetic parameters of theophylline were unaffected by *R. rosea* extract administration, the C_{max} of warfarin in the presence of *R. rosea* was significantly higher by one-third. Nevertheless, the pharmacodynamics of warfarin were unaffected. The authors concluded that application of a product containing *R. rosea* would not "necessitate stringent precautions in order to avoid possible interactions with other drugs" if it is used as an antidepressant for the treatment of mild to moderate depression. Note that the study was carried out for only 3 days, so it did not cover the possibility of an interaction on chronic administration.

While the above is a direct PK/PD interaction study in animals, several *in vitro* and *in vivo* animal and human studies outlining potential beneficial effects and activities of *Rhodiola* have given rise to cautionary notes based on possible shared mechanisms or metabolic routes between drugs and *Rhodiola* (Natural Standards Monograph 2008). The list includes anti-anxiety agents (Perfumi and Mattioli 2007), antibiotics (Ming et al. 2005; Narimanian et al. 2005), antidepressants, especially SSRIs (Perfumi and Mattioli 2007; McGovern and McDonnell 2010), antidiabetic agents (Kim et al. 2006; Kwon et al. 2006), antihypertensives and ACE inhibitors (Kwon et al. 2006), antineoplastic agents (Dement'eva and Iaremenko 1987; Udintsev and Shakhov 1991; Udintsev et al. 1992; Bocharova et al. 1995), antioxidants (De Sanctis et al. 2004; Kim et al. 2006), CNS depressants (Kelly 2001), exercise performance enhancers (Azizov and Seifulla 1998; Abidov et al. 2003, 2004), heart rate–regulating agents (Lishmanov et al. 1993, 1997; Maimeskulova et al. 1997; Maimeskulova and

Maslov 1998, 2000 Zhang et al. 2005), hormonal agents (Scott et al. 2006), immunosuppressants (Bocharova et al. 1995; Kormosh et al. 2006), impotence agents (not referenced), neurologic agents (Saratikov et al. 1968; Petkov et al. 1986), nonsteroidal anti-inflammatory agents, cyclooxygenase 2 inhibitors (Abidov et al. 2004), opiates (Lishmanov et al. 1993, 1997; Maimeskulova et al. 1997; Maimeskulova and Maslov 1998, 2000), and pentobarbital (Ahumada 1991).

11.7 CRITICAL ASSESSMENT OF CURRENT INFORMATION

11.7.1 ASSESSMENT OF *R. ROSEA* SAFETY BASED ON CURRENT EVIDENCE

The clinical trials and animal toxicology studies reviewed earlier reinforce the popular perception that *R. rosea* is relatively safe and well tolerated in adults who are not taking any other medications, at least in short-term use. There are no long-term data to evaluate the safety of *Rhodiola* on chronic administration. The safety of *R. rosea* in combination with drugs and other herbs is, however, uncertain and over the past few years the emerging information cited above indicates that potentially dangerous toxic interactions may occur.

The screening study by Scott et al. (2006) showed that extracts containing over 400 µg/mL of salidroside significantly inhibited CYP 3A4, a major drug-metabolizing enzyme. This was reinforced by the study by Hellum et al. (2010) in which Norwegian *R. rosea* extracts inhibited the CYP 3A4 metabolism with an IC_{50} value of 1.7–3.1 µg/mL and P-gp efflux in Caco-2 cells with IC_{50} values of 16.7–51.7 µg/mL. Comparing these concentrations with the C_{max} values of *Rhodiola* ingredients, salidroside and rosavin, alone, which were approximately 1 µg/mL in the human pharmacokinetic studies by Panossian et al. (2010a), it is apparent that the *in vitro* IC_{50}s are in the clinically relevant range. Comparisons with IC_{50} values of known drugs brought more relevance. Therefore, metabolic enzyme inhibition is being observed *in vitro* at similar concentrations to those reached *in vivo* in rats and humans after clinical doses. A problem with the *in vitro* studies, however, is that the identity and bioavailability of the inhibitor(s) have not been determined. Therefore, the clinical relevance of the *in vitro* data remains unknown. Panossian et al. (2010a) have conducted a single study in rats with CYP 1A2 substrates in support of their contention that *R. rosea* can safely be administered with other drugs. Clearly, this evidence is insufficient. The single case report of an interaction between a *Rhodiola* product and an antidepressant drug serves as a warning that we do not yet have sufficient information to declare whether *R. rosea* can be safely be coadministered with drugs.

11.7.2 GAPS AND TOPICS FOR FURTHER STUDY

1. Individual ingredients causing inhibition of CYP and P-gp have not been established. Follow-up *in vitro* studies on individual *R. rosea* ingredients, including a search for more potential active or toxic ingredients, is warranted.
2. Only three *in vitro* studies and a single case report give evidence of possible interactions of *R. rosea* with CYP 3A4, P-gp, and MAO. The interactions with these enzymes require further characterization, and other enzymes

important in drug metabolism, such as CYP 1A2, CYP 2B6, CYP 2C8, CYP 2C9, CYP 2C19, CYP 2D6, and CYPs 3A5 and 7, also require attention.

3. *In vitro* interaction findings require follow-up with *in vivo* studies in animals and humans in order to determine their relevance.

4. The list of potential *R. rosea*–drug interactions by Natural Standards (2008) indicates the possibility of mechanisms not yet explored. Follow-up studies are warranted.

5. The rosavins and salidroside are all glycosides (e.g., salidroside of tyrosol, rosin of caffeic acid, and rosavin of cinnamic alcohol). The studies considered in this review did not relate information on *R. rosea* to the aglycones presumably formed on hydrolysis of the glycosides. The contribution of these metabolites to the activities of *R. rosea* is worthy of further evaluation.

11.8 CONCLUSION

Although *R. rosea* appears to have an excellent safety profile when administered alone, its potential for dangerous interactions with drugs has only begun to be explored. *In vitro* studies and a case report indicate the potential for *R. rosea* to cause toxicity of some drugs that are CYP 3A4 and/or MAO substrates, but many other potential mechanisms of interaction remain unexplored. Much more study is required before *R. rosea* can be considered safe.

REFERENCES

Abidov, M., Crendal, F., Grachev, S., Seifulla, R., and Ziegenfuss, T. 2003. Effect of extracts from *Rhodiola rosea Rhodiola crenulata* (Crassulaceae) roots on ATP content in mitochondria of skeletal muscles. *Bull. Exp. Biol. Med.* 136(6): 585–587.

Abidov, M., Grachev, S., Seifulla, R. D., and Ziegenfuss, T. N. 2004. Extract of *Rhodiola rosea* radix reduces the level of C-reactive protein and creatinine kinase in the blood. *Bull. Exp. Biol. Med.* 138(1): 63–64.

Abrahamyan, H., Hovhannisyan, A., Panossian, A., and Gabrielyan, E. 2005. The bioavailability of salidroside and rosavin, active principles of *Rhodiola rosea* extract SHR-5 in rats. *Med. Sci. Armenia.* 45(1): 24–29.

Ahumada, F. 1991. Effect of certain adaptogenic plant extracts on drug-induced narcosis in female and male mice. *Phytother. Res.* 5(1): 29–31.

Azizov, A. P. and Seifulla, R. D. 1998. The effect of elton, leveton, fitoton and adapton on the work capacity of experimental animals. *Eksp. Klin. Farmakol.* 61(3): 61–63.

Blomkvist, J., Taube, A., and Larhammar, D. 2009. Perspective on Roseroot (*Rhodiola rosea*) studies. *Planta Med.* 75(11): 1187–1190.

Bocharova, O. A., Matveev, B. P., Baryshnikov, A. I., Figurin, K. M., Serebriakova, R. V., and Bodrova, N. B. 1995. The effect of a *Rhodiola rosea* extract on the incidence of recurrences of a superficial bladder cancer (experimental clinical research). *Urol. Nefrol. (Mosk)* (2): 46–47.

Brown, R. P., Gerbarg, P. L., and Ramazanov, Z. 2002. *Rhodiola rosea*: A phytomedicinal overview. *Herbalgram.* 56: 40–52.

De Bock, K., Eijnde, B. O., Ramaekers, M., and Hespel, P. 2004. Acute *Rhodiola rosea* intake can improve endurance exercise performance. *Int. J. Sport Nutr. Exerc. Metab.* 14(3): 298–307.

Dement'eva, L. A. and Iaremenko, K. V. 1987. Effect of a *Rhodiola* extract on the tumor process in an experiment. *Vopr. Onkol.* 33(7): 57–60.

De Sanctis, R., De Bellis, R., Scesa, C., Mancini, U., Cucchiarini, L., and Dacha, M. 2004. *In vitro* protective effect of *Rhodiola rosea* extract against hypochlorous acid-induced oxidative damage in human erythrocytes. *Biofactors.* 20(3): 147–159.

Fintelmann, V. and Gruenwald, J. 2007. Efficacy and tolerability of a *Rhodiola rosea* extract in adults with physical and cognitive deficiencies. *Adv. Ther.* 24(4): 929–939.

Hellum, B. H., Tosse, A., Hoybakk, K., Thomsen, M., Rohloff, J., and Nilsen, G. O. 2010. Potent *in vitro* inhibition of CYP3A4 and P-glycoprotein by *Rhodiola rosea*. *Planta Med.* 76(4): 331–338.

Hung, S. K., Perry, R., and Ernst, E. 2011. The effectiveness and efficacy of *Rhodiola rosea* L.: A systematic review of randomized clinical trials. *Phytomedicine.* 18(4): 235–244.

Ishaque, S., Shamseer, L., Bukutu, C., and Vohra, S. 2012. *Rhodiola rosea* for physical and mental fatigue: A systematic review. *BMC Complement Altern. Med.* 12(1):70.

Kelly, G. S. 2001. *Rhodiola rosea*: A possible plant adaptogen. *Altern. Med. Rev.* 6(3): 293–302.

Kim, S. H., Hyun, S. H., and Choung, S. Y. 2006. Antioxidative effects of *Cinnamomi cassiae Rhodiola rosea* extracts in liver of diabetic mice. *Biofactors.* 26(3): 209–219.

Kormosh, N., Laktionov, K., and Antoshechkina, M. 2006. Effect of a combination of extract from several plants on cell-mediated and humoral immunity of patients with advanced ovarian cancer. *Phytother. Res.* 20(5): 424–425.

Kurkin, V. A. and Zapesochnaya, G. G. 1985. Chemical composition and pharmacological characteristics of *Rhodiola rosea*. *J. Med. Plants.* 10: 1231–1245.

Kwon, Y. I., Jang, H. D., and Shetty, K. 2006. Evaluation of *Rhodiola crenulata Rhodiola rosea* for management of type II diabetes and hypertension. *Asia Pac. J. Clin. Nutr.* 15(3): 425–432.

Li, Z.-H., Zhu, S.-Y. and Du, G.-H. 2006. Comparison of the pharmacokinetics of salidroside and salidroside in the extracts of *Rhodiola rosea* L. in rats. *Asian J. Pharmacodyn. Pharmacokinet.* 6(2): 224–226.

Lishmanov, I. B., Maslova, L. V., Maslov, L. N., and Dan'shina, E. N. 1993. The anti-arrhythmia effect of *Rhodiola rosea* and its possible mechanism. *Biull. Eksp. Biol. Med.* 116(8): 175–176.

Lishmanov, I. B, Naumova, A. V., Afanas'ev, S. A., and Maslov, L. N. 1997. Contribution of the opioid system to realization of inotropic effects of *Rhodiola rosea* extracts in ischemic and reperfusion heart damage *in vitro*. *Eksp. Klin. Farmakol.* 60(3): 34–36.

Maimeskulova, L. A. and Maslov, L. N. 1998. The anti-arrhythmia action of an extract of *Rhodiola rosea* and of n-tyrosol in models of experimental arrhythmias. *Eksp. Klin. Farmakol.* 61(2): 37–40.

Maimeskulova, L. A. and Maslov, L. N. 2000. Anti-arrhythmic effect of phytoadaptogens. *Eksp. Klin. Farmakol.* 63(4): 29–31.

Maimeskulova, L. A., Maslov, L. N., Lishmanov, IuB, and Krasnov, E. A. 1997. The participation of the mu-, delta- and kappa-opioid receptors in the realization of the anti-arrhythmia effect of *Rhodiola rosea*. *Eksp. Klin. Farmakol.* 60(1): 38–39.

McGovern, E. and McDonnell, T.J. 2010. Herbal medicine—sets the heart racing! *Ir. Med. J.* 103(7): 219.

Ming, D. S., Hillhouse, B. J., Guns, E. S., Eberding, A., Xie, S., Vimalanathan, S., and Towers, G. H. 2005. Bioactive compounds from *Rhodiola rosea* (Crassulaceae). *Phytother. Res.* 19(9): 740–743.

Narimanian, M., Badalyan, M., Panosyan, V., Gabrielyan, E., Panossian, A., Wikman, G., and Wagner, H. 2005. Impact of Chisan (ADAPT-232) on the quality-of-life and its efficacy as an adjuvant in the treatment of acute non-specific pneumonia. *Phytomedicine.* 12(10): 723–729.

Panossian, A., Hovhannisyan, A., Abrahamyan, H., Gabrielyan, E., and Wikman, G. 2009. Pharmacokinetic and pharmacodynamic study of interaction of *Rhodiola rosea* SHR-5 extract with warfarin and theophylline in rats. *Phytother. Res.* 23(3): 351–357.

Panossian, A., Hovhannisyan, A., Abrahamyan, H., Gabrielyan, E., and Wikman, G. 2010a. Pharmacokinetics of active constituents of *Rhodiola rosea* SHR-5 extract. In *Comprehensive Bioactive Natural Products—Efficacy, Safety & Clinical Evaluation I*, Vol. 2. Ed. Gupta, V. K., pp. 307–329. Stadium Press LLC, Texas, USA.

Panossian, A., Wikman, G., and Sarris, J. 2010b. Rosenroot (*Rhodiola rosea*): Traditional use, chemical composition, pharmacology and clinical efficacy. *Phytomedicine.* 17(7): 481–493.

Perfumi, M. and Mattioli, L. 2007. Adaptogenic and central nervous system effects of single doses of 3% rosavin and 1% salidroside *Rhodiola rosea* L. extract in mice. *Phytother. Res.* 21(1): 37–43.

Petkov, V. D., Yonkov, D., Mosharoff, A., Kambourova, T., Alova, L., Petkov, V. V., and Todorov, I. 1986. Effects of alcohol aqueous extract from *Rhodiola rosea* L. roots on learning and memory. *Acta. Physiol. Pharmacol. Bulg.* 12(1): 3–16.

Rhodiola (Rhodiola rosea). 2008. Natural Standards Monograph.

Saratikov, A. S., Krasnov, E. A., Chnikina, L. A., et al. 1968. Rhodiolosid, a new glycoside from *Rhodiola rosea* and its pharmacological properties. *Pharmazie.* 23(7): 392–395.

Scott, I. M., Leduc, R. I., Burt, A. J., Marles, R. J., Arnason, J. T., and Foster, B. C. 2006. The inhibition of human cytochrome P450 by ethanol extracts of North American botanicals. *Pharm. Biol.* 44(5): 315–327.

Semple, H. A. 2010. Toxicology studies on *Rhodiola rosea* extract. *Pharm. Biol.* 48(S1): 25–32.

Shevtsov, V. A., Zholus, B. I., Shervarly, V. I., et al. 2003. A randomized trial of two different doses of a SHR-5 *Rhodiola rosea* extract versus placebo and control of capacity for mental work. *Phytomedicine.* 10(2–3): 95–105.

Udintsev, S. N., Krylova, S. G., and Fomina, T. I. 1992. The enhancement of the efficacy of adriamycin by using hepatoprotectors of plant origin in metastases of Ehrlich's adeno-carcinoma to the liver in mice. *Vopr. Onkol.* 38(10): 1217–1222.

Udintsev, S. N. and Shakhov, V. P. 1991. The role of humoral factors of regenerating liver in the development of experimental tumors and the effect of *Rhodiola rosea* extract on this process. *Neoplasma.* 38(3): 323–331.

Ulbricht, C., Chao, W., Tanguay-Colucci, S., et al. 2011. *Rhodiola* Rhodiola spp.: An evidence-based systematic review by the natural standard research collaboration. *Alternat. Complement. Ther.* 17(2): 110–119.

van Diermen, D., Marston, A., Bravo, J., Reist, M., Carrupt, P. A., and Hostettmann, K. 2009. Monoamine oxidase inhibition by *Rhodiola rosea* L. roots. *J. Ethnopharmacol.* 122(2): 397–401.

Walker, T. B., Altobelli, S. A., Caprihan, A., and Robergs, R. A. 2007. Failure of *Rhodiola rosea* to alter skeletal muscle phosphate kinetics in trained men. *Metabolism.* 56(8): 1111–1117.

Wing, S. L., Askew, E. W., Luetkemeier, M. J., Ryujin, D. T., Kamimori, G. H., and Grissom, C. K. 2003. Lack of effect of *Rhodiola* or oxygenated water supplementation on hypoxemia and oxidative stress. *Wilderness Environ. Med.* 14(1): 9–16.

Zhang, Z. H., Liu, J. S., Chu, J. N., et al. 2005. The effect of Hongjingtian (Gadol) injection on cardiac hemodynamics and myocardial oxygen consumption of dogs. *Zhongguo. Zhong. Yao Za Zhi.* 30(13): 1001–1005.

Zhu, J., Wan, X., Zhu, Y., Ma, X., Zheng, Y., and Zhang, T. 2010. Evaluation of salidroside *in vitro* and *in vivo* genotoxicity. *Drug Chem. Toxicol.* 33(2): 220–226.

12 Commercialization of *Rhodiola rosea*

Nav Sharma and Raimar Loebenberg

CONTENTS

12.1 Introduction and Background 265
12.2 Commercialization of New Products...266
12.3 Commercialization of Natural Health Products 267
    12.3.1 Natural Health Product Development...267
    12.3.2 Regulatory Framework for Natural Health Products 268
    12.3.3 Natural Product Number..269
12.4 *Rhodiola rosea*: Product Development..269
    12.4.1 *Rhodiola rosea*: Roots to Extract 269
    12.4.2 Processing of the *Rhodiola* Roots 270
    12.4.3 Quality Control Testing of the Dried Roots 270
    12.4.4 Development of *Rhodiola rosea* Powdered Extract........................270
    12.4.5 Active Ingredients in *Rhodiola rosea*...271
    12.4.6 Development of the Final Dosage Form 271
12.5 Specifications and Quality Control for the Finished Product.....................271
    12.5.1 Results of Analysis Conducted on Final Capsules Manufactured
            at the University of Alberta 271
    12.5.2 *In Vitro* Release Test..272
12.6 Commercialization of *Rhodiola Rosea*—Field to Medicine Cabinet 272
12.7 Conclusion 273
References...273

12.1 INTRODUCTION AND BACKGROUND

Rhodiola rosea is a biennial subarctic plant that is native to northern areas of Russia, Scandinavia, and eastern Canada. In Canada, the plant normally grows along the Newfoundland coastline, Northern Québec, and Nunavut. One of the species, *Rhodiola intergrifolia*, also grows in the Rocky Mountains in Alberta. *R. rosea* can be used in many forms: fresh root, dried root, and extracts. It can be used in tea infusion, single-ingredient supplement in capsules, or directly in an energy drink, bar, or powder for reconstitution. The extraction process starts with harvesting, washing, and drying the roots followed by several physico-chemical processes such as maceration, filtration, vacuum evaporation, spray drying and analytical testing to result into development of the standardized extract. Increasing worldwide demand has led to the production of *R. rosea* as a medicinal crop. The fact that the Rhodiola plant can grow in the arctic region and is frost resistant makes it suitable for cultivation in Alberta and other cold regions of Canada.

12.2 COMMERCIALIZATION OF NEW PRODUCTS

Commercialization of a new product or technology consists of several technical, market, and business development processes that finally lead to the introduction of the product into the marketplace. In case of *R. rosea*, the path to commercialization in Alberta was made easier due to the involvement of companies from the initial stages of product development. What is unique about the commercialization of *R. rosea* in Alberta is the collaboration between public organizations, university centers and departments and industry partners to develop a product with benefits to agriculture and industry sectors (Nebraska Business Development Centre 2014).

The commercialization pathway is represented by various models; as the Goldsmith model. This model divides activities related to commercialization into three domains: technical, market, and business. The technical development of a product/technology consists of technology analysis followed by feasibility studies and development of a prototype. Once the prototype has been developed, production is usually feasible although numerous challenges may appear at the scale-up/commercial production stage, depending on the product.

One of the first steps is to conduct the market assessment to ensure that there is demand for the product and a clear path to profit exists. This is followed by the technology/product development stages. As depicted in the model, the initial stages of market and business development include development of an elaborate business plan with details on a market study, strategic marketing plan, economic feasibility, details on plans to conduct market validation, and sales and distribution. The processes in the market, business, and technical domains are usually iterative and do not follow a clear order, but it is advisable to involve business experts in the initial stages unless there is a clear indication of interest from a commercial partner.

Commercialization of a product/technology can be achieved through various options such as licensing, creation of a start-up company, corporate partnering etc. In some cases, the licensing of technology is one of the more feasible options especially when the path to market is not clear and the costs of product development and risks are higher. This is usually the case with commercialization of drugs and medical devices that face several barriers such as, more strict regulatory process; higher product development costs; and longer path to market. In some cases, strategic partnerships and alliances with investors/corporations at various stages of development may be possible. In such cases, the investors require a clear exit strategy, which may be possible if the risks involved are lower, such as in the case of natural health products (NHPs). Commercialization of *R. rosea*, in Alberta was relatively smooth due to the collaboration between corporate entities, public organizations and research institutions and an existing demand.

12.3 COMMERCIALIZATION OF NATURAL HEALTH PRODUCTS

Commercialization pathway for NHPs proceeds through various steps, including feasibility analysis, analytical development, product development and refinement, regulatory approvals and finally the introduction of the product into the market. The actual strategy adopted for the commercialization of the product depends on the product itself and its intended application and labeled use. The commercialization of a product with an existing application that is in accordance with the traditional literature is much simpler than the commercialization of a new product with a novel application or a new label claims that is not mentioned in the traditional literature. In many cases, depending on the label claim and the product itself, the regulatory process may necessitate clinical trials to provide a proof for the safety and the efficacy of the product. Once the barriers in product development and regulatory approval have been cleared, the next step is the marketing of the product and establishing a sound business model. Marketing and business development require establishing a business entity and/or exploring other means of commercialization, such as licensing the product to an established company. In some cases, the codevelopment of the product by an established entity, such as a company with a market presence, and a researcher or even a start-up organization is an option worth exploring. In the case of *R. rosea*, the project team led by Alberta Agriculture and Rural Development and consisting of Alberta Rhodiola Growers Organization, University of Alberta, and other Researchers successfully established a complementary relationship with an NHPs companies leading to successful commercialization of the *R. rosea,* and ensuring a steady supply chain for the *R. rosea* growers in Alberta.

12.3.1 NATURAL HEALTH PRODUCT DEVELOPMENT

Development of most NHPs consists of separation of the plant part/organism containing the active ingredient, extraction, purification, and standardization followed by formulation into the required dosage form. Development of the final product, analytical development and development of the regulatory strategy happen simultaneously. The product development is accompanied by the required quality control and testing to ensure that the process and the final product meet the required Health Canada standards. The quality control includes testing for the targeted biomarkers at various stages of extraction, purification, and standardization before and after formulation. The tests also include microbiological analysis and contaminants such as pesticides and heavy metals. Once the active ingredient from the relevant plant part/organism has been extracted, purified, and standardized, it has to be formulated into a dosage form that is suitable for the intended use. The choice of formulation ranges from a solid dosage form such as tablet, capsule, or powder to a specialized dosage form such as ointment for topical application or tonic for oral use. The Health Canada guidance documents recommend the type of quality control measures that are required for each dosage form and have to be fulfilled to procure the product license (indicated by the Natural Product Number [NPN]).

NHPs face various challenges ranging from product homogeneity and standardization to maintaining the potency of the active ingredients (biomarkers) in the formulation process and their characterization in the intermediate and final products. Unlike drug development, where a drug product is built around an active compound. In addition, the therapeutic effects of most NHPs are not typically attributable to a single compound. Instead, a plethora of compounds, called active constituents, may be responsible for an NHP's effect.

Section 12.4 describes the various stages in the development of a *R. rosea* extract in a solid dosage form (capsule). The process for production of *R. rosea* capsules consisted of harvesting of roots (rhizomes), processing them into coarse pieces, and extraction of compounds of interest from the Rhodiola root powder. Quality control tests were performed at each stage to ensure adherence to pharmacopoeial and other required standards, where applicable. Quality control tests performed on the extract and dried roots indicate that the product complies with the prescribed quality standards in terms of microbial and chemical purity and the indicated quantity/potency of the active compounds.

12.3.2 REGULATORY FRAMEWORK FOR NATURAL HEALTH PRODUCTS

In Canada, food and NHPs are governed by the Food and Drugs Act and Regulations and Natural Health Product Regulations. Health Canada is the federal agency that oversees these regulations and is primarily responsible for "helping Canadians to maintain and improve their health, while respecting their individual choices and circumstances" (Health Canada 2013). The Natural Health Products Directorate (NHPD), a part of the Health Products and food Branch at Health Canada, was established in 1999 with the mandate to *"ensure that all Canadians have ready access to natural health products that are safe, effective, and of high quality, while respecting freedom of choice and philosophical and cultural diversity."* NHPD implemented the regulatory framework for NHPs in 2004 to fulfill that mandate.

According to the Natural Health Products Regulations (Canada Gazette, Part I, December 22, 2001, p. 4939), "natural health product" refers to a substance or combination of substances, a homeopathic preparation, or a traditional medicine that is manufactured, sold, or represented for use in (Natural Health Products Regulations 2013):

- The diagnosis, treatment, mitigation, or prevention of a disease, disorder, or abnormal physical state or its symptoms in humans.
- Restoring or correcting organic functions in humans
- Maintaining or promoting health or otherwise modifying organic functions in humans

The NHP regulations affect manufacturers, packagers, labelers, distributors, and importers of NHPs. The NHPs that are intended solely for export purposes are unaffected by the regulations. Persons growing and/or performing postharvest operations (e.g., cleaning, drying, and/or freezing) are also unaffected by the regulations.

However, there is a realization that postharvest operations do have an imminent impact on the quality and standards of the NHPs, and therefore the operators do tend to follow operational procedures that can reduce the chances of microbial contamination in the end product while maintaining the required quantities of the active ingredients that are responsible for the intended beneficial effects. Standardization of the operational procedures also increases the efficiency of the process, and therefore postharvest operators tend to follow those standards.

The major components of the NHP regulations pertain to site licensing, good manufacturing practices (GMPs), product licensing, packaging/labeling, clinical trials and adverse reaction reporting. The provision of a site license for NHPs is contingent upon demonstration of compliance with the GMP standards by the site and mentiones the activities that the site may undertake. A site license is another mechanism to ensure that the site meets the quality standards and assures consumers of the quality and safety of NHPs.

12.3.3 Natural Product Number

In order to be legally sold in Canada, NHPs require a product license. To obtain a product license, detailed information on: medicinal ingredients; source; potency; nonmedicinal ingredients; recommended use and purpose (i.e., health claims); labeling content; safety and efficacy evidence; risks associated with the product (i.e., cautions, warnings, contraindications and known adverse reactions); and quality of the finished product must be submitted to the NHPD. The approved NHPs receive a Natural Product Number and bear the letters "NPN" followed by an eight-digit number on the product label. For homeopathic medicines (HM), the letters DIN-HM are followed by an eight-digit number. The acronym DIN stands for Drug Identification Number and applies to drugs. The process to obtain a product license is generally easier for products that are already in the market with a traditional label claim, i.e., claim that is supported by traditional monographs and traditional usage documentation in Canada or other jurisdictions. The NHPs that aim to have a label claim that is different from the traditional use claim have to provide evidence that supports (1) the desired claim (efficacy of the medicinal ingredients in the proposed dose) either through a clinical study or through studies that have been conducted in the past and (2) safety of the medicinal ingredients in the proposed dose. In addition to the product license requirement, the Food and Drug Act and regulations also require the NHPs to be manufactured in a GMP compliant facility. This also applies to the products for the clinical study.

12.4 *RHODIOLA ROSEA*: PRODUCT DEVELOPMENT

12.4.1 *Rhodiola rosea*: Roots to Extract

Most active ingredients that are responsible for the beneficial applications of *R. rosea* are contained in the roots. Therefore, treatment of the roots before processing for extraction is a very important step in the product development of *R. rosea*.

12.4.2   PROCESSING OF THE *RHODIOLA* ROOTS

The Rhodiola roots are stored at an average temperature of 4°C. It is important to store roots at that temperature to maintain the quality of the extract and prevent any contamination of the plant parts. A standardized procedure is followed for receiving and storage of the Rhodiola roots to ensure high quality of the raw material.

Processing starts with washing of the roots to effectively remove any foreign materials and contaminants. The roots are soaked in water and the washing is repeated to ensure that foreign material is completely removed from the roots. After washing, the roots are air-dried for about 15 minutes at room temperature and then transferred to a slicer. The sliced roots are dried in a tray dryer at 60°C to a final moisture content of less than 10%. The dried roots are cooled and packaged in plastic bags and stored for further processing.

12.4.3   QUALITY CONTROL TESTING OF THE DRIED ROOTS

The dried roots of *R. rosea* were tested for the quality and content of the salidrosides and rosavins (rosarin, rosavin, and rosin). Rosavins and salidrosides are considered to be the standard biomarkers and active ingredients for the Rhodiola plant. The dried roots were also tested for microbial content including Aerobic Plate Count, Mold Count, Yeast Count, *Escherichia coli*, *Salmonella,*and *Staphylococcus aureus.* In addition to the microbial testing, the dried roots were also tested for heavy metals (Arsenic, Cadmium, Lead, and Mercury) and for various pesticides recommended by Health Canada. The pesticide tests for included fenitrothion, fenvalerate, fonofos, heptachlor + heptachlor epoxide, hexachlorobenzene, hexachlorocyclohexane, isomers, lindane (gamma BHC), malathion, methidation, parathion-ethyl, parathion-methyl, permethrin, phosalone, piperonyl butoxide, pirimiphos-methyl, pyrethrins (total), and quintozene (total).

12.4.4   DEVELOPMENT OF *RHODIOLA ROSEA* POWDERED EXTRACT

The final stage of the processing is the development of the powdered extract that can be formulated into a solid dosage form. The production of the extract is a very critical process as it should have the required amount of rosavins and salidrosides and in the right proportion.

The flowchart below explains the various processes involved in the preparation of the *R. rosea* powdered extract from the dried roots of *R. rosea*. The cleaned dried roots were macerated for 24 hours in 75% USP grade grain alcohol. The alcohol was then vacuum evaporated and the resulting filtrate was spray dried followed by grinding, sifting, and homogenizing. The extract was then subjected to high-performance liquid chromatography (HPLC) analysis and the potency of rosavins was adjusted to 3.5%. The powder extract was also analyzed for pesticide and microbial content to ensure conformity with Health Canada standards.

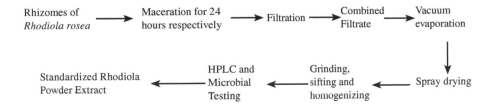

12.4.5 ACTIVE INGREDIENTS IN *RHODIOLA ROSEA*

The main ingredients of *R. rosea* extract are salidroside, rosarin, rosavin, rosin, rosiridin and tyrosol. These ingredients were characterized for the Rhodiola extract and their quantities determined for the powder extract. While it is currently unclear which specific compound(s) in *R. rosea* are active constituents, most preparations of *R. rosea* are standardized to specific levels of the marker compound rosavin, salidroside, or both.

The details on the specifications and quality control methods for the active ingredients are mentioned in Section 12.5.1.

12.4.6 DEVELOPMENT OF THE FINAL DOSAGE FORM

The production of the final dosage form (capsules) for *R. rosea* was performed at the University of Alberta's Drug Development and Innovation Center (DDIC). The figure below presents the various stages in the manufacturing of the Rhodiola capsules. The Rhodiola capsules produced were then subjected to quality control tests for content, microbiology, and *in vitro* release.

| Milling and mixing of *Rhodiola rosea* extract | Mixing with inert excipients | Encapsulation of the mixture | Packaging of capsules |

12.5 SPECIFICATIONS AND QUALITY CONTROL FOR THE FINISHED PRODUCT

12.5.1 RESULTS OF ANALYSIS CONDUCTED ON FINAL CAPSULES MANUFACTURED AT THE UNIVERSITY OF ALBERTA

Rhodiola and placebo capsules were tested for conformance to pharmacopoeial standards in terms of the microbiological content, including Total Plate count, Yeast enumeration, Mold enumeration, *Staphylococcus aureus, E. coli*, and *Salmonella* as per the USP 31<61> method of analysis. The results of the analysis performed by an Edmonton-based analytical laboratory indicated that the Rhodiola as well as the placebo capsules conformed to the quality standards.

The weight variation tests on the Rhodiola and placebo capsules were conducted by the University of Alberta's Drug Development and Innovation Center (DDIC). The results indicated that both the Rhodiola and the placebo capsules conformed to the quality standards.

12.5.2 IN VITRO RELEASE TEST

In vitro performance tests were performed on the Rhodiola and placebo capsules by the DDIC. The disintegration test was used to determine the release behavior of the ingredients from the capsule formulation. Rhodiola and placebo capsules conformed to the quality standards as per the United States Pharmacopoeia for satisfactory release of the ingredients from the dosage form.

12.6 COMMERCIALIZATION OF *RHODIOLA ROSEA*—FIELD TO MEDICINE CABINET

In Canada, approximately 70% of the population has used natural health products for various purposes. *R. rosea* is an adaptogen and can be cultivated in various parts of Alberta and this presented an opportunity to the Alberta farmers to become suppliers of the medicinal plant. This opportunity was explored by Alberta Natural Health and Agriculture Network (ANHAN), a not-for profit network of Alberta growers interested in growing crops for rural economic development and to provide healthy products to consumers, and Alberta Agriculture and Rural Development (ARD) in 2004. Early on the need to optimize the growing conditions to produce high yields of bioactive compound was realized, and ANHAN needed research to achieve this goal. The other barrier was the lack of a commercialization model for a medicinal crop grown in Alberta to provide the link to NHP companies, and necessary information on value transfer from commercial product to farmers. Hence there was a need to develop a complete value chain for *R. rosea* to demonstrate the value to farmers.

The Institute for Food and Agricultural Sciences Alberta (IFASA) was created in 2004 by three stakeholders - University of Alberta, Alberta Agriculture and Rural Development (ARD) and Alberta Research Council (ARC). The creation of IFASA provided the framework that would allow development of *R. rosea* as a crop, extract the bioactive compounds, create consumer product prototypes, obtain regulatory licenses, and commercialize the product in the marketplace.

ANHAN in collaboration with IFASA launched a promotional campaign in various parts of Alberta and actively recruited growers to the project. This campaign to promote *Rhodiola* as a crop led to about 160 farmers joining the Alberta *Rhodiola Rosea* Growers Organization (ARRGO).

A provincial venture capital organization and a member of the Alberta Funding Consortium, AVAC Ltd., provided initial funding to the project based on "strong scientific leadership." As the partnership became established and demonstrated promising results, other provincial and national funding organizations provided additional funding to bring the total cash contributions to $1.37 million.

One of the most important aspects of the commercialization process is the product's access to market through a pharmaceutical/natural health products company.

Pharmaceutical companies expressed interest in the project as a result of the coverage that the project received in the media. ARRGO developed a collaborative relationship with a German company and entered into a four-year contract to provide *R. rosea* crop to the company.

R. rosea commercialization project had several positive outcomes including, creation of a collaborative multidisciplinary team with members ranging from crop scientists to clinicians; enabling the farmers and community members to diversify their income source; diversification of the rural economy through creation of the processing facility (in Thorsby) and establishment of a steady supply chain; and creation of alliances and partnerships with industry, farmers, researchers and international stakeholders that can lead to more success stories in future.

The *Rhodiola rosea* Commercialization Project has been recognized for excellence in innovation and project management by a Gold Premier's Award of Excellence (June 2010), a Canada Award for Excellence/National Quality Institute (July 2010), and Alberta Science and Technology Foundation honoree (November 2010), and President's Award "Connecting Communities" (2012).

12.7 CONCLUSION

The *Rhodiola rosea* project had several positive outcomes including creation of a framework that can lead to better agronomic practices, postharvest handling, bioactive extraction, product development, toxicology, pharmacology, clinical trial development, regulatory license process, marketing, and value chain development. In coming years the success of *R. rosea* growers will inspire and motivate more farmers across Alberta to adopt *R. rosea* as one of the high value crops. The growers will have to embrace advanced production systems to stay competitive in the international markets. Overall, the demand for *R. rosea* crop and its value to farmers will depend on the global demand for the natural health product(s) containing the plant extracts. Expertise of researchers at Alberta Agriculture and Rural Development can help with the development of crop production practices that lower the cost of production, increase yield, and improve the quality of the active compounds, both to strengthen the global competitiveness of ARRGO and to increase the profits for ARRGO members. The *R. rosea* commercialization project is an example of the role that different academic/ research, public and not for profit organizations can play in the commercialization of such a crop. One of the most important reasons for the success of commercialization of the *R. rosea* was the collaboration with international companies with an established access to markets. These networks and linkages developed in the *R. rosea* project can be leveraged for future projects and therefore need to be strengthened.

REFERENCES

Health Canada, 2013. About Health Canada. http://www.hc-sc.gc.ca/ahc-asc/index-eng.php.
Natural Health Products Regulations, 2013. http://laws-lois.justice.gc.ca/eng/regulations/SOR-2003-196/page-1.html#h-1.
Nebraska Business Development Centre, 2014. http://nbdc.unomaha.edu/technology-commercialization/techventure/home.cfm; accessed on August 14, 2014.

Index

A

Active constituents, 268
Active ingredients in *Rhodiola rosea*, 267, 269, 271
Acute mountain sickness (AMS), 236–237
Acute toxicity study, 255
AD, *see* Alzheimer's disease
ADAPT-232, 232, 234
Adapters, used in AFLP analyses, 15
Adaptogens, 189
   augment chemotherapy, 235–236
   pretreatment with, 219
ADHD, *see* Attention deficit disorder
Adjuvant therapy, 234
AdMax™, 196, 197
Adrenal hormones, 216
Adriamycin, 235
Adverse events, 234
Aerospace medicine
   aircrew stress and fatigue, 237–238
   AMS and hypoxia, 236–237
   jet lag, 237
   space exploration, 238–239
Affective mood disorders, 217
AFLP, *see* Amplified fragment length polymorphism
AFLP dendrogram of *Rhodiola rosea*, 22, 28
Aging-associated disorders, 217–219
Agricultural production, biotechnologies for, 56
Agrobacterium-mediated transformation, 184
Airborne spores of *Sclerotinia sclerotiorum*, 165
Aircrew stress, 237–238
Alaska, climate normal for, 127
Alberta Agriculture and Rural Development (ARD), 267, 272
Alberta, climate normal for, 127
Alberta Funding Consortium, AVAC Ltd., 272
Alberta-grown *Rhodiola rosea*, 255
Alberta Natural Health and Agriculture Network (ANHAN), 272
Alberta *Rhodiola Rosea* Growers Organization (ARRGO), 272, 273
Alpine countries, research on *Rhodiola rosea*, 96–97
Alternaria leaf spot, 163–164
Alzheimer's disease (AD), 193
Amenorrhea, 233
American ginseng, 54
AMOVA, *see* Analysis of molecular variance

Amplified fragment length polymorphism (AFLP), 95
   analyses, 15–17
   results, 18–21
   Rhodiola rosea samples for, 10–14
   technique, 5
AMS, *see* Acute mountain sickness
Amyloid-β oligomers, 76
Analysis of molecular variance (AMOVA), 17, 20–21
ANHAN, *see* Alberta Natural Health and Agriculture Network
Animals, experiments in, 214–217
Anti-apoptotic mechanisms, 216
Antibacterial effects of *Rhodiola rosea*, 234–235, 245
Anticancer effects of *Rhodiola rosea*, 197
Anticholinergic medications, 243
Antidepressant activity of *Rhodiola rosea*, 192–193
Antidepressant augmentation, 228
Antidepressant drug, 260
Antidepressant effects of SHR-5, 214, 215
Antidepressive drugs, pharmacological activity of, 214
Antidiabetic effects of *Rhodiola rosea*, 196
Anti-fatigue properties, 235
Antioxidant effects of *Rhodiola rosea*, 195
Antipsychotic medications, 243
Antistress (adaptogenic) effects of *Rhodiola rosea*, 194
Anxiety, 228–230, 244–245
Anxiolytic effects of *Rhodiola rosea*, 194–195
Aphids, 145–146
Appresoria, 161
Arctic Root®, 39, 205
ARD, *see* Alberta Agriculture and Rural Development
Argonautica, The, 226
ARRGO, *see* Alberta *Rhodiola Rosea* Growers Organization
Ascospores, 162
Asian medicinal systems, 42
Aster yellows, 165–169
   disease management of, 171
Asthenia, 242
Attention deficit disorder (ADHD), 231
Autofermentation, 93
Autonomic stress–response system, 228–229

B

"Barents herbs," 95
Bare soil production, 139
Bazedoxifene, 234
Beck Depression Index (BDI), 213
Benign prostatic hypertrophy, symptoms of, 242
Bioassay-guided fractionation of *Rhodiola rosea*
    roots, 78
Bioavailability of *Rhodiola rosea,* 197–198
Biocontrol agents, 170
Biological activities of *Rhodiola rosea,* 197
Biomass partitioning in *Rhodiola* plant, 127–128
Biotechnological methods of *Rhodiola rosea,*
    177–181
Biotechnologies, 50–51, 56
Black vine weevil, 147
Botrytis cinerea, 163
Broad-spectrum pharmacological activities, 190
Bulgaria, *Rhodiola rosea* research in, 98

C

CAD, *see* Cinnamyl alcohol dehydrogenase
Caffeic acid, 72
Callus cultures
    rosavins production in, 180–181
    salidroside production in, 178–179
Callus induction, 176–177
CAM, *see* Crassulacean acid metabolism
cAMP
    deregulation of, 212
    downregulation of, 211
cAMP-dependent protein kinase A, 211
Canada thistle, 142
Canadian park system, 54
Canadian (Nunavik) populations of *Rhodiola
    rosea,* 73–74
Cancer, 235–236
Carboxy-*o*-methyl transferase (COMT), 231
Carpathian countries, *Rhodiola rosea* research
    in, 98–99
Catecholamines, 214
CCAs, *see* Compact callus aggregates
CCR, *see* Cinnamoyl-CoA oxidoreductase
Cellular oxidative stress, 76
Central nervous system (CNS), *Rhodiola rosea*
    effects on, 192–195
Cercopidae, 147
Cerebral ischemic injury, 230
CETP gene, 219
Chemotherapy, 235–236
Chisan group, 234
Chlorogenic acid, 72
Cholesteryl ester transfer protein, 219
Chromatographic techniques, 197
Chronic/intermittent hypoxia, 237
Chronic stress, 238

Cinnamaldehyde, 76
Cinnamoyl-CoA oxidoreductase (CCR),
    182
Cinnamyl alcohol, 214
Cinnamyl alcohol dehydrogenase (CAD), 75,
    182
Cinnamyl alcohol glycosides, 180–181
    biosynthesis of, 181–182
Cinnamyl-CoA reductase (CCR), 76
Cisplatin, 236
Cleistothecia, 162
Climatic effects, on *Rhodiola* growth, 128–132
Clonal propagation, 55
CNS, *see* Central nervous system
CoA ligase (4CL), 76
Cognition, impairment of, 243
Cognitive function, 230–231
    Rhodiola rosea effects on, 193–194
Coleoptera, 147
Commercialization, *Rhodiola rosea,* 265–266
    field to medicine cabinet, 272–273
    of natural health products, 267–269
    of new products, 266
    product development, 269–271
    product, specifications and quality control
        for, 271–272
Committee on Herbal Medicinal Products
    (HMPC), assessment report, 100–101
Compact callus aggregates (CCAs), 177
    use of, 178–179
COMT, *see* Carboxy-*o*-methyl transferase
Conductive tissues, phloem and xylem, 127
Conidia, 162
Conservation of *Rhodiola* species, 45–46
    in new world, 51–54
    in old world, 46–51
Container systems, 139–140
Cortex, 127
Corticotropin-releasing factor (CRF), 194
Crassulaceae, 65, 89
Crassulacean acid metabolism (CAM), 132, 133
Crassulacean plants, 132–133
CRF, *see* Corticotropin-releasing factor
Critical period of weed competition, 142
Crop nutrition, 135–138
Crop rotation, 170
Crown rot, 159–161
Cultivation
    of medicinal plants, 54–56
    of *Rhodiola rosea,* 101–103
Curculionidae, 147
Cyanogenic glucosides, 69
Cyclophosphamide, 235, 236
CYP, *see* Cytochrome P450
CYP19, 254
CYP3A4, 241, 254, 260
CYP1A2 substrates, 259
CYP2C19, 254

Cytochrome P450 (CYP), 240, 254
    enzymes, 197
    metabolism, 198
Czech Republic, *Rhodiola rosea* research
        in, 99

D

DAF-16 nuclear transcription factor, 217
Dandelion seeds, 142
Dand jari, 39
Day length, *Rhodiola* growth, 128–130
DDIC, *see* Drug Development and Innovation
        Center
Dementia, cognitive function in, 231
Dentistry, 245
Depression, 213–214
    adjunctive treatment for, 239
    pathophysiology of, 192
    Rhodiola rosea for, 227–228
Dermatology, 245
Development of *Rhodiola rosea,* 126–128
Dhodlli, 38
Diagnostic and Statistical Manual
        of Mental Disorders,
        1994 (DSM-IV), 227
Diet-induced model of obesity, 193
Digoxin, 254
DIN, *see* Drug Identification Number
Diseases, 107
Distribution of *Rhodiola rosea,* 89
DNA extractions, 9
Dorvolson mugez, 38
Dosages of *Rhodiola rosea,* 239
Dose–response correlation, 214
Dried roots, quality control testing of, 270
Drug Development and Innovation Center
        (DDIC), 271, 272
Drug Identification Number (DIN), 269
Drying, postharvest, 147–149
Drying temperature, Root harvest, 112
Dryocoetes krivolutzkajae Mandelshtam, 146
DSM-IV, *see Diagnostic and Statistical*
        Manual of Mental Disorders,
        1994 (DSM-IV)

E

Ecophysiology, *Rhodiola rosea,* 132–135
Edmonton-based analytical laboratory, 271
Efficacy, data on
        nematode *Caenorhabditis elegans,*
                206–213
        stress-induced mental and behavioral
                disorders, 206
EFSA, *see* European Food Safety Authority
Encyclopedia of Traditional Chinese
        Medicines, 36

Environmental requirements for cultivation,
        27–28, 103
Epidermis, 127
Epinephrine, 234
ERα, *see* Estrogen receptor alpha
Essential oil of *Rhodiola rosea* roots, 73
Estonia, *Rhodiola rosea* research in, 100
Estrogen receptor alpha (ERα), 217, 219
    downregulation of, 215
Estrogen-sensitive breast cancer cells, 233
Ethnobotany of *Rhodiola* species
    Central Asia, *Rhodiola* species in, 36–39
    Rhodiola rosea and *Rhodiola integrifolia,*
            circumpolar, 39–44
    Rhodiola species in modern day, 44–45
Eurasia, 41–43
Europe, *Rhodiola rosea* status in, 100–101
European Food Safety Authority (EFSA), 193,
        205–206
Europe, roseroot, 90
Explants for *in vitro* culture, 174–175
Ex situ protection, 56

F

Fatigue, 242
    aircrew stress and, 237–238
    cancer-related, 235
    Rhodiola preparations on mental
            performance related to, 207–209
    stress and, 232
Fertility management, 135–138, 232–234
Fertilization, *Rhodiola rosea,* 103–104
Fertilizers, use and placement of, 137–138
Field choice, *Rhodiola rosea,* 103
Field experiment in Nunavik, 55
Final dosage form, development of, 271
Flavonoids, 68–69, 72
Flavonols, 78
Foliar diseases, management of, 170
Folkloric tradition, 44
Food and Drug Act, 269
4-coumarate-CoA ligase (4CL), 181–182
FOXO transcription factor, 216
Fusarium avenaceum, 159
Fusarium equiseti, 159
Fusarium oxysporum, 159
Fusarium solani, 159
Fusarium species, 159, 160

G

GABA$_A$-benzodiazepine receptor agonist, 194
GAD, *see* Generalized anxiety disorder
Gas chromatography (GC), 66
Gene expression profiling, 211
Generalized anxiety disorder (GAD), 228
Genes, *Rhodiola* effects on, 218

Genetic diveristy, 49–50
Genomic testing, 246
Germany, *Rhodiola rosea* research in, 97
Germination, 55
Glucocorticoid receptors (GR), 216
　　NPY-Hsp70-mediated effects on, 215
Glutamate (GLU)-induced cell, 230
Glycosides, 69
Glycosyltransferases (GTs), 183
Goldsmith model, 266
Gossypetin-7-*O*-L-rhamnopyranoside, 78
Government-sponsored space programs, 239
GPCR, *see* GTP-binding protein coupled
　　receptors
G-protein-coupled receptor (GPCR)-mediated
　　signal transduction, 211
G-protein coupled receptors, downregulation
　　of, 215
GR, *see* Glucocorticoid receptors
Gray mold, 163
Great Britain, *Rhodiola rosea* research in, 100
Growth of *Rhodiola rosea,* 126–128
GTP-binding protein coupled receptors (GPCR),
　　211, 212
GTs, *see* Glycosyltransferases

H

HACE, *see* High altitude cerebral edema
Hail storms, on *Rhodiola* growth, 131
Haisainai, 36
HAMD, *see* Hamilton Depression Scale
Hamilton Anxiety Rating Scale (HARS), 228
Hamilton Depression Scale (HAMD), 213, 227
HAPE, *see* High altitude pulmonary edema
HARS, *see* Hamilton Anxiety Rating Scale
Harvesting process, 107–108
Harvest time for *Rhodiola* plants cultivation,
　　143–145
Haustoria, 161
HDL, *see* High-density lipoproteins
Health Canada guidance, 267
Heat-shock-proteins, 217
Hemiptera, 147
Herbal Medicinal Products, 214
Herb–drug interactions, 197–198, 240–241
High altitude cerebral edema (HACE), 236
High altitude pulmonary edema (HAPE), 236
High altitude sickness, *see* Acute mountain
　　sickness (AMS)
High-density lipoproteins (HDL), 219
High-performance liquid chromatography
　　(HPLC), 66, 73
　　analysis, 270
Hippocampal neurons, 229
HM, *see* Homeopathic medicines
HMPC, *see* Committee on Herbal Medicinal
　　Products

Homeopathic medicines (HM), 269
Hormonal, 240
　　changes, 243
Hormone replacement therapy, 233
HPA, *see* Hypothalamic–pituitary–adrenal gland
HPLC, *see* High-performance liquid
　　chromatography
HPLC-UV-based technique, 74
Human clinical trials for *Rhodiola* rosea,
　　256–257
Human cortical neuron cell line (HCN-A), 230
Human sperm, quality of, 233
Hungary, *Rhodiola rosea* research in, 99–100
Hydroalcohol extract, 96
Hydrogen peroxide (H_2O_2), 230
Hydroxycinnamic acids, 72
Hyperlipidemia, 242–243
Hypothalamic–pituitary–adrenal gland
　　(HPA), 192
Hypothetical neuroendocrine mechanism, 216
Hypothetic molecular mechanisms, 211
Hypoxia, 236–237

I

IFASA, *see* Institute for Food and Agricultural
　　Sciences Alberta
IMBP, *see* Institute of Medical and Biological
　　Problems
Imigarmît, 44
Immune functions, 234–235
Infections, 234–235
Infectious diseases, *Rhodiola*
　　alternaria leaf spot, 163–164
　　aster yellows, 165–169
　　powdery mildew, 161–163
　　root rot and winter kill, 159–161
　　rusts, *Puccinia* species, 169
　　sclerotinia stem rot, 164–165
　　seed decay and seedling blight, 156–159
Infraspecific taxonomy, *Rhodiola rosea,* 25–26
INR, *see* International Normalizing Ratio
In situ protection, 56
Institute for Food and Agricultural Sciences
　　Alberta (IFASA), 272
Institute of Medical and Biological Problems
　　(IMBP), 238
Insulin resistance, 242–243
Internal transcribed spacer (ITS) regions
　　parsimony analysis, 25
　　Rhodiola rosea, 9, 15, 17–18
　　Rhodiola samples sequenced for, 6–7
International Normalizing Ratio (INR), 241
Interspecific sequence repeat (ISSR) marker, 22
Inuit communities, 54
In vitro cultures of *Rhodiola rosea,* 174–177
　　metabolite production in, 177–181
In vitro propagated plants, 138

In vitro release test, 272
In vitro Rhodiola rosea, 235
In vitro studies, 253–255
In vitro testing, 240
In vivo animal toxicity studies, 255–256
IP3, upregulation of, 212
Isolated cells, experiments in, 214–217
ISSR marker, *see* Interspecific sequence repeat
    marker
ITS regions, *see* Internal transcribed spacer regions
IUCN Red List, 46

J

Jet lag, 237

K

Kalvegror, 43

L

Labor, demand for, 140, 141
Land preparation, 138–139
LC-ESI-TOF analyses, 73
LDL, *see* Low-density lipoproteins
Leaf blight, *Botrytis cinerea,* 163
Leafhoppers, 169
Lecithin, 233
Life cycle of *Rhodiola rosea,* 101–103
Ligand-bound estrogen receptor, 219
Lipid plasma protein, 219
Low-density lipoproteins (LDL), 219

M

Mammalian target of rapamycin (mTOR)
    protein, 197
Mantel tests, 17, 20, 27, 28
MAO-A bioassay, 214
MAO inhibition, *see* Monoamine oxidase inhibition
Mass disasters, 244–245
Mass seedling production, 55
Materia Medica, 41
Mattmark, 104
Meadow spittlebug, 147
Medication interactions of *Rhodiola rosea,*
    239–241
Medicinal implications, 21–24
Medicinal plants, cultivation of, 54–56
Menopause, 232–234
Metabolic engineering of *Rhodiola rosea,*
    181–184
Metabolic enzyme inhibition, 253, 260
Metabolic profiling of *Rhodiola rosea,* 74
Methanol, 254
Micropropagated plants, adaptation of, 176
Micropropagation of *Rhodiola rosea,* 175

Mingleshi, 36
Moldova, *Rhodiola rosea* research in, 99
Mongolia, *Rhodiola rosea* research in, 100
Mongrhoside, 73
Monoamine oxidase (MAO) inhibition, 214, 254
    oxicological implications of, 255
Monoterpene glycoside rosiridin, 214
Monoterpenes, 69
Morphology of *Rhodiola rosea,* 126–128
Mosott, 41
mTOR protein, *see* Mammalian target of
    rapamycin protein

N

Natural health products (NHPs), 266
    development of, 267–268
    regulatory framework for, 268–269
Natural Product Number (NPN), 269
Neurodegenerative diseases, 193
Neuroendocrine system, impairments in, 214
Neurological side effects, 243–244
Neuropeptide Y (NPY), 215, 216
Neuroprotection, 230–231
Ng, *see Nisseria gonorrhoeae*
NHPs, *see* Natural health products
Nisseria gonorrhoeae (Ng), 235
Nitric oxide (NO), production of, 237
Non-estrogen-sensitive breast cancer cells, 233
Norepinephrine, 234
Norepinephrine-sensitive lipase, 242
North America, 43
    Rhodiola integrifolia, 43–44
    Rhodiola rosea, 8, 44
Northern Scandinavia, traditional knowledge of
    Rhodiola rosea, 89–90
Northern Urals, Russia, 90
NPN, *see* Natural Product Number
NPY, *see* Neuropeptide Y
Nuclear ribosomal ITS regions, *Rhodiola*
    samples sequenced for, 6–7
Nunavik (Québec), *Rhodiola rosea* in, 5
Nutrients, uptake and removal of, 137

O

Obligate plant-parasitic fungus, 161
OECD guidelines, *see* Organisation for
    Economic Cooperation and
    Development guidelines
Oligomeric/olymeric proanthocyanidins, 69
Organic fertilization, effect of, 104
Organic herbicides, 143
Organisation for Economic Cooperation and
    Development (OECD) guidelines, 255
Otiorhynchus sulcatus, 147
Oxidative stress, 76
Oxidative stress markers, 237

P

PAL, *see* Phenylalanine ammonia-lyase
"Parkinsonian" symptoms, 243
Parsimony analysis, 15
    of nuclear ribosomal ITS regions and *trn*L-F
        chloroplast regions, 25
PCR, *see* Polymerase chain reaction
PDE5I, *see* Phosphodiesterase-5 inhibitors
Perennial weeds, 142
Perilipin, 242
Perkin–Elmer GeneAmp PCR System 9700
        thermocycler (ABI), 16
Pesticide tests, 270
Pests of *Rhodiola*, 107, 145–147
P-glycoprotein (P-gp), 241, 254
Pharmacokinetics, 197–198
    pharmacodynamic interactions, 259–260
    properties, 220
    of *Rhodiola* active constituents, 257–259
Pharmacological activities of *Rhodiola rosea*,
        189–191
Phenolic acids, 68
Phenolic compounds, 72
Phenylalanine ammonia-lyase (PAL), 75, 181
Phenylethanoids, 67
Phenylethanol derivatives, biosynthesis of, 74–76
Phenylmethanoids, 67
Phenylpropanoid derivatives, biosynthesis of,
        74–76
Phenylpropanoids, 67, 149
Philaenus spumarius, 147
Phosphodiesterase-5 inhibitors (PDE5I), 232
Photoperiod, *Rhodiola rosea* effects on, 131, 132
Photosynthesis, 132–133
Phylogenetic tree for *Rhodiola* species, 26
Phylogeography, 27–28
Physical endurance, *Rhodiola rosea* effects on,
        195–196
Physico-chemical processes, 265
Phytochemical analyses of *Rhodiola rosea*, 23
Phytochemical constituents, *Rhodiola rosea*,
        66–73
Phytochemicals of *Rhodiola rosea*, bioactivity
        of, 76–78
Phytochrome, 129
Phytoplasma infection of *Rhodiola*, 166–169
Phytoplasmas, 166, 169
PI3K, upregulation of, 212
Piracetam, 231
Placebo capsules, 271, 272
Plasticulture, 139
PLCB1 gene, 212
Plus-maze test, 194
Poland, *Rhodiola rosea* research in, 98
Polymerase chain reaction (PCR) for *trn*L-F
        region, 9, 15
Pond snail *Lymnaea stagnalis*, 206–213

Population structure, 27–28
Porsolt's depression test, 214
Postdrying process, 112–113
Postharvest, 147–149
Post-stroke, cognitive function in, 231
Post traumatic stress disorder (PTSD), 228–230,
        244–245
Powdery mildew, 161–163, 170
Precipitation, 126
Predrying treatments, *Rhodiola rosea*, 109–112
Prescription medications, counteracting side
        effects of, 241–244
Primers, used in AFLP analyses, 15, 16
Production economics, postharvest, 149
Production systems, *Rhodiola rosea*, 138–140
Product/technology, technical development of,
        266
Propagation methods, *Rhodiola rosea*, 105–106,
        138
PTSD, *see* Post traumatic stress disorder
Puccinia species, 169
Pythium species, 157

Q

Qinghai-Tibetan plateau, 36
Quality control testing, 268
    of dried roots, 270

R

Radiation, 235–236
Raloxifene, 234
Reactive oxygen species (ROS)
        accumulation, 193
Research priorities, 149–150
Respiration, 133–134
Rhizoctonia crown rot, 160
Rhizome, anatomical structure, 126–127
Rhodalean-400 group, 242
Rhodioflavonoside, 78
Rhodiola algida, 39
Rhodiola crenulata, 36, 38, 45
Rhodiola fastigiata, 38
Rhodiola gelida, 39
Rhodiola heterodonta, 39
Rhodiola himalensis, 38
Rhodiola imbricata, 38, 39
Rhodiola integrifolia, 4, 25, 26, 41, 265
    in North America, 43–44
    synonyms for, 2
Rhodiola kirilowii, 39
Rhodiola pamiroalaica, 39
Rhodiola pinnatifida, 39
Rhodiola plants, 55
    biomass partitioning in, 127–128
Rhodiola quadrifida, 38, 45
Rhodiola rhodantha, 4, 25, 26

Rhodiola rosea powdered extract, development of, 270–271
Rhodiola rosea supply, 91–92
Rhodiola sachalinensis, 45, 92, 178–179
Rhodiola sacra, 45
Rhodiola scolytine, 146
Rhodiola semenovii, 39
Rhodiola SHR-5, 206, 211, 214, 217, 218, 259
    HPLC-profile of, 215
Rhodiola species, 155–156, 217
    beneficial effects of, 212
    capsules, 271, 272
    in Central Asia, 36–39
    conservation of, *see* Conservation of
        Rhodiola species
    cultivation, 54–56
    ethnobotany of, *see* Ethnobotany of *Rhodiola*
        species
    growth, climatic effects on, 128–132
    infectious diseases, *see* Infectious diseases,
        Rhodiola
    neuroprotective effect of, 219
    origin, distribution and description, 3–5
    pest of, 145–147
    preparations, effects of, 207–209
    roots, processing of, 270
Rhodiola–weed competition, 140–142
Rhodiola yunnanensis, 36
Rhodioloside, 70, 73
Rhododendron caucasicum, 242, 243
Rholo mukpo, 38
Romania, *Rhodiola rosea* research in, 98
Root division, 105
Root harvest
    harvesting process, 107–108
    male–female plants, 109
    postdrying process, 112–113
    predrying treatments, 109–112
    timing, 108–109
Root rot, 159–161
    disease management of, 169–170
ROS accumulation, *see* Reactive oxygen species
        accumulation
Rosarin
    drying air temperature on, 149
    variations in concentration, 145
Rosavin kinetics in rats, 258
Rosavins, 72, 78, 214, 220, 261, 270
    biosynthesis of, 76
    drying air temperature on, 149
    preliminary storage treatments, 111
    production in callus cultures, 180–181
    variations in concentration, 145
Roseroot, 39, 42–43, 46, 89
    cultivation, 54–55
Rosin, 76
    drying air temperature on, 149
    variations in concentration, 145

Rosiridin, 214
Rural economy, diversification of, 273
Russia
    research, plant size and weight, 92–93
    roseroot, 90–91
Rusts, *Puccinia* species, 169

S

Salidroside, 193–194, 246, 256–261, 270
    bioactivity of phytochemicals, 76–77
    biosynthesis of, 74–75, 182
    drying air temperature on, 149
    formation of, 183–184
    pharmacokinetic parameters of, 197, 220
    phytochemical constituents, 70–72
    preliminary storage treatments, 111
    production in callus cultures, 178–179
    variations in concentration, 145
Salidroside metabolite tyrosol, kinetics of, 258
SAPK, *see* Stress-activated protein kinases
Scandinavia, research on *Rhodiola rosea,* 94–96
Sclerotinia sclerotiorum, 164, 165, 170
Sclerotinia stem rot, 164–165
Secondary metabolites, biosynthesis of, 93
Sedum telephium, 129–130
Seed decay, 156–159
    disease management of, 169–170
Seed harvest, 114
Seeding methods, 138
Seedling age, 105–106
    and weeding system, 107
Seedling blight, 156–159
    disease management of, 169–170
Seedling damping-off, 156
Seed phase, 101
Seed propagation, 105
    plantation, 109
Selective estrogen receptor modulators (SERMS),
        233, 234
Selective serotonin reuptake inhibitor (SSRI),
        230, 243
SERMS, *see* Selective estrogen receptor
        modulators
Serotonin 5-HT3 GPCR, downregulation of, 213
Serotonin receptors, activation of, 213
SERPINI1 gene, upregulation of, 212
7- and 28-day repeated oral dose studies, 255
17-Ketosteroids, 233
Sexual dysfunction, 243
Sexual function, 232–234
SHR-5 proprietary, 205
Side effects of *Rhodiola rosea,* 239–241
Sildenafil (Viagra), 232
Slicing effect, on *Rhodiola rosea,* 111
Slovak Republic, *Rhodiola rosea* research in, 99
Sodium-dependent glucose transporter-1
        protein, 197

Soil pH, effect of, 136
Soils, 135–136
Somnolence, 242
Sow rigpa, 38
Space exploration, 238–239
Sports medicine, 244
Sprague–Dawley rats, 255
SSRI, *see* Selective serotonin reuptake inhibitor
Statistic software R, 17
Stem, anatomical structure, 126–127
Stem rot, 164–165
Steroid receptor function, 219
Streptozotocin-induced diabetic rats, 196
Stress, 228–230, 244–245
    cancer-related, 235
    and fatigue, 232
Stress-activated protein kinases (SAPK), 216, 219
Stress-induced catabolic transformations, 211
Stress–response system, 228, 231
Strict consensus tree, 18, 19
Suo-luo-ma-bu, 36
Swedish Herbal Institute, 232
Switzerland, *Rhodiola rosea* research in, 97
Sympatho-adrenal system, 192
Synthetic hormone treatments, 233

T

Tamoxifen, 234
Tantur, 39
Taxonomic distinction, 21–24
TBR branch, *see* Tree-bisection-reconnection
    branch
TDZ, *see* Thidiazuron
Temperatures on *Rhodiola* growth, 126, 130, 132
Terpenes, 72
Testosterone metabolism, 254
Theophylline, 241, 259
Thidiazuron (TDZ), 175
Tibetan medicine, 36
Tissue culture of *Rhodiola rosea,* 177
Tonizid, 237
Toxicity assessment, 253
Traditional Herbal Medicinal Product, 205
Traditional knowledge of *Rhodiola rosea,* 89–91
Transmission electron microscopy, 169
Transpiration, 133
Transplanting seedlings, 106
    weed seedlings and, 142
Transplantation of *Rhodiola rosea,* 140
Traumatic brain injury, cognitive function in, 231
Tree-bisection-reconnection (TBR) branch, 15
Triterpenes, 69
*trn*L-F chloroplast regions
    parsimony analysis, 25
    Rhodiola rosea, 9, 15, 17–18
    Rhodiola samples sequenced for, 6–7

Tsan, 38
Tyramine, 75
Tyrosine decarboxylase (TyrDC), 75,
    182–183
Tyrosol, 78, 178, 214
    biosynthesis of, 182–183

U

UCLA Anxiety Disorders Program, 228
UDP-glycosyltransferase, 183–184
Uildyn/uildyun, 42
Unrooted UPGMA dendrogram, 24
UPGMA dendrogram of *Rhodiola*
    rosea, 23
Ural Mountains, 42

V

VAM fungi, *see* Vesicular-arbuscular
    mycorrhizal fungi
Variety of *Rhodiola rosea,* 104–105
Vegetative propagation, 138
Vesicular-arbuscular mycorrhizal (VAM)
    fungi, 170
Veterinary medicine, 245–246
Vigo dana®, 194
Vine weevil, 147
Vitamin C, roseroot, 89

W

Ward's algorithm, *Rhodiola rosea* clustering
    using, 20
Warfarin, 259
Water, use of, 134–135
Weed competition, critical period of, 142
Weed control, 106–107, 140–143
Weeding system, seedling age and, 107
Weed seedlings, 142
Weight gain, 242–243
Whetzel, 163
Wind-dispersed conidia, 162
Winter kill, 159–161
    disease management of, 169–170
Word-finding problems, 243

Y

Yield gap, 149–150
Yield potential of roseroot plantation, 113

Z

Zeatin, 175
Zerleg mugez, 38
Zyrian, 42